Human Movement

D0077463

For Churchill Livingstone

Editorial Director (Nursing and Allied Health): Mary Law
Project Manager: Valerie Burgess
Project Development Editor: Dinah Thom
Copy Editor: James Dale
Sales Promotion Executive: Maria O'Connor

Human Movement

An introductory text

Marion Trew BA MSc DipTP MCSP
Head of the Department of Occupational Therapy and Physiotherapy, University of Brighton, Brighton

Tony Everett BA MEd DipTP MCSP
Lecturer and Clinical Coordinator, Department of Physiotherapy Education,
School of Healthcare Studies, University of Wales College of Medicine, Cardiff

THIRD EDITION

NEW YORK EDINBURGH LONDON MADRID MELBOURNE SAN FRANCISCO AND TOKYO 1997

CHURCHILL LIVINGSTONE
Medical Division of Pearson Professional Limited

Distributed in the United States of America by Churchill Livingstone Inc., 650 Avenue of the Americas, New York, N.Y. 10011, and by associated companies, branches and representatives throughout the world.

© Longman Group Limited 1982
© Longman Group UK Limited 1987
© Pearson Professional Limited 1997

All rights reserved. No part of this publication may be reproduced, stored in a retrieval system, or transmitted in any form or by any means, electronic, mechanical, photocopying, recording or otherwise, without either the prior permission of the publishers (Churchill Livingstone, Robert Stevenson House, 1–3 Baxter's Place, Leith Walk, Edinburgh EH1 3AF), or a licence permitting restricted copying in the United Kingdom issued by the Copyright Licensing Agency Ltd, 90 Tottenham Court Road, London, W1P 9HE.

First Edition 1981
Second Edition 1987
Third Edition 1997

ISBN 0 443 04441 4

British Library of Cataloguing in Publication Data
A catalogue record for this book is available from the British Library.

Library of Congress Cataloging in Publication Data
A catalogue record for this book is available from the Library of Congress.

Medical knowledge is constantly changing. As new information becomes available, changes in treatment, procedures, equipment and the use of drugs become necessary. The editors, contributors and publishers have, as far as it is possible, taken care to ensure that the information given in this text is accurate and up to date. However, readers are strongly advised to confirm that the information, especially with regard to drug usage, complies with the latest legislation and standards of practice.

The author and publishers have made every effort to trace the copyright holders for borrowed material. If they have inadvertently overlooked any, they will be pleased to make the necessary arrangements at the first opportunity.

The publisher's policy is to use **paper manufactured from sustainable forests**

Produced through Longman Malaysia, PP.

Contents

Contributors

Anne-Marie Ainscough-Potts MSc MCSP
Temporary Lecturer, Physiotherapy Group, Biomedical Sciences Division, King's College London; Advisor in Orthopaedic Medicine and Ergonomics for Milligan and Hill Physiotherapy Services, London

Robert A. Charman DipTP MCSP FCSP
Former Senior Lecturer, Department of Physiotherapy Education, School of Healthcare Studies, University of Wales College of Medicine, Cardiff

J. Lesley Crow MSc GradDipPhys CertEd(FE/HE) DipTP MCSP
Former Senior Lecturer, School of Physiotherapy, University of Brighton, Brighton

Tony Everett BA MEd DipTP MCSP
Lecturer and Clinical Coordinator, Department of Physiotherapy Education, School of Healthcare Studies, University of Wales College of Medicine, Cardiff

Allen Hinde BA MA DipTP MCSP
Senior Lecturer, School of Health and Life Sciences, Nene College of Higher Education, Northampton; Visiting Lecturer, Disability Design Research Unit, London Guildhall University, London

Tracey Howe MSc PhD CertEd MCSP
Research Associate, School of Nursing Studies, University of Manchester, Manchester

Neil Messenger BSc PhD
Lecturer in Biomechanics, Department of Physiology, Centre for Studies in Physical Education and Sports Science, University of Leeds, Leeds

Ann Moore PhD GradDipPhys CertEd MCSP MMACP
Deputy Head, Department of Occupational Therapy and Physiotherapy, University of Brighton, Brighton

Di J. Newham MPhil PhD MCSP
Professor and Head of Physiotherapy, Biomedical Sciences Division, King's College London, London

Jacqueline Ann Oldham BSc(Hons) PhD RGN
Reader, School of Nursing Studies, University of Manchester, Manchester

Nicola J. Petty MSc GradDipManipTh MCSP MMACP
Senior Lecturer, Department of Occupational Therapy and Physiotherapy, University of Brighton, Brighton

Marion Trew MSc BA DipTP MCSP
Head of the Department of Occupational Therapy and Physiotherapy, University of Brighton, Brighton

Preface to the third edition

When the first edition of *Human Movement* was published in 1982 it represented a radical step forward in textbooks for physiotherapy students. It was one of the first texts in this area to provide referenced evidence for its material and represented the growing awareness that physiotherapy had to be founded on fact not hearsay.

This third edition continues the tradition of evidence-based information, but the book has been radically revised to reflect its title more accurately. Although some of the chapter headings remain the same, the content is much more biased towards human beings moving and functioning within their environment. There is a significant reduction in material relating to the treatment of patients, but it is hoped that the new content will provide a basis which the reader will use when formulating rehabilitation or training approaches. It is also hoped that the more lively style, with case studies, clinical considerations and tasks for students, will stimulate an active approach to study and a feeling of excitement for the subject.

The book is still intended as a basic text for students in the early stages of their education, and it is anticipated that they will use it in conjunction with their other scientific books. It should whet the student's appetite for the subject area and give them sufficient information to facilitate further study.

Over the years it has become clear that in addition to being used by student physiotherapists this text has proved of value for students of occupational therapy, podiatry and sports science; as a consequence the text has been modified to take this into account.

Brighton and Cardiff, 1997

M.T.
T.E.

CHAPTER CONTENTS

Introduction

T. Everett M. Trew

OBJECTIVES

When you have completed this chapter you should be:

1. **Enthusiastic about the study of human movement**
2. **Aware of the range of factors that influence the initiation, production and control of human movement**
3. **Conscious of the difficulties of simultaneously considering all aspects of human movement, yet at the same time be aware of the need to maintain a holistic approach**
4. **Beginning to develop awareness of your own body at rest and when moving**
5. **Starting to develop the skills of observation of human movement.**

THE STUDY OF HUMAN MOVEMENT

The study of human movement is fascinating for two main reasons. Firstly, because it is about ourselves and how we are able to go about our everyday lives performing a vast range of functional activities, sporting activities and other pastimes. It is natural to have a curiosity about oneself and the study of human movement inevitably leads to a sense of amazement at the wide variety of intricate tasks that we are able to perform with ease, and often without thought. The second fascination lies in the complexity of human movement and the challenges that arise out of this. There are still a surprising number of gaps in our knowledge not only about the fine details of how movement is initiated and controlled but even in the apparently simple areas of exactly what happens when we perform basic everyday tasks. Whilst there have been a substantial number of scientific studies of walking and running, there are only a few research papers on other major lower limb functions and very few detailed studies of the everyday activities we perform with our upper limbs.

Observation of human movement reveals a complex and seemingly infinite variety of positional change which involves or is controlled by a wide range of internal and external factors. To begin to understand how the systems of the body interact to produce finely controlled and purposeful movement it is essential that some order is introduced into the study. It is necessary to know how human movement is initiated, performed and controlled and such knowledge forms the basis of those professions working in this area.

Human movement can be viewed from a number of different standpoints:

- Anatomical: describing the structure of the body, the relationship between the various parts and its potential for movement. Incorrect alignment or disruption of anatomical structures will clearly affect movement.
- Physiological: concerned with the way in which the systems of the human body function and the initiation and control of movement. In many cases incorrect functioning or failure of integration between systems will lead to movement abnormalities.
- Mechanical: involving the force, time and distance relationships in movement.
- Psychological: examining the sensations, perceptions and motivations that stimulate movement and the neurological and chemical/hormonal mechanisms which control them.
- Sociological: considering the meanings given to various movements in different human settings and the influence of social settings on the movements produced.
- Environmental: considering the influence of the environment on the way in which movement occurs.

The following chapters expand the anatomical, physiological and mechanical basis of human movement. This does not lessen the importance of the other approaches and readers should familiarise themselves with this information which can be obtained from other well recognised sources.

By studying the musculoskeletal basis of movement it is possible to have an anatomical framework on which movement can be referenced and described in an unambiguous manner. An intact musculoskeletal system is essential for correct movement to take place. Joints need to possess sufficient feedom of movement to perform the required activities and to be able to move in a smooth and unrestricted manner. Muscles provide the means of achieving this movement and they must possess the necessary strength, power and endurance to carry out this function. They must also be controlled in an extremely delicate and sensitive way and the efficient functioning and correct integration of the central and peripheral nervous system is essential.

Whilst some movements arise out of a conscious decision and require active thought processes, most have been previously learned and are automatic. The neural processes that store, adapt and use these learned movements are complex and demonstrate well the interdependence of all the systems of the body in the production of movement.

Movement does not take place in isolation from the external environment. There is a complex interaction of forces acting on the body including the constant force of gravity and the changing frictional forces. The force of gravity may have the effect of initiating or producing actual movements but it is always a force which the body has to counter to achieve and maintain an upright position. Friction, on the other hand, changes as the body comes into contact with a variety of surfaces, thus causing different reactions of the body's internal environment in response to these changing conditions.

Forces from many other sources such as wind, water, animate and inanimate objects all have their effect on the way movement is carried out. It is important that the physical laws of the external environment are understood so that the prediction of, and compensation for, these forces can be implemented by the controlling systems.

It is obvious that each of these aspects is interrelated and that between them they give a framework and a direction for the study of movement. However, any attempt to consider all possible factors simultaneously would result in a very lengthy and complex process and this inevitably means that research into movement rarely encompasses more than one or two aspects. In this book the initial chapters examine some of the different theoretical backgrounds to human movement and in the later chapters there is a more holistic evaluation of some common everyday activities and consideration of how movement can be affected by factors such as ageing or stress. The final chapter addresses some of the ways in which human movement is measured and emphasises the difficulties which arise from attempting to measure and evaluate complex, multidimensional activities.

It is essential for the reader to remember that, in the practical or clinical situation, most approaches to considering human movement will look at only one or two of the many aspects. This is obviously a limited approach and may give a false or distorted picture of the individual's ability. Sadly, it is not possible to achieve the ideal situation and take simultaneous measurements of all components of a complex functional activity, but providing there is an awareness of the limitations of the way in which human movement is measured and assessed, the right conclusions about how an individual's performance may be corrected or improved may still be reached.

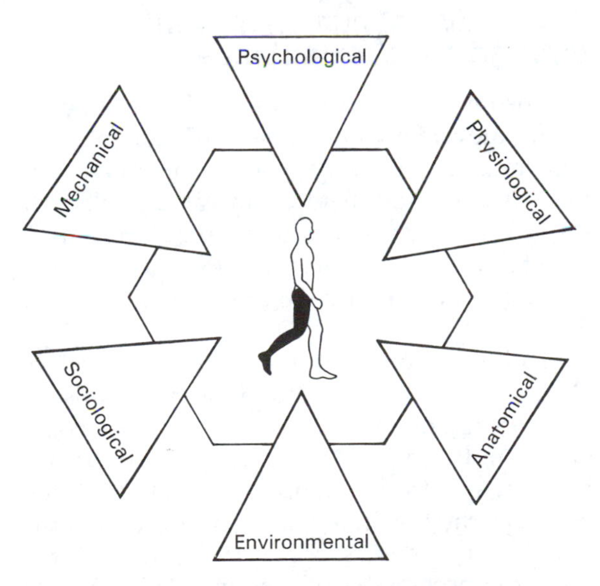

Figure 1.1 There are a number of ways in which the study of human movement can be approached; each approach is valid in its own right but, on its own, limited. For a holistic understanding of how the human body moves and why the component parts work as they do, a multidimensional approach has to be taken.

Box 1.1 Case Study 1

A premier division soccer team were undergoing a series of tests of 'fitness' which, it was hoped, would provide information which could be used to build an improved training programme. These tests included anthropometric measurements, tests for elasticity of soft tissue, muscle length and the strength and endurance capacity of their lower limb muscles. Towards the end of the day one of the players, Mr F., was having the peak torque ratio of his knee and hip flexors and extensors tested on the isokinetic dynamometer.

When the results for the team were analysed, it was found that the torque generated by the non-dominant limb was always greater or equal to that of the dominant limb, except in the case of Mr F., where his dominant limb appeared to have generated substantially more torque than expected. There was concern that either there had been an error in the data collection procedure or that Mr F. had an injury to his non-dominant limb which he had not mentioned to the laboratory staff and which was being reflected in abnormally low torque levels. It was decided to call him back in for review.

On reattendance, discussion with Mr F. revealed that he had begun to lose interest in the testing procedures by the end of the day when his isokinetic test was scheduled. He speculated that he had not been putting as much effort as he might into the test until, just as the peak torque in his dominant limb was being measured, his team manager and coach walked into the laboratory. He clearly remembered trying to impress them with his keenness and fitness by working as hard as he could and as a consequence had distorted the data.

In this case the laboratory staff were looking for physiological or mechanical reasons why the data for Mr F. should be aberrant. In the end, the reason was psychological.

Self-awareness and observational skills

Case study 1 was used to illustrate how movement and performance can be affected by a number of factors, which may not always be those that are most obvious. Another interesting fact that became apparent during the testing procedures undertaken on these professional soccer players was that their balance and spatial awareness was surprisingly poor. This information came to light inadvertently during a plyometric testing procedure, when the players found the test hard to complete and lost their balance repeatedly. Despite the fact that their profession required a high level of physical ability, none of the players had a good level of awareness of the way in which their body moved and they all found it difficult to work out how to modify their approach to the plyometric test in order to remain in balance. This lack of consciousness of movement is common in the general population, with most people never considering how they are able to undertake everyday tasks until they lose that ability.

Take yourself: do you know exactly what movements occurred in the joints of your upper limb as you picked up this book and opened it? Were you even aware of the process or did it happen automatically? All people, during their waking hours, are constantly moving and yet they rarely stop to analyse these movements and have no idea how complicated most of them are.

In the later chapters of this book, common activities such as walking or getting out of a chair will be considered and their complexity will become apparent. However, most people are oblivious to the way in which they perform tasks, and many daily activities occur at a subconscious, reflex level. This frees the brain to undertake other tasks at the same time, for example, it becomes possible for a musician to play a guitar, sing and move about the stage simultaneously.

If a detailed analysis of everyday activities is carried out, it is clear that there are patterns of joint movement and muscle action which are common to several activities. For example, going up stairs uses the same basic pattern of movement as standing up from a chair or walking up a slope. Swinging the upper limbs in walking is similar to taking food to the mouth, though the range of movement is different. The brain works in terms of patterns of movement rather than movement of individual joints and contractions of individual muscles. It is probably because of this that tasks performed frequently become reflex. As long as the ability to perform tasks automatically exists, most familiar movements can be undertaken in a smooth and efficient manner. When this ability is lost, perhaps through injury or disease, movement becomes noticeably slower and less coordinated.

Anyone whose work involves the moving human body needs to become aware of the way normal movement occurs and which patterns of movement are frequently used. Once an understanding and awareness of normal movement is acquired, it becomes possible to recognise deviations from normal and to plan rehabilitation or training programmes with precision.

Knowledge of human movement through direct experience

Few people consciously explore their full potential for movement, but students of human movement must become very aware of themselves and the way they move before they can consider others. It is as important to be aware of movement and to 'feel' or consciously experience joints moving and muscles contracting as it is to observe others.

In addition to developing self-awareness it is also essential to learn observational skills, as these are the mainstay of clinical practice. Every opportunity should be taken to observe the movements of other people. Look at the different ways they walk or stand and try to identify exactly what makes one person move differently to another. Be precise in this, observing not only which joint moves, but by how much or how fast. As you develop the skills of observation you should try to compare groups of people.

Box 1.2 Task 1

You need to develop personal self-awareness and the skills of observation if you are to have a full understanding of normal movement. This can be done in a number of ways, all of which require you to put in some effort.

1. Try to become aware of all parts of your body. For example, think about the position of your shoulder girdle: is it elevated or depressed? Be aware of your vertebral column: are the various components flexed, extended or laterally flexed? Constantly re-evaluate how your body is aligned and notice how your body changes the alignment of its parts for different activities. Become aware of the differing ranges of joint movement that can occur between individuals; compare yourself with others to see if your joints are more or less mobile. Notice what it feels like when you reach the limit of a movement. Is it the same feeling for all joints?
2. Think about what your body feels like when it moves in contrast to when it is still. If some movements cause discomfort, ask yourself why and try to work out exactly which structures are involved. Is it because you have moved too near to the limit of your normal range of movement or because you are working a muscle particularly hard?
3. Notice the difference in feeling when a muscle is contracted or relaxed by making a very tight fist and holding it tight for 30 seconds. Then relax and notice the changing sensations as relaxation occurs: is the process of relaxation instantaneous?
4. Try to become aware of the way in which your body weight is distributed during activities which require balance ability. Is your weight equally distributed between both feet, or is it more on one foot than the other? Consider whether your weight is distributed across the whole foot evenly or if there is more taken through the ball of the foot than the heel. What advantages come from different alignments of body weight across one or both feet?

Box 1.3 Task 2

Do elderly people move differently from the young, or women from men? If you think the answer is yes, then you should try to identify the differences.

UNDERSTANDING HUMAN MOVEMENT AND ITS CLINICAL APPLICATION

To develop the skills of human movement analysis it is first important to become more self-aware and this, combined with a knowledge of relevant research, will lead gradually to a firsthand understanding of many of the factors of 'normal' movement. This needs to be combined with the ability to observe, in a structured and purposeful manner, the way other people perform everyday activities. In the professional setting it is possible to use the senses of hearing, sight and touch to collect information about an individual and their problems. The skills of interviewing, listening with understanding, looking and seeing, palpating and testing, all contribute towards a pool of knowledge and modern measurement techniques will enable some quantifiable data to be collected.

When working with patients, the skills they require to perform activities of daily living need to be identified and the way in which they actually try to undertake these activities should be analysed. With this knowledge it becomes possible to consider how certain tasks might be made more efficient or how a person with a disability might be helped towards greater independence. Specific problems can be identified, goals set and a realistic programme designed. Finally, the patient's progress will need to be regularly evaluated and goals altered when necessary. By systematically approaching each individual's movement problems in this way, clinical judgement can be developed and clinical practice becomes more effective.

USING THIS BOOK

Use the early chapters, in conjunction with other specialised text books, to gain a grounding in the theories underpinning how human movement is planned, initiated and controlled. This basic knowledge is essential to an understanding of what is happening when movement occurs and why it happens. Use the later chapters as introductions to the way in which various parts of the body contribute to functional movement and as an introduction to how complex movements are analysed. These chapters will also consider some of the common factors that lead to deviations from normal movement patterns.

Throughout the book there are case studies which form the link between theory and actuality, illustrating why the acquisition of knowledge and understanding will lead to better results. If you are already working with patients or clients, then you should try to see how the content of the chapters relates to your experience.

In all the chapters there are tasks which are designed to encourage thought or help you develop skills. If you are to benefit from the learning process you should undertake each task, attempting to fulfil all its requirements. Some of the tasks are short, but a number will develop into skills which you will use for the rest of your working life. Most of these tasks are easy and do not require answers to be provided within the book; if you are unsure of any of the answers then reread the relevant parts of the chapter, discuss the problem with your colleagues and talk to more experienced staff.

When you use this book you must be aware that it is a basic text designed as an introduction to the study of human movement. It is not a definitive repository of all knowledge in the subject area but should help in the understanding of more advanced research. If you are to be excellent in your work then you must constantly strive to further your knowledge through reading and enquiry. This should not be a burdensome task because human movement is a fascinating subject, intriguing in its complexity and of direct interest to every one of us.

CONTENTS

2

Biomechanics

N. Messenger

OBJECTIVES

At the end of this chapter you should be able to:

1. **Define mass, matter, inertia and weight**
2. **Explain what is meant by gravity and centre of gravity**
3. **Discuss the concept of force and its analysis**
4. **Describe levers and their analysis**
5. **Explain equilibrium, stability and their components**
6. **Discuss the classification of motion (kinematics)**
7. **Discuss the relationship between force and motion (kinetics) with reference to Newton's laws of motion**
8. **Define friction and explain the concepts of static and dynamic friction**
9. **Discuss the differences between work, power and energy**
10. **Discuss the principles of fluid mechanics relevant to physiotherapy**
11. **Discuss the concepts of mechanical stress and strain and the relationship between these and the behaviour of biological materials.**

INTRODUCTION

Mechanics is the study of forces and their effects.

These effects are a change in motion of an object or a deformation of the material from which the object is made. Biomechanics is the study of mechanics applied to living bodies. Since the human body may be acted upon by a variety of forces that can lead to it moving, remaining at rest or being placed under harmful stress and strain, an understanding of the basic principles of mechanics and particularly of biomechanics is essential to the physiotherapist.

SCALAR AND VECTOR QUANTITIES

In mechanics, physical quantities are classified as being either scalar or vector. *Scalar* quantities are those which are described completely by a measure of their magnitude alone, whereas for a full description of a *vector* quantity, not only is the magnitude of the quantity required but also the direction in which it acts. Examples of scalar quantities are: distance, area, volume, mass and speed. Examples of vector quantities are: displacement, velocity, force, weight and mechanical work.

It is necessary to classify a quantity correctly as being either vector or scalar in order to be able to treat it mathematically. If we wish to examine the combined effect of scalar quantities, we simply add their numerical magnitudes: for example, combining a 20 kg mass with a 30 kg mass produces a 50 kg mass. However, the combined effect of a 20 N force and a 30 N force, because force is a vector quantity, will be 50 N only if the forces are acting in the same direction. How we treat vectors whose directions do not coincide will be dealt with later when force is considered in more detail.

MATTER, MASS, INERTIA, WEIGHT

Matter, which may be gas, liquid or solid, is defined as the material or substance that occupies space. The mass of an object is simply the quantity of matter that makes up that object. Objects, including the human body or any of its segments, are often referred to as masses. Mass is a scalar quantity and the unit of mass is the kilogram (kg).

In mechanics, objects are often referred to as bodies. When used in this text, therefore, the term *body* does not necessarily imply reference to the human body.

Everyday experience indicates that it is usually more difficult to change the motion of a massive body than that of a less massive one. This reluctance of matter to change its state of motion is termed *inertia* and is stated in Newton's law of inertia. Mass is also, therefore, a measure of inertia. In common usage the terms *mass* and *weight* are often used synonymously. However, the correct definitions of the two terms are quite different. The weight of a body is the force exerted on the body by gravity. Unlike mass, therefore, weight is a vector quantity with its direction always acting towards the centre of the earth, i.e. vertically downwards.

As weight is a force, the correct unit of weight is the newton (N) and not the kilogram.

The relationship between the mass, m, of an object and its weight, W, is expressed in the equation:

$$W = m.g$$

where $g = 9.81\ m/s^2$ and is the acceleration due to gravity.

For all practical purposes, for objects close to the surface of the earth, g is considered to be constant irrespective of the mass of the object. This means that, ignoring air resistance, two objects of differing mass dropped from the same height will accelerate towards the ground at the same rate and hit the ground at the same time.

CENTRE OF GRAVITY

It is often convenient in mechanics to describe the motion of a body as if it were a single point in space. The *centre of gravity* is usually used for this purpose. This is the point in a body at which the whole weight of the body may be considered to act. Knowledge of the position of the centre of gravity is, therefore, also necessary when examining the effects of the weight of the object, for example, on its motion. Sometimes the reader may come across the term *centre of mass* in this context. This is the point about which the mass of an object is evenly distributed and therefore has a quite different meaning. However, in most practical cases, the positions of the centre of mass and centre of gravity will coincide.

The centre of gravity of a body has no anatomical reality. For some positions adopted by the human body the centre of gravity of the whole body may even be located outside its physical boundaries (Fig. 2.1). The position of the centre of gravity depends upon the arrangement of the body segments.

Studies have been undertaken to investigate the position of the centre of gravity of the human body in a number of poses, particularly in upright stance. In this position it is generally accepted that the centre of gravity lies within the pelvis at approximately the upper sacral region, anterior to the second sacral vertebra (Hellebrandt et al 1938). However, the exact location differs between individuals and may vary with such factors as stature, build and body proportion (Reid & Jensen 1990).

The position of the centre of gravity changes with every change in body position. For example, in walking the centre of gravity usually remains within the pelvis but it will oscillate both laterally and anterioposteriorly as the limbs swing with each stride (Inman et al 1980). Similarly, when bending forwards to pick an object from the ground, the centre of gravity will move anteriorly. In both of the above cases, the total body weight will remain the same.

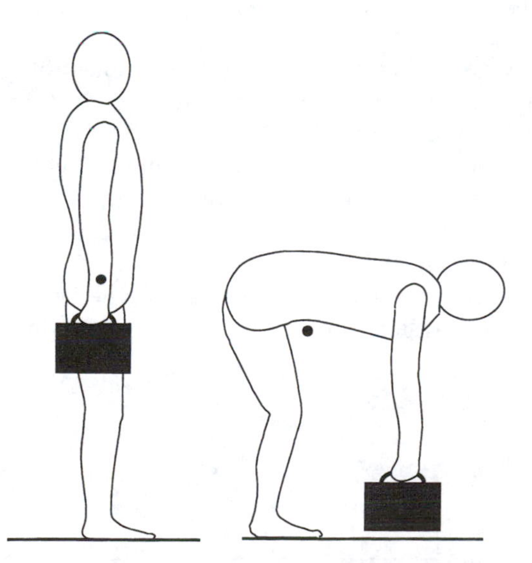

Figure 2.1 The position of the centre of gravity (●) of the body and carried object combined. Note that an incorrect lifting position increases the displacement of the centre of gravity relative to the free-standing position.

If weights are added to the body, as in the case when carrying a backpack or lifting an object, the centre of gravity of the total system will need to be reconsidered. Likewise, loss of a body part through amputation not only alters the total body weight but the distribution of that weight, so the location of the centre of gravity will be altered. This has important implications for posture and balance control.

Unlike the whole body, the position of the centre of gravity of a body segment may usually be assumed to be fixed. A number of studies investigating the positions of the segmental centres of gravity have been undertaken and the results of such studies are readily available (Gowitzke & Milner 1988, Reid & Jensen 1990). But, by definition, the centre of gravity will always be towards the heavier end of a segment, approximately $\frac{4}{7}$ths (57%) of the segmental length from the distal end, the precise position differing between individuals and between body segments.

Knowledge of the position of a segment's centre of gravity, as opposed to total body centre of gravity, is useful to the physiotherapist when it is necessary to examine the leverage effect of gravity around a joint. For example, in Figure 2.2, the lever effect produced by the weight of the arm about the shoulder increases as the length, a, increases and is, therefore, greatest in this case, when the arm is horizontal, i.e. a_2 is greater than both a_1 and a_3. Note that the weight of the arm always acts vertically downwards.

FORCE

Force is a characteristic of the interaction between masses. It is a quantity that we cannot see directly but we know it is there because we can observe its effects. We are generally aware of force when we push or pull an object in an attempt to move or deform it. Force may, therefore, be defined as a quantity which changes or tends to change the motion of a mass to which it is applied. If a change in motion does not occur, it is because the applied force is opposed by an

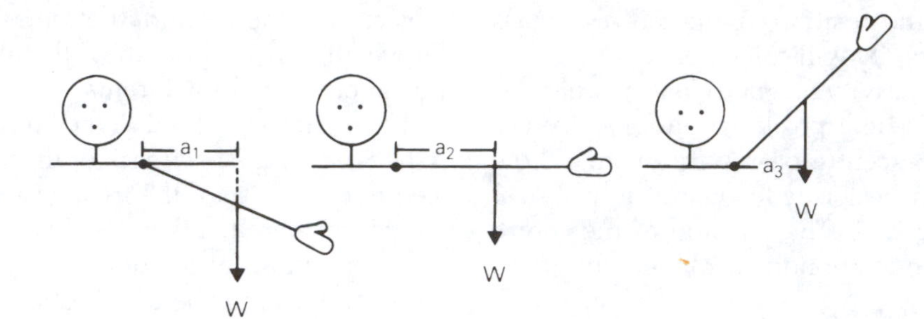

Figure 2.2 The change in position of the line of action of the weight of the arm, W, relative to the shoulder as the arm is elevated or lowered. Reproduced with permission from LeVeau B 1992 Williams and Lissner: Biomechanics of Human Motion. W B Saunders, Philadelphia.

equal force acting on the mass in the opposite direction. Forces may be classified as contact forces (e.g. friction) or non-contact forces (e.g. gravity), but for a force to be generated, one mass must act upon another. Force is a vector quantity — its direction is as important as its magnitude — and the unit of force is the newton (N).

The human body moves under the influence of, and is affected by, both internal and external forces. Within the body, forces are produced principally by the contraction of muscles. Examples of external forces acting on the body are those of gravity (weight), forces generated by interaction with external objects, and the forces applied either manually, or with the aid of equipment, by a physiotherapist exercising a patient.

To fully describe a force, because it is a vector, it is necessary to know its magnitude, line of action, direction and the point at which it is applied. It is often convenient to represent forces graphically. This allows the nature and effects of the forces applied to an object to be more easily visualised. The conventional method of representing any vector quantity graphically is to use an arrow: the length of the arrow being proportional to the magnitude of the vector; the arrow head indicating its direction; and the position and orientation of the arrow on the diagram indicating the vector's origin and line of action. This can be seen in Figure 2.2 where the weight of the arm is represented by a vertical arrow pointing downwards and drawn from the centre of gravity of the segment.

FORCE SYSTEMS

Any group of two or more forces is known as a force system. Within a force system two or more forces may be considered:

- colinear — if all forces act along the same line of action
- coplanar — if all forces act in the same plane
- concurrent — if the line of action of all forces passes through the same point.

Force systems may also be classified as being linear, parallel, concurrent or general force systems.

Linear force systems are the simplest of force systems in which all forces are colinear (Fig. 2.3 and Fig. 2.4).

Parallel force systems are systems in which the forces are coplanar and parallel but not all colinear. These systems also tend to cause rotary effects (Fig. 2.5).

Concurrent force systems occur when all forces in the system are concurrent. An anatomical

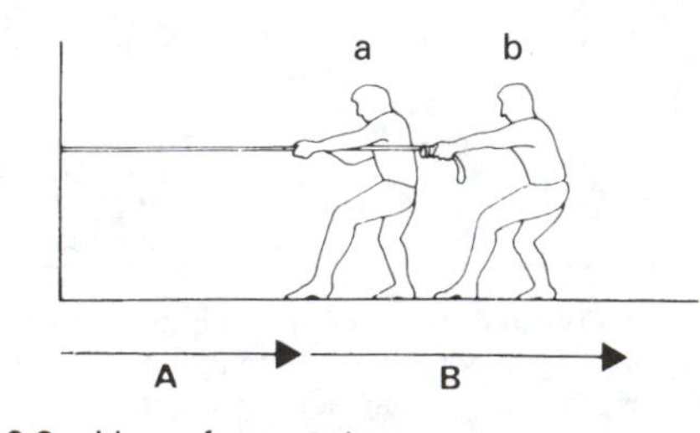

Figure 2.3 Linear force system.

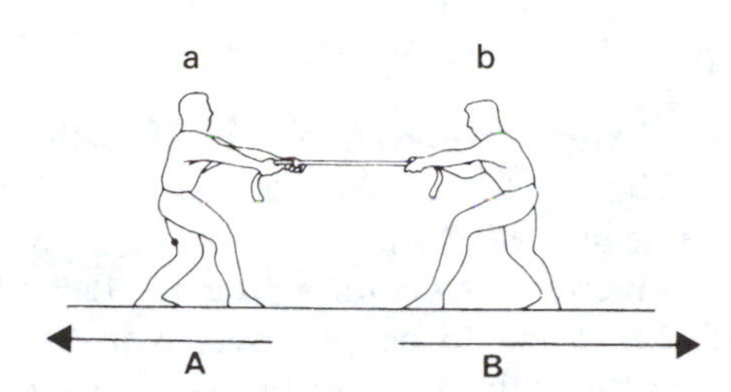

Figure 2.4 Linear force system.

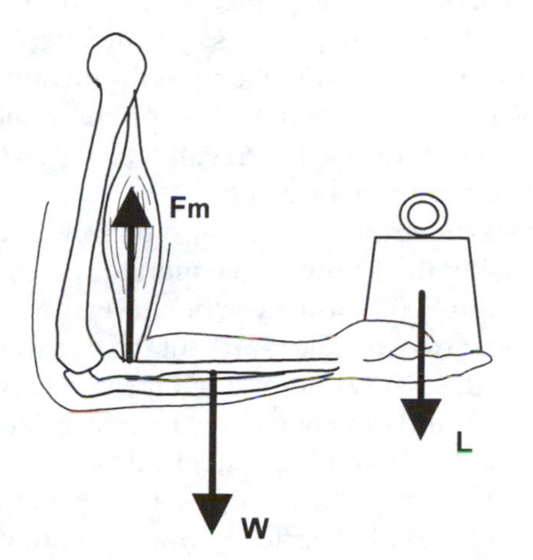

Figure 2.5 Example of a parallel force system. In the case shown, the biceps are pulling vertically upwards with force Fm in opposition to the weights of the foream (W) and the object carried in the hand (L).

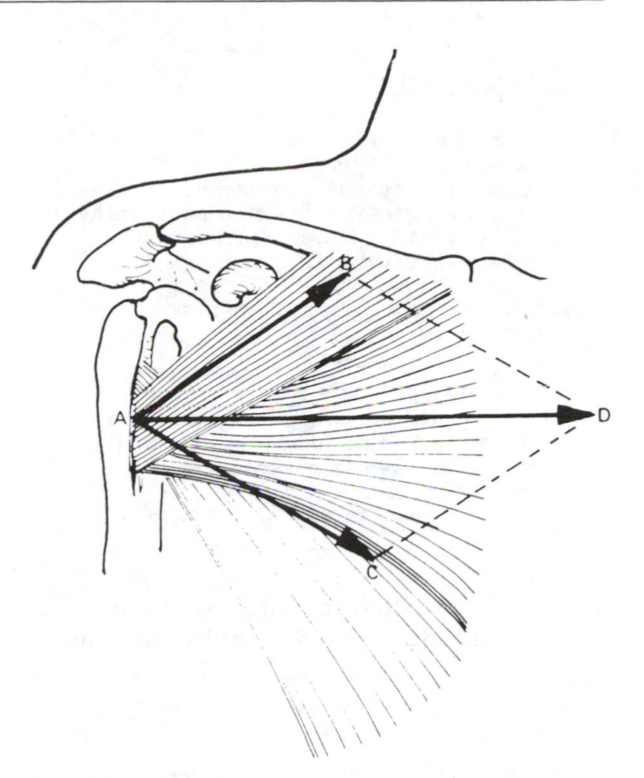

Figure 2.6 Example of a concurrent force system. The resultant force of the clavicular head of the pectoralis major, AB, is concurrent with the resultant force of the sternal head, AC. The force vector AD represents the combined effects of AB and AC, i.e. the resultant muscle force. Reproduced with permission from Gowitzke B A, Milner M 1988 Scientific Bases of Human Movement, 3rd edn. Williams and Wilkins, Baltimore.

example of such a system would be the sternal and clavicular heads of the pectoralis major (Fig. 2.6). The forces produced by these two components of the muscle pull in differing directions, but are considered to act from the same origin and are, therefore, obviously concurrent. It should be noted, however, that the physical origins of the forces need not necessarily coincide for a system to be concurrent, only that their lines of action intersect at a common point. Therefore, it can be seen that any system of two non-parallel coplanar forces will be concurrent.

Concurrent forces alone will not produce angular motion. In Figure 2.6, adduction occurs because of an additional reaction force at the shoulder joint. The total force system describing motion about the shoulder is, therefore, a general force system.

General force systems are force systems that cannot be classified according to the above definitions. The majority of force systems encountered will fall into this group and they may cause both linear and angular motions. However, it is often convenient in the analysis of these systems to break them down into component systems that conform to the above classifications.

VECTOR ANALYSIS OF FORCES

In real life many forces may be acting on a body simultaneously, each separately tending to change the motion of or deform the body. It is often necessary to know the final effect of these forces.

It is always possible to define a single force

Box 2.1 Task 1

Force systems of the body
Get into a side lying position and flex and extend your top limb at the knee. Ignoring the weight of the limb, analyse the force system that is used to produce flexion. How does this differ from the force system that is in operation during extension? If you consider the weight of the leg, how would we now classify the force system operating in both cases?

Using fibre direction as a guide to the line of action of a muscle body, classify the following muscle groups into the appropriate force system: biceps; trapezius; hip adductors (brevis, longus, magnus): gracilis, semitendinosus and semimembranosus.

that would have the same effect as a force system acting on a body. This force is called the resultant or vector sum of the forces in the system. The resultant, therefore, is the simplest single force that would produce the same effect as all the forces acting together.

Addition or composition of forces

Adding vectors to find the resultant is called composition. Two basic methods of composition may be used. They are the algebraic method and the graphic method.

Consider the linear force system illustrated in Figure 2.3. In this example, two people are pulling on a single rope. A is pulling with a force of 600 N and B with a force of 800 N. The problem is to determine the resultant force generated by them both.

As they are both pulling in the same direction, intuitively we can see that the resultant is simply the sum of the two forces. That is:

$$600\text{ N} + 800\text{ N} = 1400\text{ N}$$

Figure 2.4 shows the same two people, again A pulling with a force of 600 N and B with a force of 800 N, but this time involved in a tug of war. Now the resultant is quite different. Intuitively we can say that B is likely to win because he is pulling with 200 N more force. The algebraic solution therefore is:

$$-600\text{ N} + 800\text{ N} = 200\text{ N}$$

The difference between this and the solution to Figure 2.3 is that the 600 N force applied by A is given a negative (−) sign.

By convention, vectors acting to the left are usually assumed to be negative, whilst vectors acting to the right are positive — similarly vectors acting upwards are positive and vectors acting downwards negative. Although this is a general convention, not a strict rule, vectors acting in opposite directions will always have opposing signs. This rule also applies to the solution. Therefore, as the resultant is positive, it must be acting to the right.

These two examples can also be solved graphically using the arrow convention to represent the 600 N and 800 N force vectors. In Figure 2.3 two arrows drawn to scale represent the two forces. By placing the arrows nose to tail it can be seen that the length and direction of the resulting line represents the resultant force magnitude and direction. Using the same principles to solve the tug of war problem; the tail of vector arrow B, pointing to the right, is drawn from the head of vector arrow A, pointing to the left. The resultant is then represented by an arrow drawn from the tail of vector arrow A to the head of vector arrow B (Fig. 2.7).

Analysis of non-linear force systems

If the forces to be analysed are acting at angles to each other, i.e. the force system is non-linear, the same basic approaches described so far may still be used, but some modification is necessary to the algebraic approach. This is because simple algebraic summation does not take into account the precise direction of the force vectors.

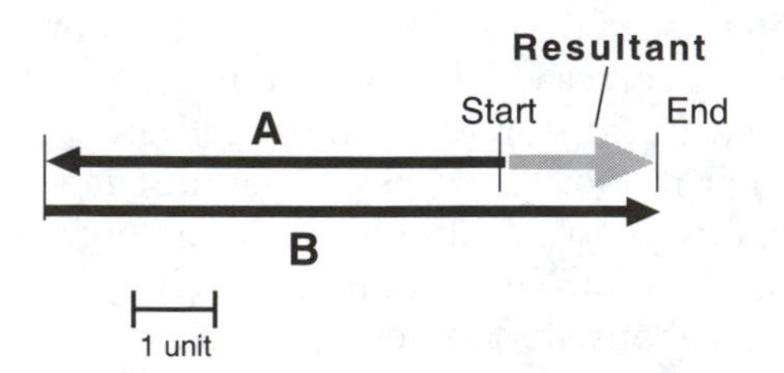

Figure 2.7 Graphical composition of two opposing linear forces (1 unit = 100 N).

Consider first the graphic method: two basic approaches are available; the parallelogram of forces and the triangle or polygon of forces.

The parallelogram of forces

This approach is used to find the resultant when any two forces are acting at the same point. As an example, Figure 2.8 illustrates two forces applied to a patient's leg in traction. To find the resultant, it is simply necessary to draw two vectors to a suitable scale and from the same origin or starting point and at the appropriate angles as in Figure 2.8b. Two further lines are then drawn parallel to these, the resultant being represented by the diagonal of the resulting parallelogram.

In this case, as the scale of the diagram is one division = 10 N and the diagonal of the parallelogram is 66 units long, the resultant is 66 N. To complete the analysis, it would also be necessary to use a protractor to measure the angle of the diagonal to determine the orientation of the resultant. The direction of the resultant vector, and therefore the direction in which the vector arrow will point, will always be away from the origin of the diagram.

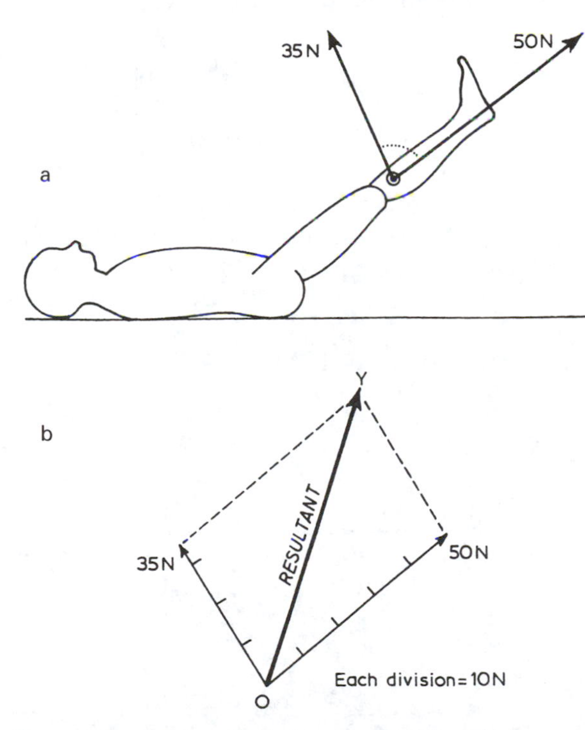

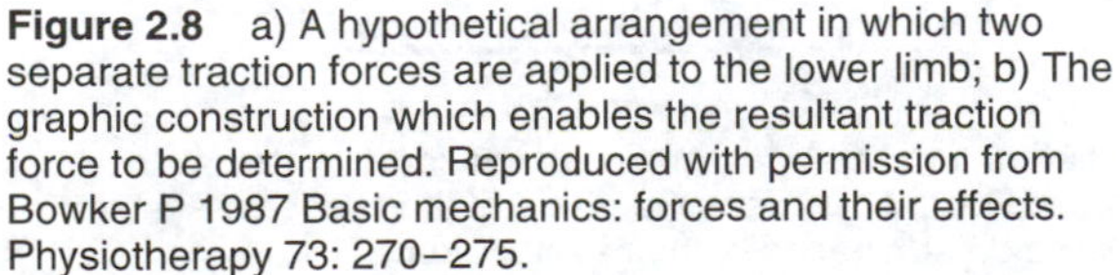

Figure 2.8 a) A hypothetical arrangement in which two separate traction forces are applied to the lower limb; b) The graphic construction which enables the resultant traction force to be determined. Reproduced with permission from Bowker P 1987 Basic mechanics: forces and their effects. Physiotherapy 73: 270–275.

The triangle and polygon of forces

An alternative method of analysing the problem in Figure 2.8a is to follow the method described previously for the linear system in Figure 2.4. Here one of the forces is chosen, say the 50 N force, and drawn to scale parallel to the direction of the force applied to the patient. The second force is then drawn, again to scale and in the appropriate direction, but starting from the end of the first vector as in Figure 2.9. The resultant is then represented by the vector drawn from the beginning, or tail, of the first vector to the end, or head, of the second. The resulting triangle is called the triangle of forces.

As can be seen, this method produces exactly the same result as the parallelogram of forces. In fact, the triangle of forces is simply half of the parallelogram of forces found in Figure 2.8b. Note, also, that the same result would have been obtained irrespective of which force had been drawn first.

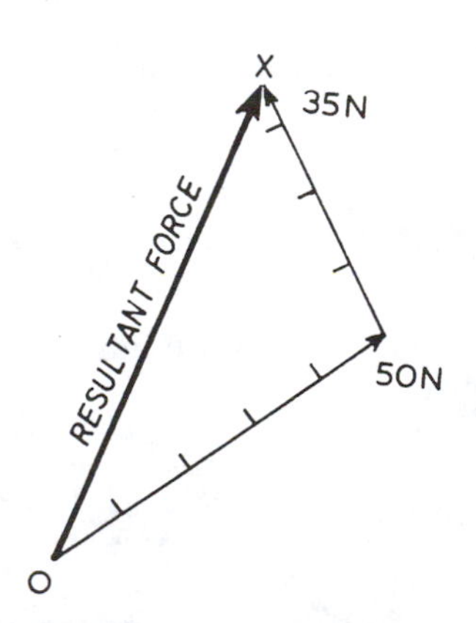

Figure 2.9 The triangle of forces for the system of Fig. 2.8a. Reproduced with permission from Bowker P 1987 Basic mechanics: forces and their effects. Physiotherapy 73: 270–275.

The advantage of the triangle of forces method is that it can be easily extended to obtain the resultant of force systems containing more than two forces. Consider the four force system illustrated in Figure 2.10. To solve this we choose any force and draw it to scale, starting at any suitable origin and in the correct direction. Then take the remaining forces in turn, drawing each from the head of the previous force in the appropriate direction. The resultant is the vector which is required to complete the diagram, drawn from the tail of the first vector to the head of the last vector. Note that the order in which the forces are drawn has no effect on the result. The polygon of forces method, then, allows the determination of the resultant of any combination of uniplanar forces.

The resultant by calculation

The graphic methods outlined above are useful, particularly in that they help us to visualise the effects of force systems, but have limited accuracy. It is often necessary, therefore, to be able to find the resultant mathematically. Here, simple trigonometry is used to account for the directions of the forces.

The simplest situation to analyse is that when the two forces are at right angles to each other as in Figure 2.11a. First sketch the geometric solution, as in Figure 2.11b — this is always a useful first step. The result is a right angle triangle. Therefore, the resultant, which is the hypotenuse of the triangle, can be found from the Pythagoras theorem which gives the equation:

$$R = \sqrt{(P^2 + Q^2)}$$

where R is the resultant and P and Q are the two perpendicular forces being added.

Remember, it is necessary to indicate the direction and orientation of the resultant as well as its magnitude. This is often done by specifying the angle it makes with one of the original forces. These may be found from the following:

$$\tan\theta = \frac{P}{Q} \quad \text{or} \quad \theta = \tan^{-1}\frac{P}{Q}$$

$$\tan\alpha = \frac{Q}{P} \quad \text{or} \quad \alpha = \tan^{-1}\frac{Q}{P}$$

If the two forces are not perpendicular, it is necessary to use the more complex equation known as the cosine rule:

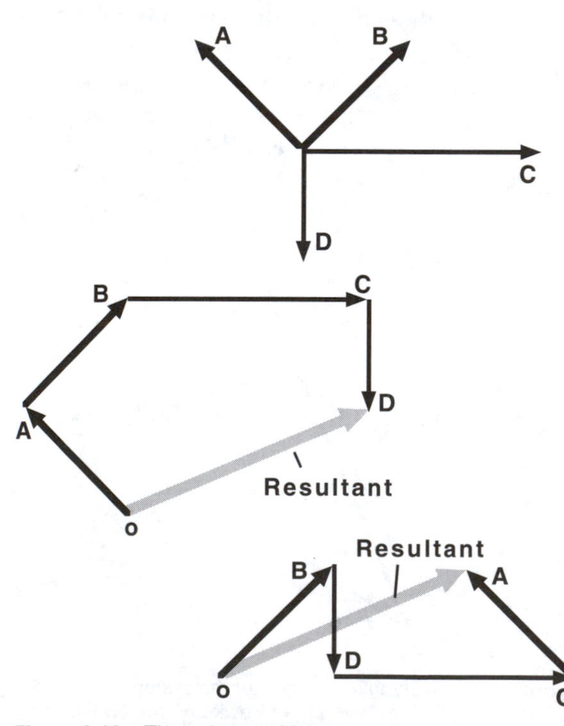

Figure 2.10 The polygon of forces solution to the composition of multiple force systems.

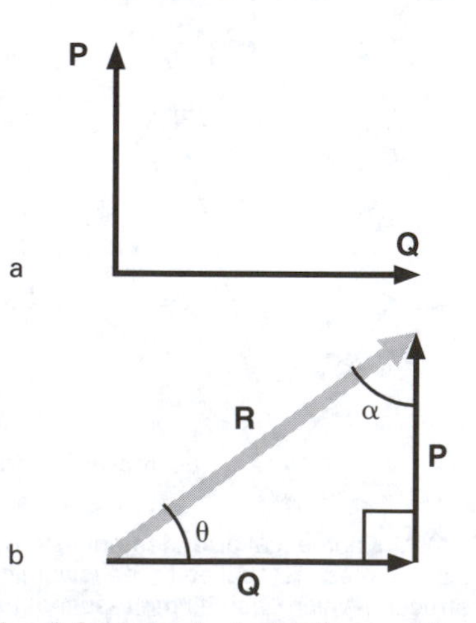

Figure 2.11 a) A perpendicular two force system with forces P and Q. b) Vector diagram of the two force system for the determination of the resultant force R.

$$R = \sqrt{(P^2 + Q^2 - 2.P.Q.\cos \beta)}$$

where R is the resultant, P and Q are the two non-perpendicular forces and β is the angle between them.

To find the resultant of force systems containing more than two forces it is necessary first to resolve the forces into vertical and horizontal force components as described below. The resultant vertical and horizontal components are then added in the usual way.

Resolution of forces

If a force is applied to an object at some angle, it is sometimes useful to know what part of that force is acting vertically and what part horizontally. This can be done by reversing the process of composition. This new process is called the *resolution of force*.

Resolution allows the determination of two or more forces that would have the same effect as a single force. In theory, any force may be replaced by an infinite variety of hypothetical force systems that have the same effect. Figure 2.12 illustrates two possible force systems that could replace the original force *F*. These are, however, hypothetical systems, which would have little practical value.

In practice, resolution is most frequently used to find two mutually perpendicular forces that may replace the known single force. These two forces are called the components of the force. In the original example, it is the vertical and horizontal components of the force that are required. In these cases, force resolution is relatively straightforward as we are always dealing with a right angle triangle.

Again, either a graphical or a mathematical approach may be used. Consider the force acting at an angle θ to the horizontal, and α to the vertical (Fig. 2.13). To find the components graphically, simply draw a vertical line from one end of the force vector and a horizontal line from the other end of the force vector, terminating the lines where they intercept. F_h is then the horizontal component, or horizontal effect, and F_v the vertical component, or vertical effect of the original force.

Because the triangle is right angled, the components can be found mathematically using the following equations:

$$F_h = F.\cos \theta$$

$$F_v = F.\sin \theta$$

where F is the magnitude of the original force and θ is the angle to the horizontal, or:

$$F_h = F.\sin \alpha$$

$$F_v = F.\cos \alpha$$

where α is the angle to the vertical.

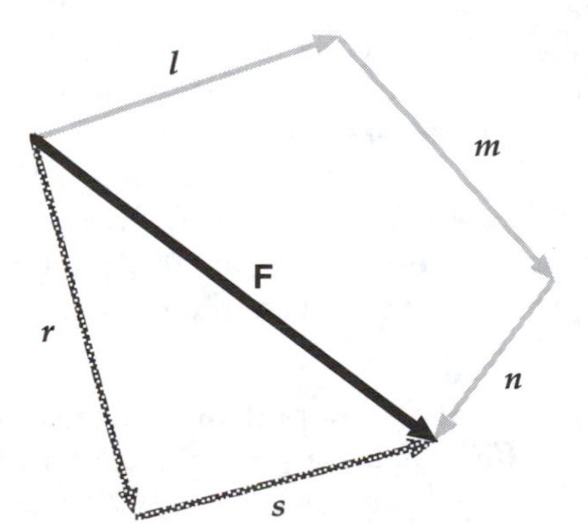

Figure 2.12 Examples of a three force system (*l,m,n*) and a two force system (*r,s*) which would have the same effect as the single force F.

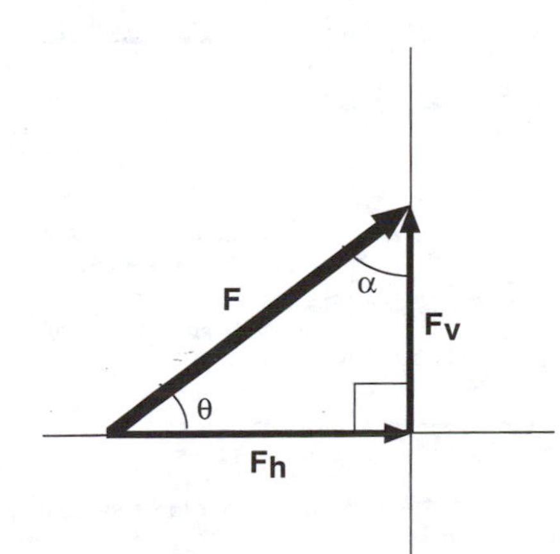

Figure 2.13 Determination of the vertical, F_v, and horizontal, F_h, components of the force F using the triangle of forces.

To determine the directions of the components, observe that the component vectors always point away from the origin or tail and towards the head of the original vector being resolved.

Resolution is used in the composition of multiple force systems and to aid in the visualisation of the effects of a force applied at an angle.

Composition of multiple force systems using force resolution

Consider, for example, the four force system in Figure 2.14. The first step is to resolve each force in turn into vertical and horizontal components. This results in four vertical components and four horizontal components. Then add all vertical components and all horizontal components, remembering to use the correct sign convention. This leaves a single resultant vertical R_v and a single resultant horizontal R_h force from which the final resultant can be found in the usual way.

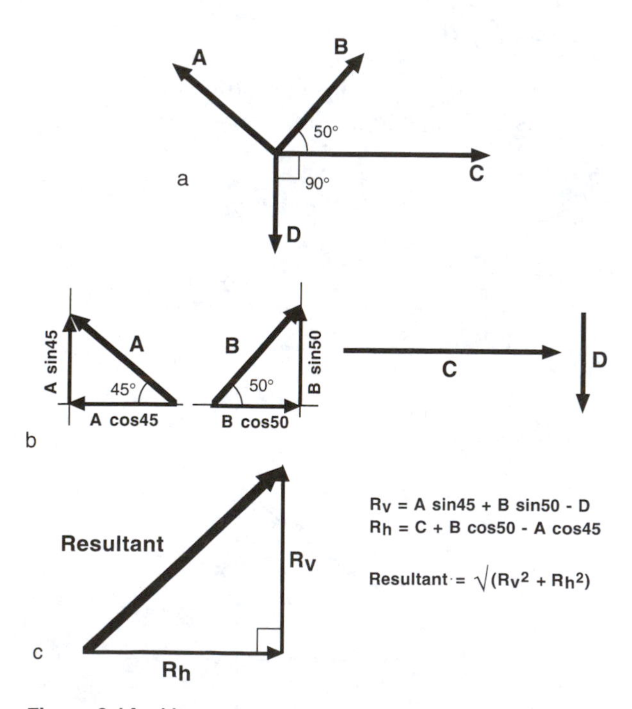

Figure 2.14 Vector resolution of a four force system.
a) Four force system with forces of magnitude A,B,C and D.
b) Resolution of each force into vertical and horizontal components. c) Resulting triangle of forces used to calculate the magnitude of the resultant.

Visualisation of force effects

One useful application of force resolution is in determining the effectiveness of a particular force in moving a bony lever or body segment. This will vary depending on the angle of pull of the muscle, which is defined as the angle between the line of pull (line of action) of the muscle and the mechanical axis of the bone or segment involved.

Figure 2.15 illustrates the effect of the biceps brachii on the forearm at three positions of elbow flexion. In *a* the elbow is flexed to approximately 90° at which point the line of action of the muscle force, F_m, is perpendicular to the segment axis. It will be seen later, in the discussions of levers and moments of force, that the turning effect of a force arises solely from that component of a force which is perpendicular to the long axis of the lever. Therefore, in this position all the muscle force is tending to cause rotation.

In contrast, in *b* and *c*, although the resultant muscle force is the same, it is no longer perpendicular to the forearm. To observe its effects, the muscle force must be resolved into a component perpendicular to the segment axis, F_p, and a component acting along the axis of the segment, F_a. The component, F_p, is called the rotary component or turning effect. It causes the forearm to rotate about the elbow.

The component, F_a, is called the non-rotary component or the axial effect. F_a may have a stabilising effect when the angle of pull is less than 90° (Fig. 2.15b) and a distracting (traction) effect at angles of pull greater than 90° (Fig. 2.15c).

When joint motion takes place, the magnitudes of the two force components F_p and F_a change, as the angle of pull changes. However, it should be noted that both F_p and F_a will always be less than the muscle force F_m except at 90° flexion when F_p will equal F_m. This leads to the following general observations:

- As the angle of pull approaches 90°, the turning effect of a muscle increases in magnitude.
- When the muscle force is applied to the bony lever at a right angle, the effect of the muscle force is entirely rotary. This is the position of

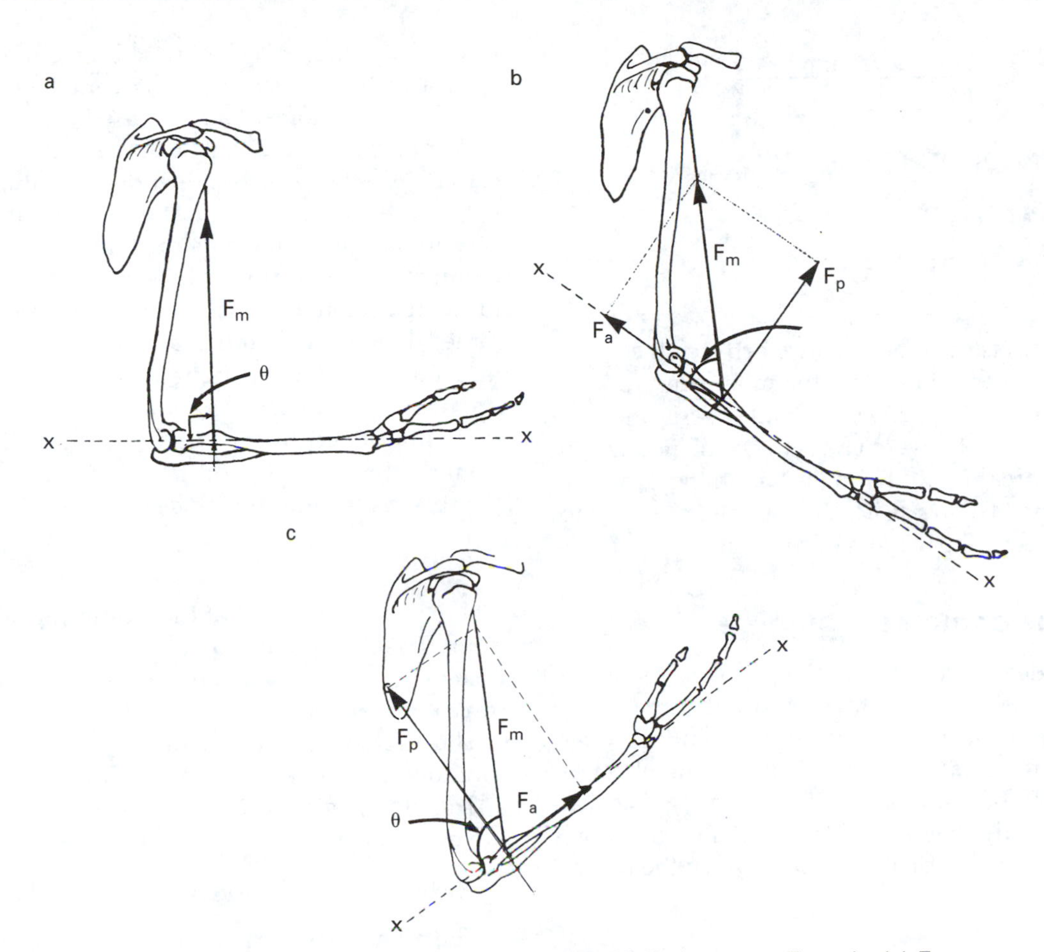

Figure 2.15 Resolution of total muscle force of biceps brachii, F_m, into rotary, F_p, and axial, F_a, components. Adapted from Gowitzke B A, Milner M 1988 Scientific bases of human movement, 3rd edn. Williams and Wilkins, Baltimore.

greatest mechanical efficiency for producing rotation.

- As full extension or flexion is approached and the angles of pull become small the mechanical efficiency of the muscle is low because of the large axial component F_a.

LEVERS AND THE MOMENT OF FORCE

Just as a force is needed to generate motion in a straight line, a force is required to produce rotation or angular motion. This is often achieved by the use of a lever and the locomotor system of the human body may be considered as comprising a series of linked levers. By definition, a *lever* is a rigid bar, or mass, which rotates around a fixed point called the axis of rotation or *fulcrum*. Rotation is then produced by a force applied to the lever at some distance, called the lever arm, from the fulcrum (Fig. 2.16). For example, the radius and ulna in the forearm can be viewed as a lever with the elbow the fulcrum.

A force which is applied to a lever to overcome resistance is called the *effort* and the distance between the point on the lever where the effort is applied and the fulcrum is called the *effort* arm. In the forearm, during elbow flexion against resistance, the effort is the force exerted by the muscle; for example, the biceps brachii (Fig. 2.17). The

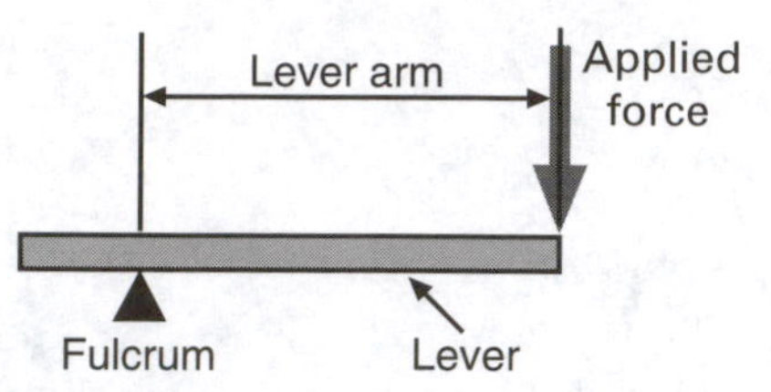

Figure 2.16 A simple lever.

effort arm is then the distance between the axis of rotation of the elbow and the muscle insertion. In this example (Fig. 2.17), the load or resistance that the effort has to overcome is the combined weight of the forearm and hand and anything carried in the hand. The *resistance* arm is then the distance from the load to the axis of rotation.

Torque or moment of force

When dealing with levers, and rotary motion in general, it is important to know the amount of rotation or turning about the fulcrum that a particular force can produce. This turning effect, or tendency of the force to cause turning, is called *torque* or the *moment of force*. Figure 2.17 represents a parallel force system in which the muscle

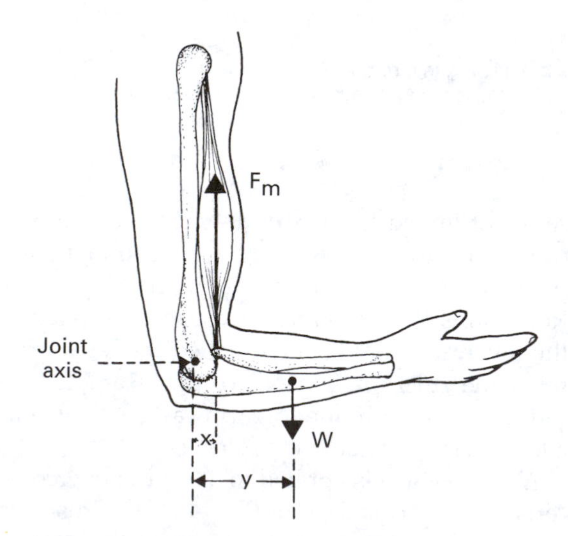

Figure 2.17 A parallel force system acting on the forearm. Where W is the weight of the forearm; F_m is the force of the biceps; x is the perpendicular distance from the line of action of F_m to the axis of rotation of the elbow – the effort arm; and y is the perpendicular distance from the line of action of W to the axis of rotation of the elbow – the resistance arm.

force, F_m, produces a clockwise moment and the weight of the forearm and hand combined, *W*, an anticlockwise moment about the elbow.

From experience, it can be appreciated that the turning effect of a force depends not only on the magnitude of the force but the distance at which it is applied from the axis of rotation, that is its moment arm. This principle is observed if we try to open a door by pushing close to the hinge rather than at the handle. Defined precisely, the magnitude of the moment of force about any point is equal to the magnitude of the applied force multiplied by the perpendicular distance from the line of action of the force to that point. Expressed algebraically, this is:

$$T = F.d$$

where *T* is the moment of force or torque about a point, *F* is the applied force and *d* is the perpendicular distance from the line of action of the force to the point.

The unit of measurement of the moment of force and of torque is the newton-meter (N.m).

Therefore, in Figure 2.17, because the forces are perpendicular to the forearm:

The moment of the force $F_m = F_m.x$

The moment of the force $W = -W._y$

Note: the moment of *W* is negative because it is clockwise. Moments and torques are vector quantities, they are either clockwise or anticlockwise. By convention, just as forces acting vertically downwards and to the left are usually considered negative, clockwise moments are also considered negative.

When considering human movement, the forces acting on body segments often act at angles other than 90°. In these cases the moment arms of the effort and resistance can now no longer be measured directly along the length of the segment. It is now necessary to consider the perpendicular distance from the line of action of the force more carefully.

Examine, for example, Figure 2.18. Here the moment arm of the muscle force about the elbow is the perpendicular distance, *x*, from the joint axis to the line of action of muscle force, F_m,

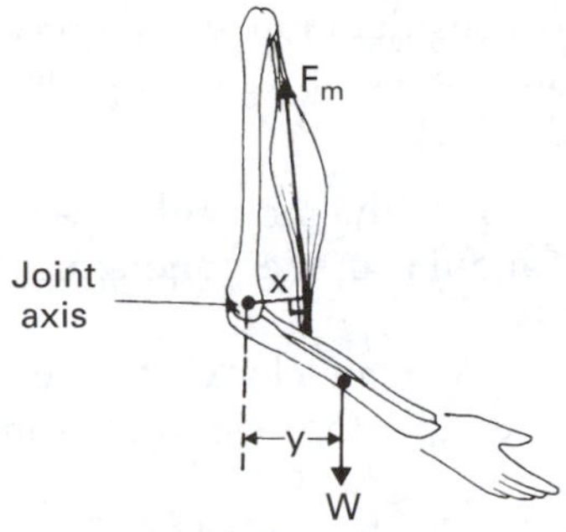

Figure 2.18 A parallel force system acting on the forearm when the angle of pull of the biceps brachii is less than 90°. Note $F_m.x$ is the moment of force produced by the muscle about the joint axis and W.y is the moment of force produced by the weight of the forearm about the joint axis.

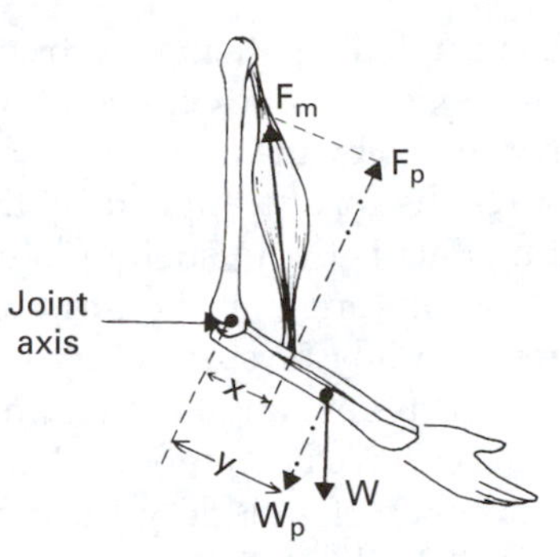

Figure 2.19 A parallel force system acting on the forearm when the angle of pull of the biceps brachii is less than 90°, where F_p is the rotary component of F_m and W_p is the rotary component of W. Note $F_p.x$ is the moment of force produced by the muscle about the joint axis and $W_p.y$ is the moment of force produced by the weight of the forearm about the joint axis.

extended from the point of muscle insertion; and the moment arm of the forearms weight about the elbow is the perpendicular distance, y, from the joint axis to the line of action of the weight, W, acting from the centre of gravity of the segment. The distances, x, and, y, would have to be found either by drawing the lever to scale or by using basic trigonometry and it should be noted that although W will always be vertical, F_m may not.

Analysing the moments in this way, it can be seen that the lengths x and y, when the forces are not perpendicular to the lever, will be reduced, hence the moments generated by the forces will also be reduced. Therefore, the moments of F_m and W in Figure 2.18 will be smaller than those in Figure 2.17.

Figure 2.19 illustrates an alternative method of viewing the problem introduced in Figure 2.18. Here the distances x and y are the same as those in Figure 2.17. This is because it is now the components of the forces F_m and W perpendicular to the segment which are used to calculated the moments of force. As mentioned earlier, in the discussion of the resolution of forces (Fig. 2.15), the component of a force perpendicular to a lever is known as the rotary component of that force. Another way of defining torque, then, is that it is the product of the force component perpendicular to the lever arm (its rotary component) and the distance from the line of action of this force to the fulcrum. Or, expressed algebraically:

$$T = F_p.d$$

where T = torque or moment of force about a point, F_p = perpendicular (rotary) component of force, and d = perpendicular distance from the line of action, F_p, to the fulcrum.

Torque measurement

The magnitude of the rotary component of muscle group contraction can be found for a variety of joint angles in a number of ways. An estimate of isometric strength may be made by the physiotherapist holding the limb in the desired position whilst the patient pushes or pulls against the resistance generated by the therapist. More accurate measurements may be obtained using a force transducer or hand-held dynamometer. This is a device which measures the force applied by the therapist to resist the patient's muscle contraction. Although these methods still only give a static measure of strength, they are widely used and provide useful information in both clinical and research applications. See, for example, Puharic & Bohannon (1993). However, more detailed dynamic measures can be obtained using an isokinetic dynamometer.

An isokinetic dynamometer is a device which allows joint motion only at a fixed and preset angular velocity (note: the accuracy and precision at which the velocity can be fixed is a limiting factor in the use of such devices). With the position of the joint axis fixed, the patient or subject attempts to maximally extend or flex the joint against the resistance provided by the dynamometer which in turn measures the torque applied by the patient. The resulting data is then usually plotted against either time or joint angle in the form of a torque curve. Often a family of curves are obtained for contractions at different angular velocities.

Torque curves have been reported in the literature for a number of muscle groups at various joints. Figure 2.20 illustrates a typical curve for the hamstrings and quadriceps at the knee.

Data of this type are used clinically, for example to monitor rehabilitation progress. Isokinetic dynamometers are often used as training apparatus both in injury rehabilitation and sports performance training. The data are also used in clinical research, particularly in the study of sports injury. For example, the ratio of peak dynamic hamstring torque to peak dynamic quadriceps torque may be related to the incidence of hamstring injury in certain sports (Read & Bellamy 1990).

Rotary equilibrium

For a lever or any other system of forces to be in equilibrium the sum of all clockwise moments must equal the sum of all anticlockwise moments. That is, the vector sum of all moments,

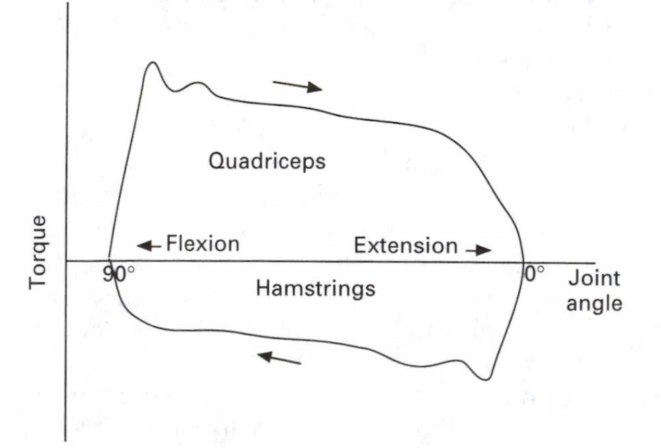

Figure 2.20 A typical torque versus joint angle curve for the quadriceps and hamstring muscle groups obtained using an isokinetic dynamometer.

taking into consideration the correct sign convention, must be equal to zero. Relating this to Figure 2.17:

- if $F_m.x = W.y$, the moments are balanced; so there is no angular motion; therefore the system is in equilibrium
- if $F_m.x < W.x$, the clockwise moment is the larger and therefore the lever rotates in the clockwise direction
- if $F_m.x > W.x$, the anticlockwise moment is the larger and therefore the lever rotates in the anticlockwise direction.

Therefore, with all classes of lever, equilibrium is achieved if the following equation is satisfied:

$$\text{effort} \times \text{effort arm} = \text{resistance} \times \text{resistance arm}$$

For example, if in Figure 2.17 the weight of the forearm W = 13 N, the position of the centre of gravity y = 180 mm from the fulcrum and the muscle insertion x = 30 mm from the fulcrum, the magnitude of the muscle force required to achieve equilibrium is found in the following way:

$$F_m \times 30 = 13 \times 180 \quad \therefore F_m = \frac{13 \times 180}{30} = 78\text{ N}$$

When gravity is the only resistance force acting on the body mass, as above, only three elements of the equation can readily be altered:

- The effort: the magnitude of the muscle force can clearly be changed at will, but as the weight of the limb is constant, any force other than that required to generate equilibrium will cause motion about the joint.
- The effort arm: this changes as the joint position alters.
- The resistance arm: the position of the centre of gravity of the body part being moved may be altered by changing the position of the segments in space, e.g. the centre of gravity of the leg is brought closer to the hip if the knee is flexed.

However, the resistance or weight of the body segments involved cannot be significantly altered in this case.

Additional forces may be added to the classical free exercise situation to resist or assist the muscle contraction or effort force. This may be done by using exercise weights, or by manually assisting or resisting the limb's motion as, for example, in Figure 2.21a and b.

The effectiveness of these forces will depend on:

- The distance of their point of application from the fulcrum, that is, their moment arm. The longer the moment arm of these forces the greater will be their effect.
- The angle at which they are applied. A resistance applied perpendicular to the segment will have a greater effect than one applied at an angle.

Mechanical advantage

An important concept in the study of levers and leverage is that of *mechanical advantage*. The mechanical advantage of a lever is simply the ratio of resistance to effort. If the effort required to move a lever is less than the resistance, the lever is said to have a mechanical advantage of greater than one. Whereas if the effort is greater than the resistance, the mechanical advantage is less than one. If the mechanical advantage equals one, the effort and resistance are the same.

Classification of levers

In mechanics levers are often classified in terms of the relative positions of load, effort and fulcrum. There are three basic classifications; first order, second order and third order.

First order levers are levers in which the fulcrum is situated between the effort and the load as illustrated in Figure 2.22.

These levers may have a mechanical advantage greater or less than one. If the mechanical advantage is less than one, the effort arm is shorter than the resistance arm (Fig. 2.22b). This lever is then said to be designed to gain speed. This is because, although the effort will be greater than the resistance and the effort and resistance move through the same angle, the resistance has to cover a greater linear distance and therefore has a greater linear velocity. If the mechanical advantage is greater than one, the effort arm is longer than the resistance arm (Fig. 2.22a). This lever is said to be designed for force because the effort force will be less than the resistance; however, the resistance will have a lower velocity than the effort and therefore the effort will have to move through a greater distance than the resistance.

Anatomical examples of first order levers include:

- The action of the muscles of the neck against

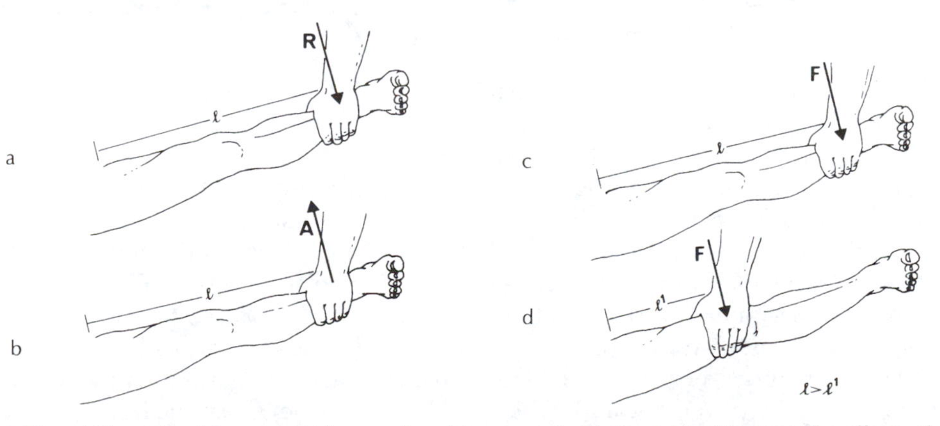

Figure 2.21 a) Manual resistance and b) manual assistance applied to the right adductors. The effects of a constant force F will vary depending on the distance it is applied from the hip joint. Therefore exercise c is much harder than d for the hip adductors. Figures a) and b) are after and Figures c) and d) are from LeVeau B 1992 Williams and Lissner: Biomechanics of Human Motion. W B Saunders, Philadelphia.

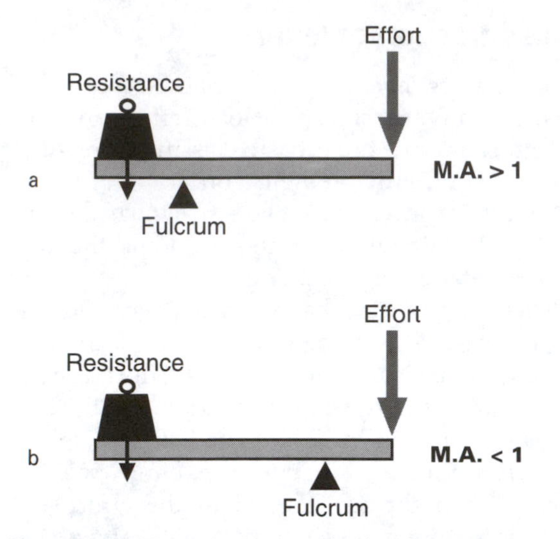

Figure 2.22 First order levers with: a) mechanical advantage greater than one, and b) mechanical advantage less than one.

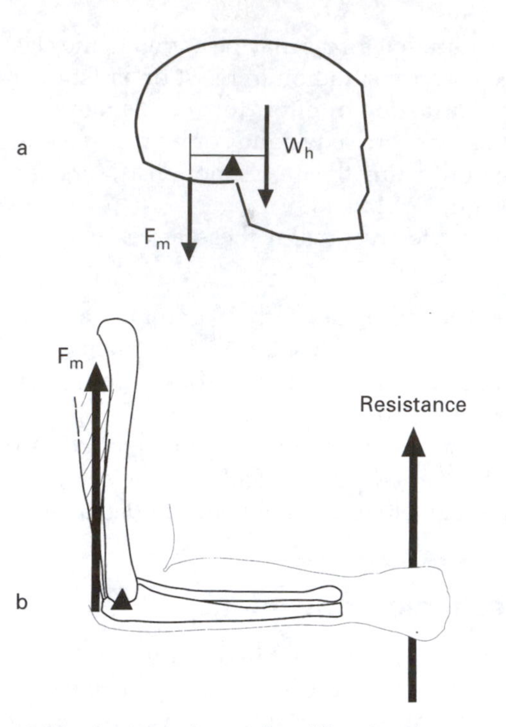

Figure 2.23 Anatomical examples of first class levers. a) The atlanto-occipital joint is the fulcrum for the weight of the head, W_h, and the muscles of the neck, F_m. b) The triceps, F_m, extending the elbow joint against resistance.

the head in the upright position, with the atlanto-occipital joint acting as the fulcrum (Fig. 2.23a).

- The action of the triceps muscle in extending the forearm against resistance as in Figure 2.23b.

Second order levers are levers where the resistance always lies between the fulcrum and the effort (Fig. 2.24). As the mechanical advantage of these levers is always greater than one, these are force levers. That is, the effort applied will always be less than the resistance to be overcome.

In the past, in an attempt to find anatomical examples of this type of lever, some authors have erroneously classified some joints as being second order. For example, it has been suggested that rising on to the ball of the foot in a weight-bearing position constitutes a second order lever. The metatarsal heads are said to be acting as the fulcrum, with the muscles acting on the heel being the effort and the weight of the body passing through the ankle the resistance. However, as argued by Gowitzke & Milner (1988), this situation is more accurately classified as a third order lever (Fig. 2.25). In reality, most joints in the human body may only be properly classified as second order when they are doing negative (eccentric) work. Here the weight of the limb, and any load applied to the limb are the effort (the force producing motion) and the muscle force is the resistance.

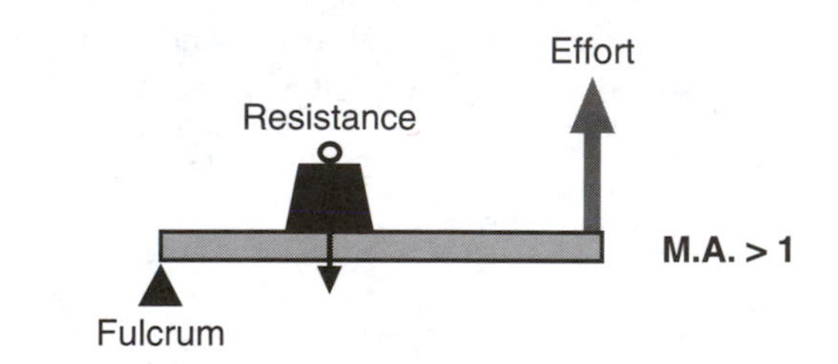

Figure 2.24 A second order lever.

Third order levers are levers in which the effort is always placed between the resistance and the fulcrum (Fig. 2.26). As the mechanical advantage of these levers is always less than one, they are used for achieving speed at the expense of force. Many of the levers in the human body are of this type. The action of the biceps brachii muscle in flexing the elbow is an example of a third order lever (Fig. 2.17).

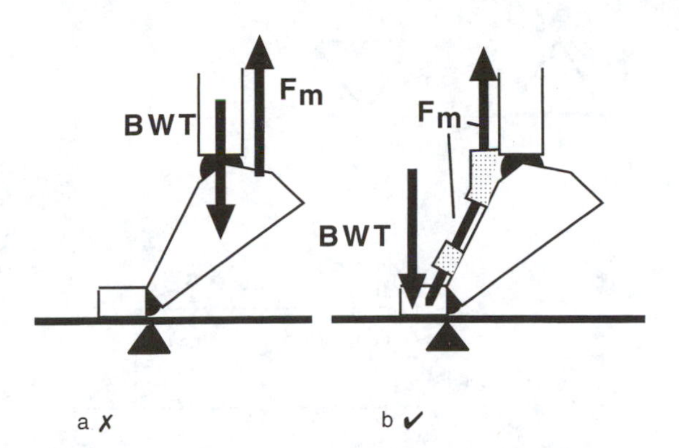

Figure 2.25 a) An erroneous example of a second order lever in which the effort is considered to be the muscle force transmitted through the tendo-achilles. b) The example correctly defined as a third order lever in which the effort, F_m, is the muscle force produced by the toe extensors and both F_m and BWT — the resistance — act anteriorly to the fulcrum. BWT = body weight.

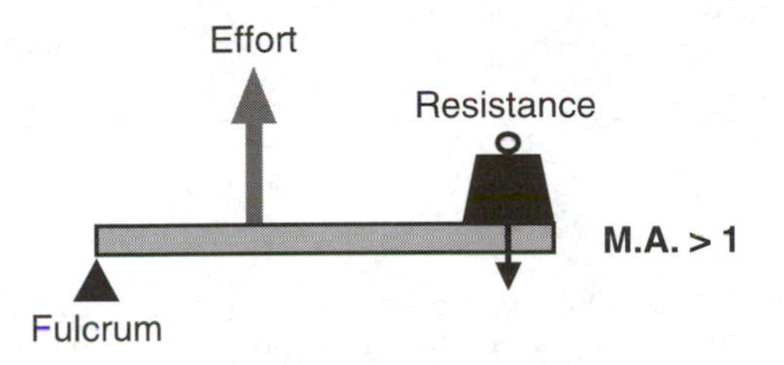

Figure 2.26 A third order lever.

Pulleys

Levers, as illustrated above, can be used to overcome a large resistance by a small effort — if the mechanical advantage of the lever is appropriate — but they may also be used to change the direction of an applied force (consider the relative directions of effort and resistance in second and third order levers). Another simple mechanical device which allows these actions is the pulley.

A pulley consists of a grooved wheel with a rope running over it. Pulley systems may comprise one or any number of individual pulleys (Fig. 2.27). In general, the ratio of load to effort is equal to the number of pulleys used in the system.

To simply change the direction of a force only one pulley is required. Although there are no true anatomical examples of pulleys, as described above, the peroneus longus can be seen to act in a very similar way (Fig. 2.28).

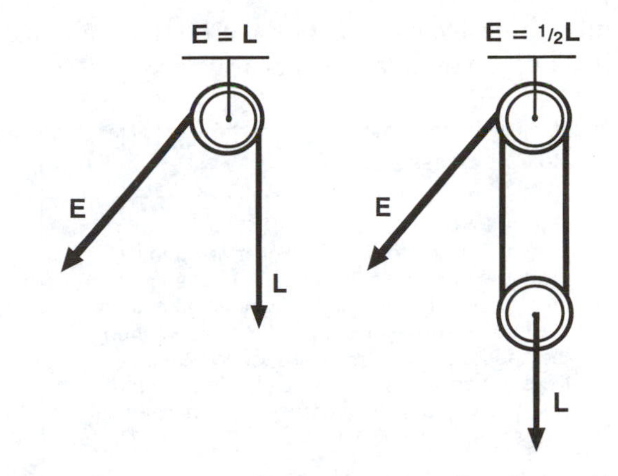

Figure 2.27 Examples of pulley systems. E = effort and L = load or resistance. The relationships between E and L given assume the pulleys to be frictionless.

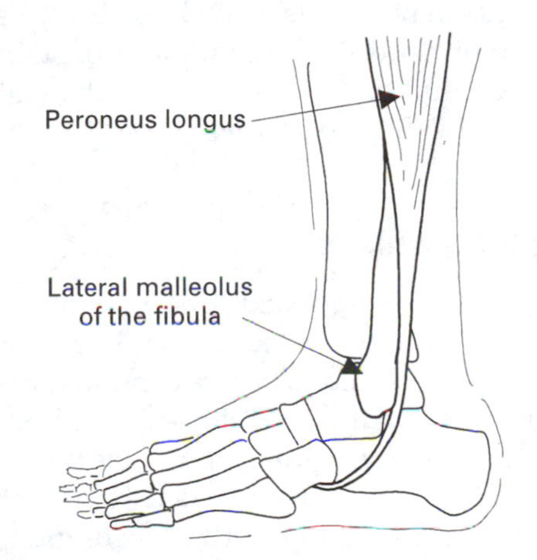

Figure 2.28 The lateral malleolus of the fibula acting as a pulley to change the line of action of the applied force of the peroneus longus.

Physiotherapists may more usually encounter pulleys in lifting or exercising equipment or in traction sets.

EQUILIBRIUM AND STABILITY

A body is said to be balanced and in *equilibrium* when the resultant of all the forces and moments acting upon it are equal to zero. If a body is in equilibrium, its state of motion remains unchanged. A body may be in a state of static or dynamic equilibrium. In *static equilibrium*, the body is stationary, whilst in *dynamic equilibrium* the body is moving with constant angular and linear velocity.

Box 2.2 Task 2

Use of levers
One of the simplest ways to alter an exercise to progress or regress its effects is to change the lever arm of the limb. If you want to strengthen the flexors and extensors of the shoulder then the sequence of progression would be elbow flexed, then extended then a club in the hand. Can you explain this progression in mechanical terms?

When a rigid body in static equilibrium is acted upon by an additional force, its state of balance will be indicated by its subsequent behaviour and may be described as being stable, unstable or neutral. The conditions of stable and unstable equilibrium are illustrated in Figures 2.29a and 2.29b, respectively.

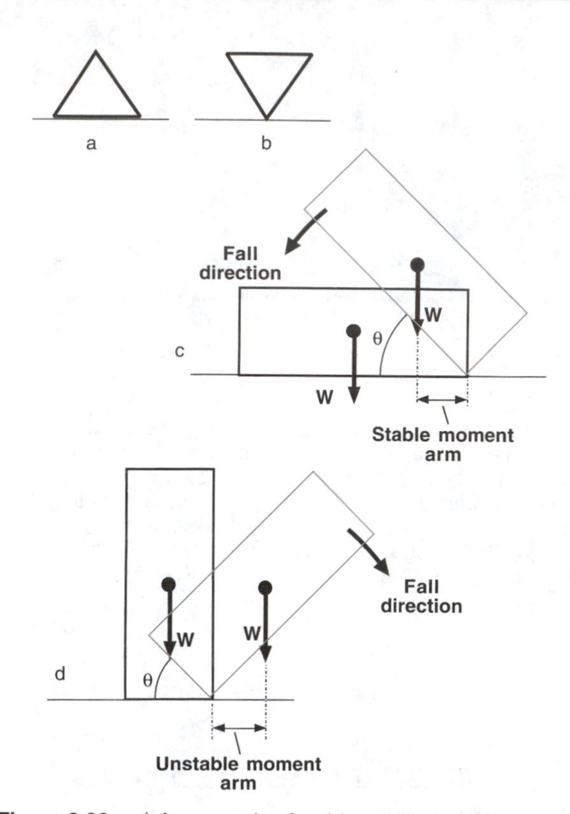

Figure 2.29 a) An example of stable equilibrium. b) An example of unstable equilibrium. c) This object is said to be stable because when it is displaced through an angle, θ, to the position indicated by the shaded line, its weight, W, produces a stabilising moment about the supporting edge. d) This object is said to be unstable because when it is displaced through an angle, θ, to the position indicated by the shaded line, its weight, W, produces a destabilising moment about the supporting edge. Notice that although the block has been moved through the same angle, θ, in conditions c) and d), the consequences, as represented by the fall direction, are diffferent.

Stable equilibrium

If, after a force is applied to a body at rest, the body tends to return to its original starting position, the body is said to be in a condition of stable equilibrium. Figure 2.29c illustrates a body in a condition of stable equilibrium. If the body is displaced to the position indicated by the shaded outline, it will tend to return to its original position. This is because the weight of the body acting from its centre of gravity creates a stabilising moment about the supporting edge. However, if the object were to be displaced so far that its centre of gravity passed over the supporting edge, the moment created by the object's weight would tend to prevent the object returning to its original position. The further the object's centre of gravity has to be displaced to reach this critical point, the more stable the object can be considered.

In conditions of stable equilibrium, displacement of the object requires the centre of gravity to be raised. The further the centre of gravity needs to be raised before it passes over the supporting edge the greater the degree of stability. The stability of an object can, therefore, be increased by lowering its centre of gravity and increasing the size of its area or base of support. Figure 2.30 illustrates how the effective base of support is affected by foot placement and the use of walking sticks in standing. In Figure 2.30a there is relatively low stability, whilst in Figure 2.30b the anterior-posterior stability is increased and in Figure 2.30c both medial-lateral and anterior-posterior stability has been increased.

Unstable equilibrium

When displaced a short distance by a force, a

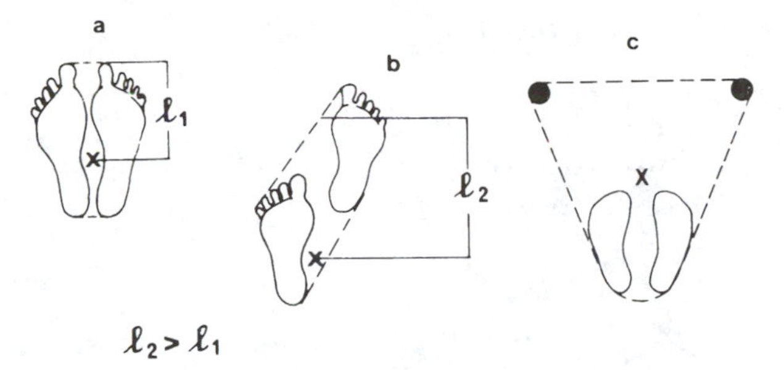

Figure 2.30 Base off support. X is the point of intersection of the line of gravity with the base, --- marks the boundaries of the base.

body is said to be in a condition of unstable equilibrium if it then tends to increase its displacement further under the influence of gravity alone. During this process the centre of gravity drops below its original position as the line of gravity falls outside the base of support. Once an object's centre of gravity moves to a point outside of its base of support, the object becomes unstable, as can be seen in Figure 2.29d.

It should be noted that although Figure 2.30 suggests that the human body is in a condition of stable equilibrium in upright stance, this is only true if the ankle and all other supporting joints are considered to be locked. In practice it is perhaps more accurate to consider the upright body to be in a condition of unstable equilibrium in which continuous adjustments have to be made to prevent the body from falling over. This can be seen by observing the slight sway of the body when somebody is asked to stand still.

Neutral equilibrium

A body is in neutral equilibrium if, when it is displaced, it remains at rest in its new position and, if displaced along a level surface, there has been no change in the vertical position of its centre of gravity. A ball having been rolled along a level surface would fulfil these conditions but, strictly speaking, there are probably no true body position examples of neutral equilibrium.

Summary of conditions for stability

The stability of a rigid body depends on:

- the area of the base of support
- the height of its centre of gravity above the base
- the position of the line of gravity relative to the base of support
- the weight of the body.

The state of balance of the human body is improved if it is made more stable. It becomes more stable when:

- the centre of gravity of the total weight supported over the base is lowered. The total weight will include any additional weights being carried
- the total weight of the body and any weight being carried is increased
- the centre of gravity is directly above the centre of the base of support
- the area of the base of support is enlarged.

The shape of the base of support is broadened in the direction of the force being applied to the body. For example, when a force is applied from behind, extending the front edge of the base forward, as in Figure 2.30b, produces greater stability when compared with the shape of the base illustrated in Figure 2.30a.

Box 2.3 Task 3

Equilibrium and stability
If someone's balance is disturbed by a musculoskeletal problem then as part of their rehabilitation they may use walking aids. A progression may be made from frame to crutches then to sticks. How can you use the principles of equilibrium and stability to explain this progression?

MOTION

Motion occurs when an object undergoes a change in place or position. More precisely, motion is a process involving continuous change in place or position of a body with respect to some agreed frame of reference. In mechanics the general term used to describe the study of motion is dynamics. Dynamics, in turn, may be subdivided into two fields: kinematics and kinetics.

Kinematics is concerned with the description of motion in precise mathematical terms without reference to the forces producing the motion. *Kinetics* is concerned with the forces affecting motion of body so that it behaves in the particular way that it does.

Classification of motion

Although there is considerable variety in the way human beings move, all motion can be described as being translatory, rotary or a combination of the two. This latter form of motion is termed general plane motion.

Translatory motion is said to occur when the orientation of a moving body remains constant. In translation, all points on the body follow parallel paths. Translation may be linear (more correctly, rectilinear), in which case the body moves along a straight line, or curvilinear, in which case the body moves along a curved path. An example of simple linear translation would be the motion undergone by a wheelchair occupant being pushed along a level floor. An example of curvilinear motion would be the motion of the trunk when jumping from a step or wall. (Fig. 2.31)

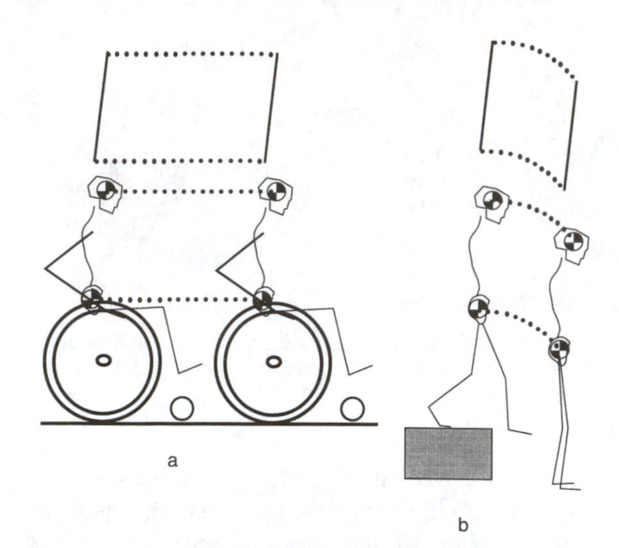

Figure 2.31 Examples of translatory motion. a) Rectilinear motion and b) curvilinear motion. •••• trajectories of points on the subject's head and pelvis. Notice the lines joining the hip and head remain parallel throughout the movement.

Rotary or angular motion occurs when all points on a rigid body describe independent circular paths about a common fixed axis. In pure rotation, all parts of the body move through the same angle in the same time. From this description it can be seen that most joint movements are rotary in nature, as can be seen in Figure 2.15.

General plane motion is the most common form of motion observed. However, we can often describe that motion in terms of the translatory and rotary components of the motion. For example, in walking, the whole body may be said to undergo translatory motion, as a result of the rotary motion occurring around the joints of the lower limbs. This is illustrated in Figure 2.32.

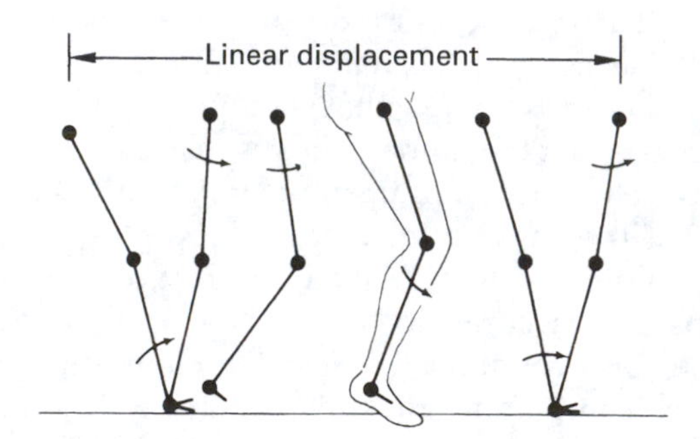

Figure 2.32 General plane motion: rotation of the limbs to obtain translation of the body. Reproduced with permission from LeVeau B 1977 Williams and Lissner: Biomechanics of Human Motion, 2nd edn. W B Saunders, Philadelphia.

Description of motion: kinematics

In the description of motion we are concerned with how far and how fast a body is moving. In everyday language the terms *distance* and *displacement* tend to be used synonymously to describe how far an object has travelled and the terms *speed* and *velocity* tend to be used synonymously to describe how fast an object is travelling. In mechanics, however, these terms have very specific meanings.

Linear kinematics

Displacement is a measure of how far a body is

moved from a starting point in a given direction. For example, when we say how far is one point on a map to another point on a map as the crow flies, we are really asking what is the displacement a body would undergo in moving between the two points. Displacement is, therefore, a vector quantity and for completeness should be specified with a direction. To manipulate measures of displacement mathematically, it is necessary to use the same rules which were earlier applied to the manipulation of forces.

Distance, in comparison to displacement, is simply a measure of how far an object has actually travelled in getting from one point to another. Distance conveys no sense of direction and is therefore a scalar quantity. However, the correct unit of both distance and displacement is the metre (m).

In Figure 2.33, in moving from point S to point F a person has walked along a path indicated by the dotted line. The distance that the person has walked is simply the length of that line, whilst the person's displacement is the length of the solid vector arrow — the shortest distance between the two points.

Walking from S to F in Figure 2.33 would obviously take some time. Whilst it is useful to be able to specify the time taken, it is often of greater value to combine the measured time with the measure of distance or displacement to describe how fast the object was moving. This leads to the two quantities: speed and velocity.

Speed is a measure of the distance covered by an object in a given time. It is, therefore, a scalar quantity and infers no sense of direction. The average speed of an object is calculated by dividing the distance travelled by the time taken.

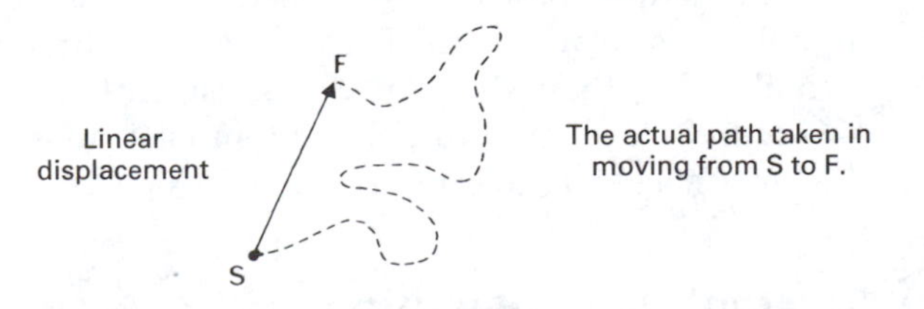

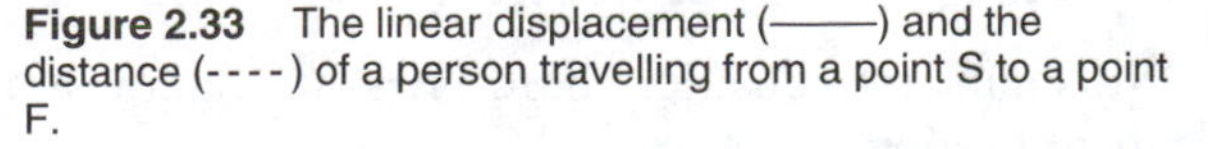

Figure 2.33 The linear displacement (——) and the distance (- - - -) of a person travelling from a point S to a point F.

Velocity is a measure of how fast an object is moving based on its displacement. As displacement is a vector quantity, then so is velocity and to specify a velocity correctly the appropriate direction must also be quoted. The average velocity of an object is the ratio of displacement to time taken.

This can be written algebraically as:

$$\text{Average velocity} = \frac{\text{Displacement}}{\text{Time taken}}$$

Both speed and velocity are expressed in the units of metres per second (m/s or ms^{-1}).

The above equation is an average velocity because it is calculated over a finite time period. However, as a body moves from one location to another, its instantaneous velocity may not remain constant. For example, an Olympic sprinter may cover 100 metres in 10 seconds, giving an average velocity of 10 m/s. However, on leaving the blocks, the sprinter's velocity would be much less than this, and in the middle of the race the sprinter's velocity would probably be in excess of this. Therefore, knowing the rate at which the velocity of an object is changing is also very useful. The quantity which describes the rate of change of velocity is termed **acceleration**.

Just as it is possible to define both instantaneous and average velocity, it is also possible to define an instantaneous and an average acceleration. Average acceleration, calculated between two points, is the ratio of the difference between the instantaneous velocities at the two points and the time taken to travel between them. Or:

$$\text{Average acceleration} = \frac{\text{Change in velocity}}{\text{Time taken}}$$

Then units of acceleration are the metre per second squared (m/s^2 or ms^{-2}).

Angular kinematics

The kinematics of rotary motion are important to the physiotherapist because so much human movement involves rotation of body segments around a joint axis. Although angular kinematics describes motion around the arcs of circles, the terms used are analogous to those used for linear kinematics. However, the units of measurement of angular motion are different to those of linear motion.

The most familiar unit of angular distance or displacement is the degree, a full circle being divided into 360 degrees (360°). In mechanics, a common and more useful alternative unit is the radian, abbreviated to rad. There are 2π radians in a full circle where $\pi = 3.142$, therefore 1 rad = $360/2\pi = 57° 17'$. Although these may seem odd units to use, they are useful because an angle of one radian generates an arc whose length is exactly equal to its radius. Measuring angular motion in radians, then, tells us something directly about the relationship between the angular and linear motion of an object.

Angular displacement and angular distance refer to angular changes in position during movement of a body about its axis of rotation. Angular displacement is a vector measure of that change whilst angular distance is a scalar measure of that change. For example, consider Figure 2.34. In this illustration a patient has been asked to abduct their arm at the shoulder from the vertical at A to a horizontal position at B. During this motion the arm has gone through an angular distance of 90° and a clockwise angular displacement also of 90°. If the patient was then asked to move their arm back through the vertical to the position of shoulder hyperadduction at C, the arm would have gone through an angular distance from A of 200° (90° clockwise + 110° anticlockwise), but an angular displacement from A of 20° anticlockwise. Note that if the arm had simply been returned to the vertical, the angular displacement would have been zero whilst the angular distance would have been 180°.

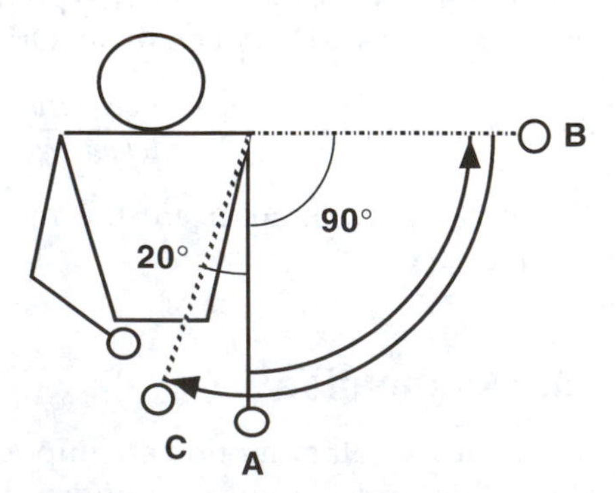

Figure 2.34 Measurement of angular distance and angular displacement. In moving from A to B the angular distance = angular displacement = 90°. In moving from A to C through B angular distance = 90° + 90° + 20° = 200° and angular displacement = 90° − 90° − 20° = −20°.

Angular velocity and angular speed are also analogous to their linear counterparts. Angular velocity is the rate of change of angular displacement and is, therefore, a vector quantity, whilst angular speed is the rate of change of angular distance and is, therefore, a scalar quantity. Both angular velocity and angular speed are measured in degrees per second (°/s or $°s^{-1}$), or radians per second (rads/s or $rad.s^{-1}$).

For example, if in Figure 2.34 the arm had been moved from position A to position C in 5 seconds, the average angular velocity would have been 4°/s whilst the average angular speed would have been 40°/s. It may be seen from this that the concept of average angular velocity has little practical value; however, the concept of instantaneous angular velocity, as with linear velocity, is extremely valuable. Angular acceleration is the rate of change of instantaneous angular velocity. It is expressed in degrees per second squared ($°/s^2$ or $°s^{-2}$), or in radians per second squared (rad/s^2 or $rad.s^{-2}$).

FORCE AND MOTION

All motion is the result of the application of one or more forces to a body. Whilst we may have an intuitive understanding of the relationship between force and motion, there are many situations in physiotherapy practice where a more fundamental understanding is essential. In classical mechanics, there are three basic laws which describe this relationship. These laws were first formulated by the 17th century scientist, Sir Isaac Newton. In order to understand motion it is therefore necessary to understand these laws.

Newton's laws of motion

We shall first consider these laws in relation to linear motion.

The law of inertia

Newton's first law states:

A body will remain at rest or continue to move with constant linear velocity unless it is acted upon by an unbalanced set of forces.

This means that to start, stop or change the motion of an object a force must be applied to it. This tendency of a body to continue in its present state of motion is called its *inertia* and is dependent upon the mass of the object. An object with a larger mass when compared to another has a larger inertia. In fact, when considering linear motion, mass is a direct measure of a body's inertia. The unit of inertia, therefore, is the kilogram.

The effects of this law are seen, for example, in whiplash injuries to the neck in motor vehicle accidents (Gowitzke & Milner 1988). Here, for example, a car is rapidly decelerated as a result of the forces at impact. The inertia of the driver's body is in turn overcome by the forces applied through the seat belt and shoulder harness. However, the driver's head tends to continue travelling forward as a result of its inertia until restrained by stretching of the posterior and compression of the anterior anatomical structures of the neck.

The law of momentum

Newton's second law states:

The rate of change in the quantity of motion possessed by a body is directly proportional to the force causing that change.

The quantity of motion possessed by a body is called momentum and is dependent not only upon the body's velocity, but also its inertia. Momentum is calculated by multiplying an object's velocity by its mass and it is a vector quantity with the derived units kilogram metre per second (kg.m/s or $kg.m.s^{-1}$).

Newton's second law, then, can be restated as: the rate of change of momentum of a body is directly proportional to the force causing that change. This can be written algebraically as:

$$F = \frac{m_2.v_2 - m_1.v_1}{t}$$

where F is the force, m_1 and v_1 are the object's initial mass and velocity and $m_1.v_1$ is therefore the object's initial momentum, m_2 and v_2 are the object's final mass and velocity and $m_2.v_2$ is, therefore, the object's final momentum and t is the time taken.

Very often, the mass of an object remains constant. In this case:

$$F = \frac{m.(v_2 - v_1)}{t}$$

where m is the constant mass of the object,

but $\frac{v_2 - v_1}{t}$ = the rate of change of velocity

= acceleration

therefore $F = m.a$

or *Force = mass × acceleration.*

Newton's second law, therefore, is also known as the law of acceleration and indicates that in order to accelerate (or decelerate) an object a force has to be applied to it, and if the mass remains constant, the acceleration produced will be directly proportional to the magnitude of the force. For example, doubling the applied force will double the acceleration. Note, however, that once the force is removed the object will stop accelerating but will continue to move at a constant velocity until another force is applied. Further, the force required to produce a given acceleration is proportional to the mass (inertia) of the object. For example, doubling the mass of an object will require a doubling of the applied force to produce the same acceleration.

The direction of the acceleration or momentum change produced by a force will always be in the direction of that force. Therefore, in order to decelerate an object the force must be applied in the opposite direction.

Impulse

The algebraic form of Newton's second law may be rearranged as follows:

$$F.t = m_2.v_2 - m_1.v_1$$

This is known as the impulse–momentum relationship, where:

Impulse = Force × Time = $F.t$

and $m_2.v_2 - m_1.v_1$ = *change in momentum.*

This impulse–momentum relationship is useful because it emphasises, among other things, the relationship between applied force and the time over which the force acts. We are often interested in producing a change in momentum rather than a specific acceleration. The impulse momentum relationship illustrates that this requires not only the application of a force, but also that the force must act for a finite time. In human activity, the muscle force available to move an object or the body may be limited; however, by contriving to apply that force for a longer period of time an effect similar to the application of a larger force over a short time can be obtained. This relationship also illustrates why the forces generated in impact, where initial contacts are very rapid, are large. In landing from a jump or a fall, a person attempts to reduce the impact forces generated as a result of the rapid change in the momentum by extending the period over which the impact forces are applied, for example, when landing feet first, by flexing their knees immediately after contact. For similar reasons, in walking and running, the impact forces at heel strike are reduced through the controlled plantar flexion of the ankle and slight flexion at the knee (Inman et al 1980).

Impulse, because it is based on force, is a vector quantity and the unit of measurement is the newton second (N.s).

The law of reaction

Newton's third law states:

For every action there is an equal and opposite reaction.

For a force to be generated, two or more masses must be in contact. Newton's third law, therefore, states that the force exerted by one mass on another produces an equal but opposite reaction force in the second object.

For example, Figure 2.35 illustrates three phases of foot to ground contact during a normal walking step. In each case the walker clearly exerts a force on the ground, in reaction to this, the ground exerts an equal but opposite reaction force on the walker's foot. It is this ground reaction force which acts to decelerate the motion of the walker's centre of gravity from heel strike to mid-stance and acts to accelerate the walker's centre of gravity during the push off phase from mid-stance to toe off.

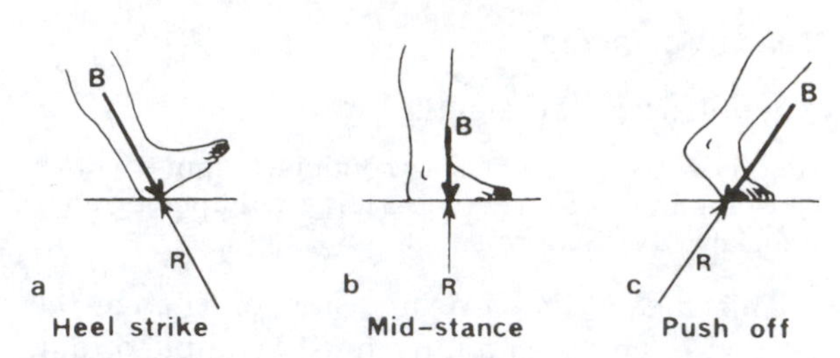

Figure 2.35 Resultant ground reaction force, R, during the stance phase of gait at, a) heel strike, b) mid-stance and c) push off. Notice that R is exactly opposite and equal in magnitude to the force applied to the ground by the walker, B.

Measurement of the ground reaction force (GRF) in walking is common in research and clinical biomechanics. This is usually done by using an instrumented device known as a *force platform*. This is a steel or alloy structure embedded in the ground which measures the components of GRF in the three cardinal planes; that is, vertical, anterior-posterior and medial-lateral reaction force components (Cunningham & Brown 1952, Whittle 1991).

Typical GRF curves obtained in normal, level ground walking are illustrated in Figure 2.36. The vertical reaction (F_z) is the force required to support the person's body weight and change their vertical momentum during the step. The anterior-posterior reaction force (F_y) is the force which changes the person's momentum in the general walking direction and the medial-lateral reaction force (F_x) produces the lateral changes in the person's momentum during the step. The horizontal components, F_y and F_x, can also be seen to be the friction forces preventing the foot from sliding along the ground.

Newton's laws and angular motion

Newton's three laws of motion can be applied equally to angular motion. Here, however, the laws relate torque to angular velocity, angular momentum and angular acceleration.

Newton's first law of angular motion then becomes:

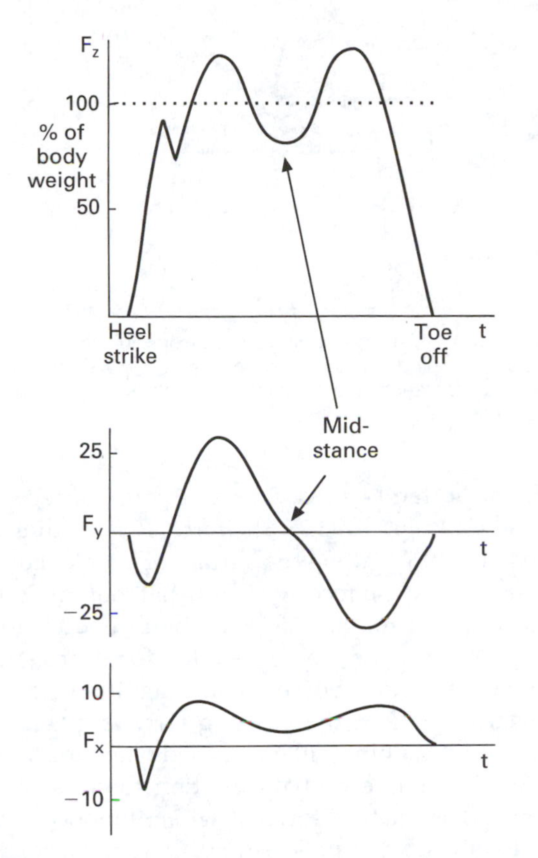

Figure 2.36 Typical ground reaction force components for normal level walking. F_z = vertical component, F_y = anterior-posterior component and F_x = medial-lateral component.

A body will remain at rest or continue to move with constant angular momentum unless it is acted upon by an unbalanced torque.

The tendency for an object to resist changes in its state of angular motion is termed its *moment of inertia*. Unlike an object's linear inertia, its moment of inertia is determined not only by its mass but also the position of that mass relative to the axis of rotation. Therefore, although a mass's linear inertia is constant, its moment of inertia will change if the position of the axis of rotation changes. The actual moment of inertia of a mass about a given axis is calculated from the following equation:

$$I = m.k^2$$

where I is the moment of inertia, m is the mass of the object and k is the radius of gyration.

The radius of gyration is the perpendicular distance from the axis of rotation to a point about which the mass of the object may be considered to act. For a very small object some distance from the axis of rotation, this point coincides with the object's centre of mass. For most real objects, however, this point and the centre of mass may not coincide.

It can be seen that as the mass of an object increases so does its moment of inertia, but more significantly, as the mass distribution moves away from the axis of rotation, the radius of gyration increases and the moment of inertia increases in proportion to the square of that increased distance — a two-fold increase in the radius of gyration will produce a four-fold increase in the moment of inertia.

In the human body, muscle bulk is concentrated towards the proximal end of many body segments so that most of it is close to the axis of joint rotation. This anatomical distribution of body mass helps to reduce the moment of inertia of the segment and, therefore, to reduce the resistance to changes in angular motion.

By changing body positions, the mass distribution around any axis can be altered, consequently changing the body's or limb's moment of inertia. In a sports context, this can be seen in gymnastics and diving when an athlete adopts a tuck position in order to more easily perform somersaults. Similarly, in running, flexing the knee during the swing phase brings the mass of the leg closer to the hip joint allowing for a more rapid swing through. The physiotherapist also utilises this principle in selecting starting position for exercises which make initiation of movement — overcoming inertia — more or less difficult.

The radii of gyration and moments of inertia of human body segments have been investigated by a number of researchers. The results of these studies are well documented, though basic data and useful reviews may be found in Gowitzke & Milner (1988) and Reid & Jensen (1990), respectively.

Moment of inertia is a scalar quantity measured in kilogram metres squared ($kg.m^2$).

Newton's second law applied to angular motion becomes:

The rate of change of angular momentum is directly proportional to the torque causing that change.

As for linear motion, this quantifies the relationship between angular momentum and torque. However, angular momentum is derived from the object's angular velocity and moment of inertia such that:

Angular momentum $= I.\omega$

where I is the moment of inertia and ω is the angular velocity.

Newton's first and second laws suggest that unless a net torque is applied to an object, the angular momentum of that object remains constant. If the linear momentum of an object is constant, it follows that its linear velocity will also be constant. However, because it is possible to alter the moment of inertia of a rotating body by moving its mass concentration closer to the axis of rotation, it is possible to alter its angular velocity without changing its angular momentum. This effect can be observed in the motion of a diver performing a somersault from a high board. Once the diver leaves the board and becomes airborne, the angular momentum about the somersault axis must remain constant until entry into the water. This is because no external torque is being exerted. Initially in an extended position, the diver's somersault velocity is low, but if the diver adopts a tuck position — thereby reducing the moment of inertia about the somersault axis — the angular velocity will increase in order to maintain a constant angular momentum. The diver's angular velocity will again decrease on returning to an extended position.

Newton's third law applied to angular motion simply states:

For every torque applied by one body on a second body the second body applies a torque on the first which is equal in magnitude but opposite in direction.

This implies that a muscle producing a torque across a joint on the distal segment will produce an equal but opposite torque on the proximal segment.

Friction

Friction is the force which arises when one object slides or tends to slide across the surface of another. In Figure 2.37, the force, P, is tending to push the block along the surface of a table. Force, F_S, is the friction force resisting that motion. If the block is not moving, it must be in equilibrium and the friction force — the force resisting motion — must be equal in magnitude to the pushing force. If the pushing force is gradually increased, a point will be reached when the block begins to move and the friction force is overcome. The friction force at which this occurs is called the *limit of static friction*. Once moving, the friction force remains constant but the force required to keep the object moving is slightly less than the force required to initiate the motion in the first place. The frictional resistance force which has to be overcome to maintain motion is called the *limit of dynamic friction*.

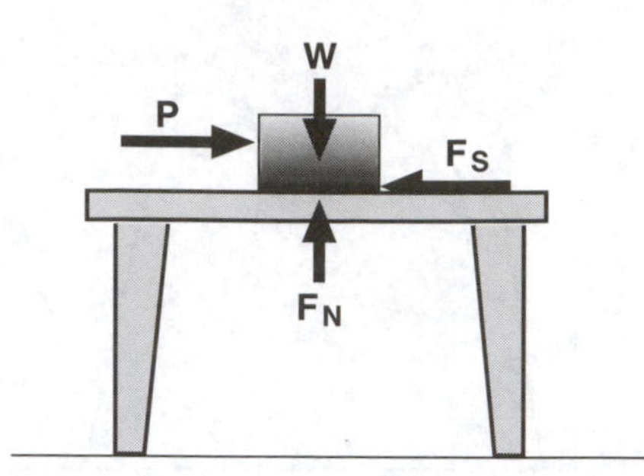

Figure 2.37 The forces on a block being slid across a table. P = sliding force, F_S = friction force, W = weight and F_N = vertical reaction force between the block and the table. Note: $W = -F_N$.

The magnitudes of both the limit of dynamic and the limit of static friction are dependent upon the same basic factors. These are: the surface roughness of the two objects, the materials from which the two objects are made and the contact force pressing the two objects together. However, contrary to what might initially be thought, limiting friction forces are not affected by the surface area of contact. In general, friction is increased if the surface roughness is increased, the contact force is increased and the materials of the two objects are chemically similar. Friction is reduced if the surface roughness is reduced, the contact force is reduced and the materials of the two objects are chemically dissimilar.

The ratio of friction force to contact force is

known as the coefficient of friction and is represented by the symbol μ. There are in fact two important coefficients of friction; the coefficient of static friction μ_s and the coefficient of dynamic friction μ_d. Both coefficients of friction are determined by the material properties and the surface roughness of contact, but for any pair of contact materials they may be considered constant.

For any two materials μ_d will be slightly less than μ_s but both will usually be between 0 and 1, with $\mu = 0$ indicating frictionless conditions. For example, the coefficient of static friction of a rubber crutch tip on a clean tile is 0.30 to 0.40; whereas the coefficient of static friction of a rubber crutch tip on unpolished wood is 0.70 to 0.75. The crutch is therefore less likely to slip on the wooden floor than on a tiled floor.

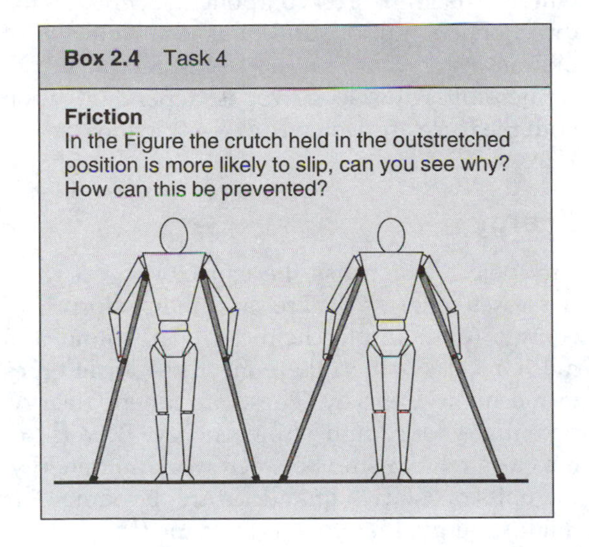

Box 2.4 Task 4

Friction
In the Figure the crutch held in the outstretched position is more likely to slip, can you see why? How can this be prevented?

Pressure

Where two bodies are in contact they exert a force on each other. The distribution of that force over the area of contact is known as *pressure*. Pressure is defined as the total force applied per unit area of force application and may be expressed by the equation:

$$P = \frac{F}{A}$$

where P is the pressure, F is the applied force and A is the area of contact.

The units of pressure are the newton per metre squared (N/m^2 or N.m^{-2}) or the pascal (Pa), where 1 Pa = 1 N/m^2.

By enlarging the area of contact the same force, for example body weight, will result in a lower contact pressure. In standing, the area of contact is much less than in lying (Fig. 2.38), hence the pressure exerted by the body weight in standing is much greater than in lying. Typically, pressure beneath the foot in standing and walking are in the ranges 80–100 kPa and 200–500 kPa respectively (Whittle 1991). Note that the area of contact is not the same as the base of support, nor is the pressure necessarily evenly distributed over the contact area. It is the increase in the area of support and the more even distribution of the pressure over the contact area which improves comfort in a softer bed. This has practical significance in the care of the bedridden where the reduction of pressure, using yielding materials which conform to body contours, is necessary in the prevention of pressure sores. Similarly, it is useful in the design of seating and wheelchair cushions.

WORK, POWER, ENERGY

A number of terms used in mechanics have more specific or subtly different meanings than the same terms have in common usage. Amongst these are the terms *work*, *power* and *energy*. These need to be defined to avoid error or ambiguity.

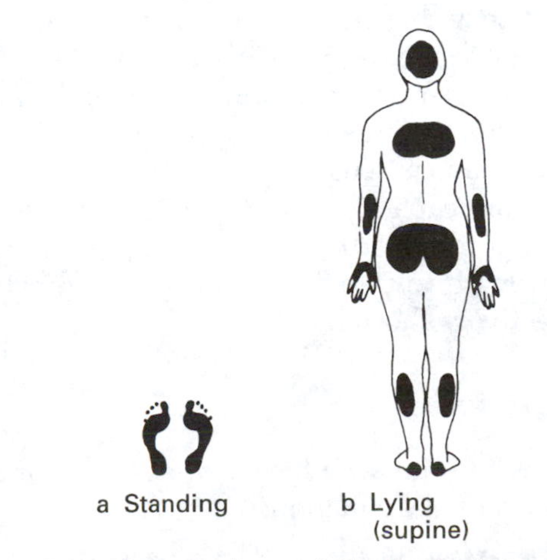

Figure 2.38 Area of pressure distribution in a) standing and b) supine lying (area of pressure shaded in black).

Work

When a force, for example that arising from a muscle contraction, is used to move a load through a distance, mechanical work is said to be done. Work is expressed by the equation:

$$W = F.d$$

where W is the symbol of work, F is the force component acting in the direction of motion and d is the distance through which the load is moved.

The units of work are the joule (J).

Work is a vector quantity. Therefore, the direction in which work is performed is important and will affect the sign given to the magnitude of the work. For example, in work done against gravity, the motion is vertically upwards. This is termed *positive work*. Conversely, work in which the motion is vertically downwards is *negative work*. Similarly, work done by a muscle may be positive or negative.

Positive work is done by a muscle when it actively shortens to move or lift a load against gravity. Concentric muscle work is, therefore, positive.

However, when muscles are actively lengthening, work is done on them by an external force, for example gravity, to produce movement. This is called negative work. Eccentric muscle work is, therefore, negative.

It is important to note that if movement does not occur at all under muscle tension, as would be the case with isometric tension, no mechanical work is done (i.e. $W = 0\ J$). That is not to say that there is no energy expenditure and that the muscle does not fatigue, but this is not work in the strict mechanical sense.

The work capacity of a muscle is related to the amount of force, or tension, it can generate, that is its strength, and to the distance through which it can actively shorten.

Power

Power is the rate at which work is done and is expressed by the equation:

$$Power = \frac{Work}{Time} = \frac{F.d}{t}$$

where F is the force component acting in the direction of motion, d is the distance through which the load is moved and t is the time taken.

The unit of power is the watt (W).

Muscle power is the rate at which work can be done by a muscle or muscle group. It is a common mistake to confuse this with muscle strength. Muscle strength is simply the maximum force that the muscle can actively produce. Therefore:

$$Muscle\ power = \frac{Work\ done\ by\ the\ muscle}{Time\ taken}$$

It can also be shown that power is related to the velocity of movement by the equation:

$$Power = F.v$$

where F is the force component acting in the direction of motion and v is the velocity of motion.

Therefore, muscle power is dependent upon both the force and velocity of contraction.

Energy

Any object which has the capacity to do work possesses energy. There are many forms of energy, for example chemical, heat, sound and mechanical. Further, there are three basic types of mechanical energy. These are kinetic energy, potential energy and strain energy. Because of the close relationship between work and energy, the units of the two quantities are the same. The unit of energy, therefore, is the joule (J).

Potential energy is the energy possessed by a body due to its position above the ground. It is a measure of the capacity of the object to do work in falling to the ground under the influence of gravity. An object's potential energy, therefore, increases as the distance it can fall on release increases. For example, a person standing on a box placed on the floor has a greater potential energy than when they are simply standing on the floor.

The amount of potential energy a body possesses is given by the equation:

$$PE = m.g.h$$

where PE is the potential energy, m is the mass of the body, g is the acceleration due to gravity and h is the height through which the body can fall.

Kinetic energy is the energy a body possesses due to its motion. Only moving bodies possess kinetic energy. The amount of energy possessed by a body depends upon the body's velocity and its inertia measured by its mass. For example, if muscles are made to contract during a movement so that the velocity of the movement is increased, the body part concerned will possess an increased capacity to do work — it has a greater kinetic energy.

The amount of kinetic energy a body possesses is given by the equation:

$$KE = \frac{m.v^2}{2}$$

where KE is the kinetic energy, m is the mass of the body and v is the body's velocity.

Strain energy is the capacity a body has to do work after being deformed from its original shape. For example, a spring which has been extended or compressed is said to possess stored strain energy which is released when the spring is released. That is because the spring does work in regaining its original length. In some texts strain energy is treated as a form of potential energy. Whilst this is not incorrect, it helps to avoid confusion if the two definitions are treated separately.

Conservation of energy

Although energy can be converted from one form to another it cannot be destroyed nor can it be created from nothing. The total sum of energy is constant. Chemical, or metabolic, energy used to produce muscle tension is transformed into mechanical and heat energy but no energy is lost. Even during isometric exercise, where no work is done, no energy is lost. This does not mean that no energy is expended, simply that it is not converted into kinetic energy.

Similarly, in standing on the ground an individual possesses greater potential energy than when lying on it, because in the former case, the person's centre of gravity is higher. If the standing person falls, the potential energy possessed in the upright position is gradually transformed into increasing kinetic energy as the person accelerates towards the floor. On reaching the floor, all potential energy will have been converted into kinetic energy, which in turn is transferred in the form of sound, heat and strain energy as the person comes to rest.

Box 2.5 Task 5

Conservation of energy
If a person's leg is being exercised in sling suspension using displaced axial suspension, then the movement that occurs is pendular. After being given an initial displacement, the energy changes gradually and cyclically from potential to kinetic and back to potential. If energy cannot be created or destroyed, why doesn't the sling continue to move indefinitely without muscular effort by the patient?

FLUID MECHANICS

Fluid mechanics is the branch of mechanics which deals specifically with the forces which result from a body's presence in, or motion through, a fluid. As hydrotherapy is increasingly used in physiotherapy practice, some understanding of the basics of fluid mechanics is therefore useful.

Buoyancy

A body floating motionless in water is in a state of equilibrium. The weight of the body acting downwards is balanced by an equal force exerted by the water on the body acting upwards (Fig. 2.39a). This upward force is called *buoyancy*. Similarly, a person standing on the bottom of a pool is also in equilibrium, but in this case there are two upwards forces; buoyancy and the reaction force between the person's feet and the pool floor. In this case the buoyancy force is tending to reduce the reaction force — or weight — taken by the person's feet. Thus, buoyancy may be seen as producing an apparent reduction in the

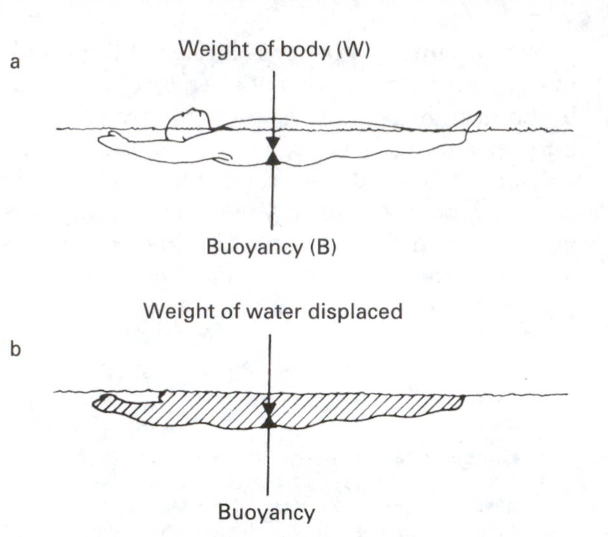

Figure 2.39 a) Forces acting in a balanced horizontal float showing body weight, W and buoyancy, B. b) Buoyancy is equal to the weight of water displaced by the floating body.

weight of the body or body part immersed in the water. This phenomenon can be used, for example, in walking rehabilitation exercises where only partial loading — weight bearing — of the lower limb is required.

In general, the deeper an object is immersed in a fluid the greater will be the magnitude of the buoyancy force. This is because of the Archimedes principle which states that when a body is wholly or partially immersed in a fluid, the magnitude of the buoyancy force exerted is equal to the weight of the fluid displaced (Fig. 2.39b). This can be used by the physiotherapist in the above example; by increasing the depth of the pool in which the walking exercise is undertaken, the degree of weight reduction obtained will be increased (Fig. 2.40).

Specific gravity and relative density

When a body is placed in water, it will sink until the weight of the fluid it displaces is equal to the weight of the body. At this point it will float, the buoyancy and weight forces being in equilibrium. If, however, the body is never able to displace a sufficient volume of water, the body will continue to sink. Whether a body is able to float depends upon the ratio:

weight of the body ÷ weight of an equal volume of water

This ratio is known as the *specific gravity* of the body. For an object to float, its specific gravity must be less than one. The lower the specific gravity, the higher in the water the object will float; the object is said to be more buoyant. Clearly the specific gravity of water = 1.

Density is a measure of the concentration of mass in a body. It is calculated by dividing the body's mass by its volume. The above ratio can,

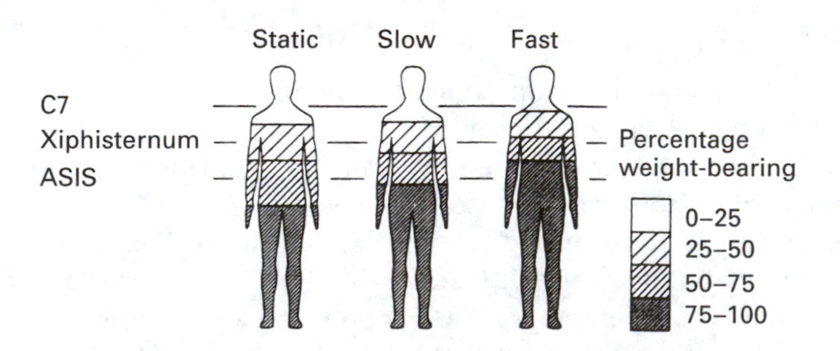

Figure 2.40 The effect of partial immersion on the percentage weight-bearing during standing and slow and fast walking exercise. (Note: the shaded areas indicate the level of immersion required in each condition to achieve the given % weight-bearing.) Reproduced from Harrison R A, Hillman M, Bulstrode S 1992 Loading of the lower limb when walking partially immersed: implications for clinical practice. Physiotherapy 78: 164–166.

therefore, be written in terms of the density of the body and the density of water. That is the ratio:

density of the body ÷ density of water

This ratio is known as the *relative density of the body*. If the body has a lower density than that of water, it will float, whereas if the body has a higher density than water, it will sink.

Note that values obtained for the specific gravity and the relative density of any object will be identical.

Different tissues in the body have different specific gravities — fat has a lower specific gravity than muscle and bone. As individuals differ in their body composition, their ability to float will vary; those individuals with a greater proportion of fatty tissue are generally better able to float.

The distribution of the tissue is also important. Body parts, such as the lower limb which contains predominantly muscle and bone, will tend to sink because the density of muscle is similar to that of water, whilst the density of bone is, on the whole, greater than that of water. The amount of air in the lungs will also determine the specific gravity of the body. During inspiration, the volume of the body is increased as the chest inflates, although the corresponding increase in body mass is negligible — the effective density of the body therefore decreases. As a result, the specific gravity is reduced and the more able the body is to float. On expiration, however, the reverse is true and the body tends to sink lower in the water.

Centre of gravity, centre of buoyancy

When a body is immersed in water its weight, as usual, acts through its centre of gravity. However, the buoyancy force acting on the body acts through a point known as the centre of buoyancy. But, as illustrated in Figure 2.41, these two points may not coincide. The relative positions of these points determine the spatial orientation of a free floating body.

The centre of buoyancy of any floating object coincides with what would have been the centre of gravity of the water displaced by the object before its displacement. As water has a uniform density, this point coincides with the centre of volume of the displaced water — that is, the centre of volume of that portion of the body which is submerged. Figure 2.39b illustrates the water displaced by the floating human body in Figure 2.39a in its original position. The centre of gravity of this volume of water, therefore, coincides with the point at which the buoyancy force is considered to act.

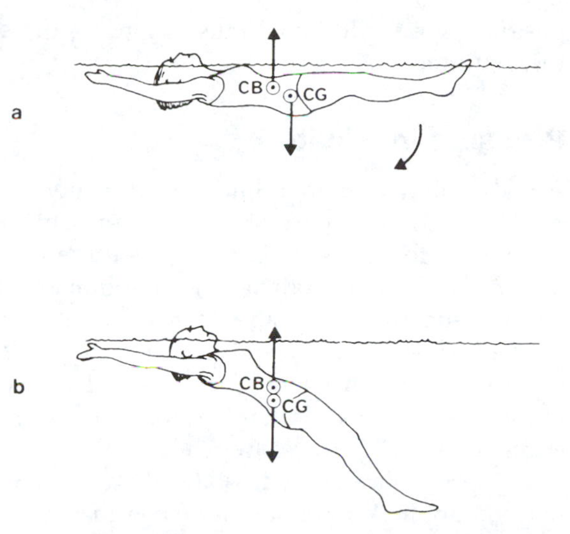

Figure 2.41 Floating positions are governed by the relative positions of the lines of action of body weight and buoyancy. Equilibrium is achieved when they are on the same vertical line. CG = body's centre of gravity; CB = centre of buoyancy.

The centre of gravity and the centre of buoyancy of a body only coincide if the body, like the fluid, is of uniform density. Since the human body is not of uniform density, due to the usual tissue distribution, the centre of buoyancy is closer to the head than the centre of gravity. This situation sets up a force couple when an individual tries to float horizontally in the water (Fig. 2.41a). Rotation of the body occurs until the centre of buoyancy is directly above the centre of gravity (Fig. 2.41b). How much rotation takes place depends upon the distance between the centres. This, in turn, largely depends upon the somatotype of the person attempting to float. If a person wishes to float in a position which is more or less horizontal to their equilibrium floating position, some additional muscle force is therefore required.

Pressure in a fluid

A body submerged in a fluid such as water will be subjected to an external pressure. This is known as the *hydrostatic fluid pressure* and is determined by the depth at which it is immersed. Hydrostatic fluid pressure is felt by all submerged surfaces, the magnitude of the pressure increasing proportionally with the depth of immersion. The net result is that the body experiences compressive forces from all sides.

The principles of hydrostatic fluid pressure may be applied, for example, when in therapy the compressive forces are utilised to help reduce oedema, which tends to accumulate distally in the lower extremities. The patients are here given walking exercises in the hydrotherapy pool. The pressure is greatest distally all around the foot and gradually lessens towards the proximal end of the limb giving a beneficial pressure gradient or differential.

MECHANICS OF MATERIALS

Mechanics of materials is the branch of biomechanics concerned with the strength, deformation and failure characteristics of biological materials. When subjected to a force (or load), all materials will undergo some deformation, even if the load is so small or the material is so stiff that the deformation is imperceptible to the naked eye. Understanding the nature of the relationship between applied force and the resulting deformation for a given material is fundamental to an understanding of the strength and failure characteristics of that material. As all injuries fundamentally result from the mechanical failure of a material, in order to fully understand injury it is necessary to understand this relationship for the biological materials involved in that injury. Fortunately, for any given material, the relationship between applied force and resulting deformation is a constant characteristic of the material, provided that the type of loading and size of the material are known.

Types of pure loading

There are three types of pure loading. These are compression, tension and shear (Fig. 2.42).

Pure compression occurs when two equal loads are applied to an object such that they act along the same line of action towards each other. This can be seen as a squeezing or pushing action. This is the principal loading in the long bones of the leg, for example, in stance: body weight acting downwards and the ground reaction force acting upwards. The resulting deformation is a shortening (with some widening) of the object.

Pure tension occurs when two equal loads are applied to an object such that they act along the same line of action away from each other. This can be seen as a stretching or pulling action. This is the principal loading in tendons and ligaments, for example, which results from their normal anatomical action. The resulting deformation is a lengthening (with some narrowing) of the object. Pure tension is the opposite of pure compression.

Pure shear occurs when two parallel and equal loads act on an object in opposite directions but not on the same line of action. This is frequently

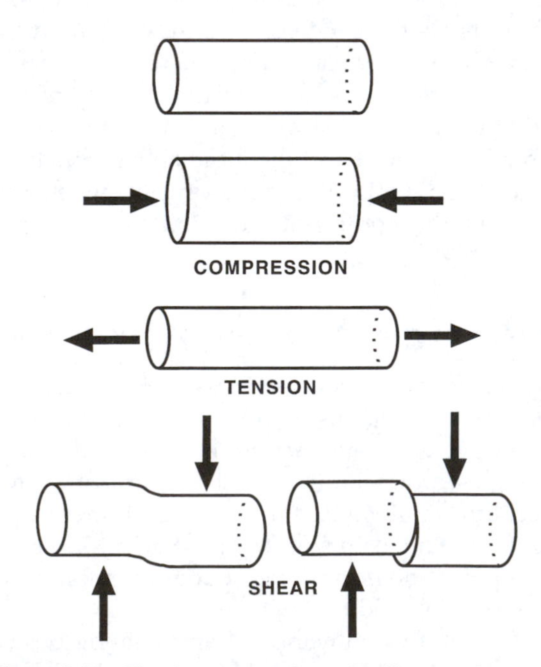

Figure 2.42 The three types of pure loading.

seen as a scissoring or sliding action. Friction is a type of shear force. Deformation at failure is a sliding of one surface over another. Prior to failure, shear deformation is angular.

Size effects: stress and strain

Experience tells us that the size of an object affects the absolute magnitude of force required to break or deform that object a given amount. In order to understand the force deformation relationship for a given material, therefore, we need to account for potential size differences. This is done by expressing the force acting on the object as stress and the deformation as a strain.

Consider Figure 2.43. If the two objects are made from identical materials but one has twice the cross-section area (CSA) of the other, it will require twice the tensile force to break or deform the object with the larger CSA. It is, therefore, possible to standardise the measure of loading on the two objects by dividing the applied load by the cross-section area of the object. This standardised measure of loading is called *stress* and has the unit newton per metre squared (N/m^2 or $N.m^{-2}$). At fracture, for example, although the loads acting on the two will be different the stress will be the same.

In Figure 2.43 the stress will be a tensile stress because the loads applied are tensile loads. Compressive loads will produce compressive stress and shear loads will produce shear stress. In each case:

$$Stress = \frac{applied\ load}{cross\text{-}section\ area} = \frac{F}{CSA}$$

In compression and tension, the cross-section area is measured perpendicular to the direction of loading; in shear the cross-section area is measured parallel to the direction of loading between the two loads.

Now consider Figure 2.44. Here two objects are made from identical materials but one has twice the original length (l) of the other. If the stress applied to each is the same, the longer object will experience twice the deformation — here seen as an increase in length. It is, therefore, possible to standardise the measure of deformation of the two objects by dividing the measured deformation by the original dimension. This standardised measure of deformation is called *strain*.

In Figure 2.44, the strain will be a tensile strain because the loads applied are tensile loads.

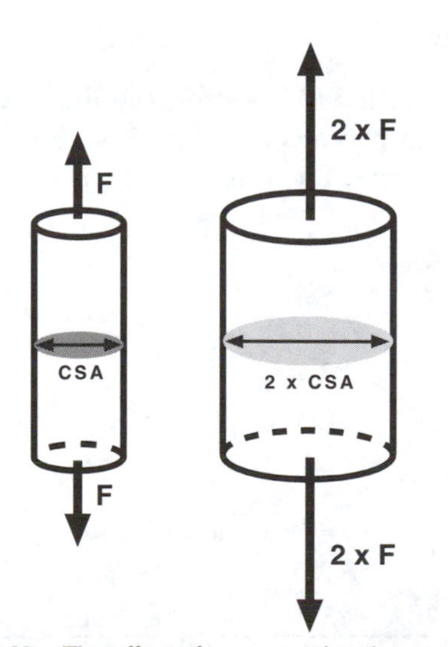

Figure 2.43 The effect of cross-sectional area on the strength of a material. These effects can be accounted for by expressing the load in terms of stress.

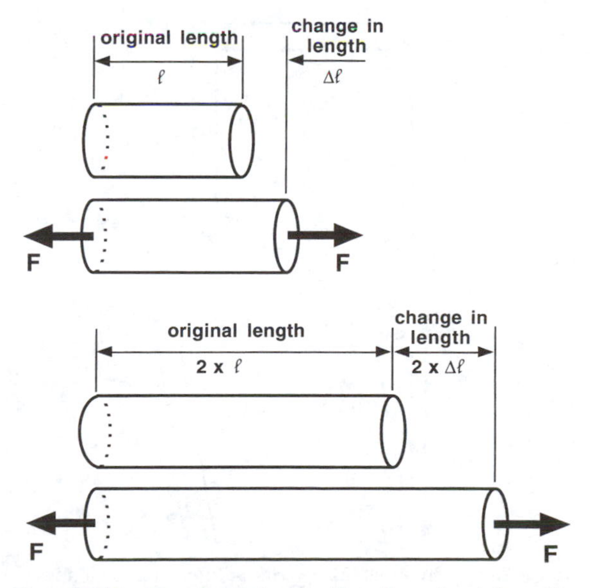

Figure 2.44 The effect of original length on the deformation of a material. These effects can be accounted for by expressing the deformation in terms of strain.

Compressive loads will produce compressive strain. In these cases:

$$Strain = \frac{change\ in\ length}{original\ length} = \frac{\Delta l}{l}$$

Shear strain is not a change in length but an angular deformation as illustrated on the block of material shown in Figure 2.45. The force *F* acting on the block produces a deformation in which the top face of the block moves a slight distance *d* relative to the bottom face. In this situation, the angle θ measures the resultant geometrical deformation and is known as the angle of shear strain. If the height of the block is *h*, the shear strain on the block is given by:

$$Shear\ strain = \frac{displacement}{height} = \frac{d}{h}$$

$$= \theta \text{ if } d \text{ is much less than } h$$

Stress/strain diagram

The relationship between stress and strain for any given material is only dependent upon the type of loading and the mechanical properties of that material. This relationship will stay the same for any object made from that material irrespective of its size or shape. The relationship between stress and strain is often illustrated in terms of a stress/strain diagram (Fig. 2.46).

A typical stress/strain diagram has two principal regions: a linear region at relatively low stresses and a curved region at relatively high stresses; the point of transition between the two is called the *elastic limit*. The linear portion of the curve up to the elastic limit is termed the *elastic region* and that after the elastic limit the *plastic region*. Any strain that occurs as a result of stresses that are placed on an object less than the stress required to achieve the elastic limit of the material from which it is made is non-permanent: the deformation is elastic. Strain occurring as a result of stresses greater than the stress required to achieve the elastic limit will have an elastic element but will also result in some permanent deformation. Any permanent deformation is termed *plastic deformation*. Although ultimate failure occurs at the breaking (fracture) point, failure is often considered to start once the elastic limit has been reached, i.e. the stress required to achieve the elastic limit of the material has been exceeded. In biological materials this is often the point at which the onset of micro-damage of the tissue is seen to occur.

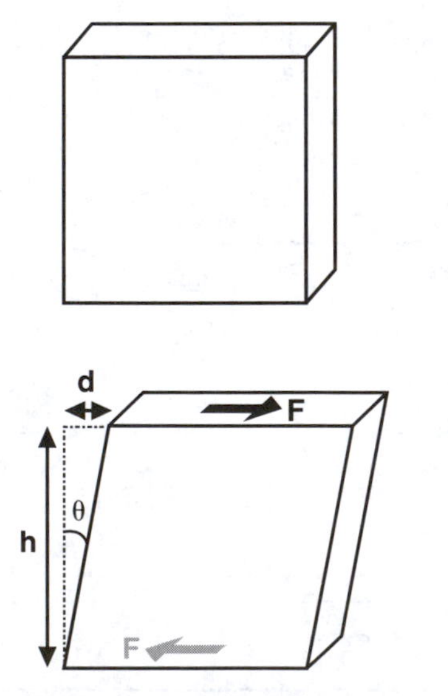

Figure 2.45 Shear strain. F = Shear force applied to the block; h = height of the block; d = the distance moved by the top surface of the block when F is applied, and θ = angle of shear strain.

Stiffness

The linear portion of the stress/strain curve up to the elastic limit is a consequence of Hooke's Law which states that for a material behaving elastic-

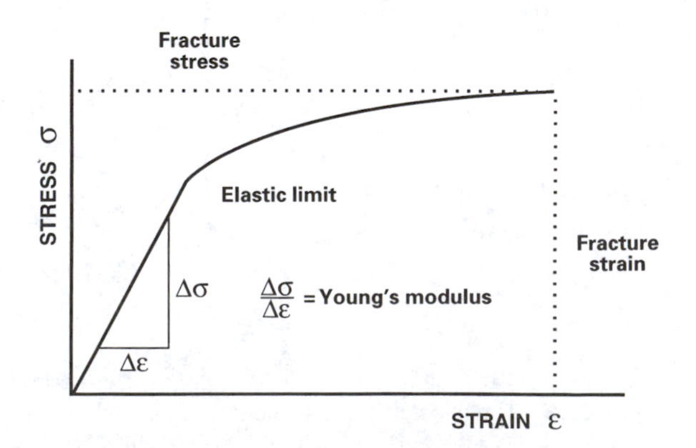

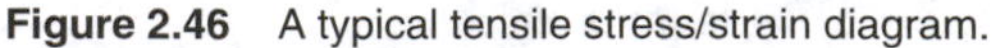

Figure 2.46 A typical tensile stress/strain diagram.

ally the deformation of the material is proportional to the applied force. The constant of proportionality is a characteristic of the material and is known as the modulus of elasticity. For linear stresses this is called *Young's modulus* and is given by:

$$\text{Young's modulus} = \frac{\text{Linear stress}}{\text{Linear strain}}$$

The constant of proportionality for shear stress is known as shear modulus and is given by:

$$\text{Shear modulus} = \frac{\text{Shear stress}}{\text{Shear strain}}$$

Stiffness is a measure of the resistance offered by a structure or object to deformation. It is dependent upon the shape and size of the object and the modulus of elasticity of the material from which the object is made. For example, materials with a relatively high modulus of elasticity, such as bone, are stiffer than those with a low modulus of elasticity such as tendon.

The architecture of trabecular bone affects the stiffness and strength characteristics of the bone when loaded in differing directions. In fact, the architecture of most skeletal bone is optimised for normal load bearing function by aligning the trabeculae along lines of internal stress (Fig. 2.47). Such a structure helps to maintain a beneficial strength to weight ratio for the bone. However, a large stress applied in an unusual loading direction may cause the bone to fracture at loads which it would normally be capable of maintaining. This characteristic of biological materials in which they exhibit different strength and stiffness characteristics when loaded in different directions is termed anisotropy. Tissues tend to be strongest in the direction of normal stress as a result of morphological adaptation.

Though Figure 2.46 represents an idealised stress/strain curve, some materials exhibit a non-linear region at stresses below the elastic limit. Tendon, skin and other connective tissue in the musculoskeletal system are examples of such materials (Fig. 2.48). This behaviour is largely due to the arrangement of the collagen fibres in the tissue. In skin, collagen is present as a dense network of fibres running in different directions, whereas in tendons and ligaments the arrangement is more orderly with the collagen having regular undulations that tend to run parallel with each other. As stress is applied, the collagen fibres first unravel producing relatively larger strains until the fibres become straight and aligned with the direction of loading. This is referred to as the *initial lax phase of elastic behaviour*. Skin has a longer initial lax phase due to the more random orientation of its collagen fibres that allows the skin to strain further before the fibres become aligned. Once aligned, the colla-

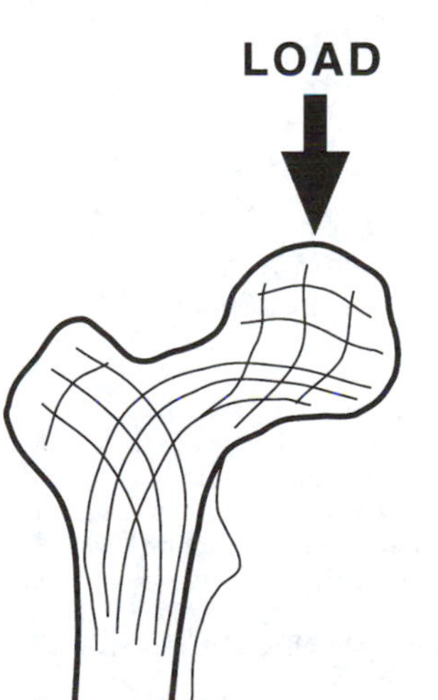

Figure 2.47 The lines of principal stress in the head and neck of the femur.

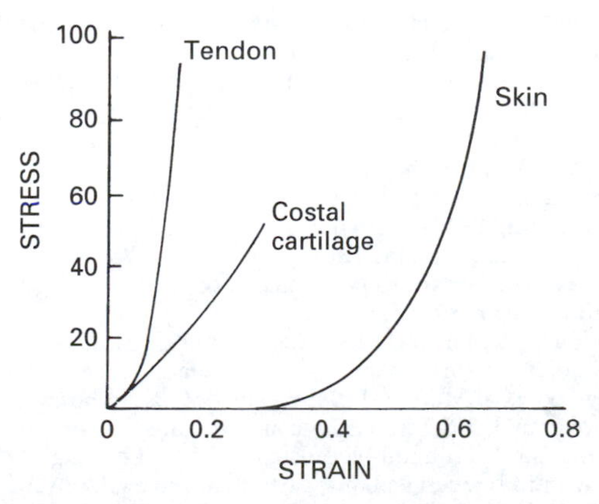

Figure 2.48 The response of some soft tissues to uniaxial loading. After Barbinel J C, Evans J N, Jordan M N 1978 Tissue mechanics. Engineering in Medicine 7: 5–9.

gen is then able to take more of the load and the tissues tend to stiffen in response to increasing stress and behave in the more expected linear fashion.

Viscoelasticity and creep

Another factor that affects the stiffness of biological materials is the rate at which the stress is applied. This property of a material to show sensitivity to rate of stress loading is called *viscoelasticity*. Viscoelastic materials are apparently stiffer when a force is applied quickly; the force meets with a higher resistance to deformation than when the force is applied more gradually. For example, joint cartilage deforms less when a rapid load is applied, as in impact situations for example, than when the load is applied more gradually. In cartilage this effect is related to the water permeability of the material. Tendons, ligaments and skin also exhibit viscoelastic characteristics to a greater or lesser extent.

When a constant force is applied to a material, there is an immediate deformation which, as we have seen, depends upon the stress applied and the stiffness of the material. Viscoelastic materials continue to deform after initial loading taking some time to reach a steady state (Fig. 2.49). This phenomenon is called *creep* and may be used to the physiotherapist's advantage in certain treatments, for example, when applying serial plasters to increase joint range of motion that has been limited by contracted soft tissues. After stretching the tightened soft tissue during a treatment session, a plaster is applied to maintain its new length until the next treatment session when it is removed and a further stretch applied. This process is repeated until the required tissue length is achieved and is able to be maintained by normal functioning of the body part.

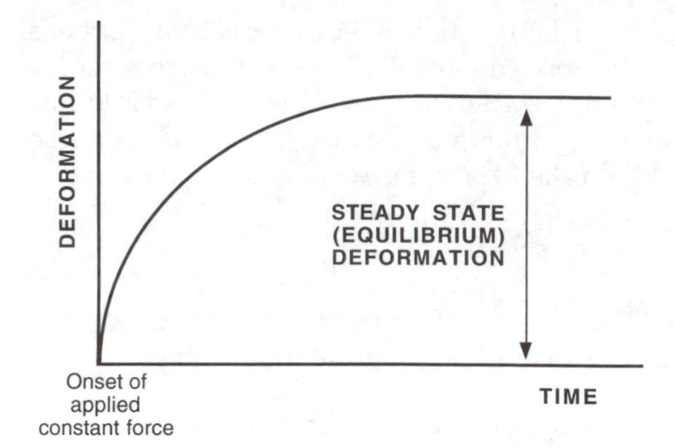

Figure 2.49 A deformation/time curve illustrating the creep phenomenon of biological materials.

REFERENCES

Barbinel J C, Evans J N, Jordan M N 1978 Tissue mechanics. Engineering in Medicine 7: 5–9
Bowker P 1987a Basic mechanics: forces and their effects. Physiotherapy 73: 264–270
Bowker P 1987b Applied mechanics: biomechanics of orthoses. Physiotherapy 73: 270–275
Cunningham D M, Brown G W 1952 Two devices for measuring the forces acting on the human body during walking. Experimental Stress Analysis IX: 75–90
Enoka R M 1988 Neuromechanical basis of kinesiology. Human Kinetics, Champaign
Frankel V H, Nordin M 1980 Basic biomechanics of the skeletal system. Lea and Febiger, Philadelphia
Gowitzke B A, Milner M 1988 Scientific basis of human movement, 3rd edn. William and Wilkins, Baltimore
Harrison R A, Hillman M, Bulstrode S 1992 Loading of the lower limb when walking partially immersed: implications for clinical practice. Physiotherapy 78: 164–166
Hellebrandt F A, Tepper R H, Braun G L, Elliot M C 1938 The location of the cardinal anatomical orientation planes passing through the centre of weight in young adult women. American Journal of Physiology 121: 465–470
Inman V T, Ralston H J, Todd F 1980 Human walking. Williams and Wilkins, Baltimore
Luttgens K, Deutsch H, Hamilton N 1992 Kinesiology: scientific basis of human motion, 8th edn. Brown, Madison, Wisconsin
Puharic T, Bohannon R W 1993 Measurement of forearm pronation and supination strength with a hand held dynamometer. Isokinetics and Exercise Science 3: 202–206
Read M T, Bellamy M J 1990 Comparison of hamstring/quadriceps isokinetic strength ratios and power in tennis, squash and track athletes. British Journal of Sports Medicine 24: 178–182
Reid J G, Jensen R K 1990 Human body segment inertia parameters: a survey and status report. Exercise and Sport Science Reviews 18: 225–242
Roberts S L, Falkenburg S A 1992 Biomechanics: problem solving for functional activity. Mosby Year Book, St Louis

Rodgers M M, Cavanagh P R 1984 Glossary of biomechanical terms, concepts and units. Physical Therapy 64: 1886–1902
Whittle M 1991 Gait analysis: an introduction. Butterworth Heinemann, Oxford
Williams M, Lisner H 1977 Biomechanics of human motion, 2nd edn. W B Saunders, Philadelphia
Winter D A 1990 Biomechanics and motor control of human movement, 2nd edn. Wiley Interscience, New York

CHAPTER CONTENTS

3

Musculoskeletal basis for movement

D. J. Newham
A. M. Ainscough-Potts

OBJECTIVES

At the end of this chapter you should be able to:

1. **Describe the structure and function of muscle**
2. **Discuss the physiological process for the different types of contraction**
3. **Discuss the different types of muscle activity**
4. **Explain the role of the muscle in different activities.**

INTRODUCTION

In order to be able to analyse movement it is essential to have a good understanding of muscle function, anatomy and biomechanics. This chapter does not attempt to reproduce the basic textbook material in these areas. Its aim is to bring together the components of knowledge necessary for movement analysis and help the reader integrate them.

The first section describes skeletal muscle structure and function. The second section incorporates this knowledge into the consideration and analysis of human movement.

MUSCLES

There are three major types of muscle: skeletal (striated), cardiac and smooth muscle. The latter is found in the walls of blood vessels and gut. Skeletal muscle is that which enables the maintenance of posture and movement and will be considered in this section. It is under both voluntary and reflex control.

Skeletal muscle structure

An understanding of how skeletal muscle works requires a knowledge of its structure from the level of gross anatomy to molecular organisation. There is a hierarchial structure seen in Table 3.1 and illustrated in Figure 3.1. Muscle structure and function is discussed in more detail in specialised works (e.g. Jones & Round 1990, Lieber 1992).

Skeletal muscle is composed of contractile and non-contractile components. The *contractile components* are the actin and myosin filaments which are responsible for the generation of active force (tension).

The *non-contractile components* are elastic structures such as tendons, connective tissue sheaths and structural proteins which contribute to the development of passive tension.

Muscle fibres

These long, multinucleated muscle cells may extend over the whole length of the muscle or may be much shorter, depending on the arrangement of fibres within the muscle. They are surrounded by a network of capillaries, many of which will be closed off at rest. During activity they become patent and muscle blood flow can be greatly increased.

The muscle cells contain many *mitochondria* which are responsible for aerobic metabolism. Their presence allows the muscle to function continuously in the presence of oxygen and their number varies with fibre type and also endurance training. The cells also contain *glycogen* and *lipid droplets*.

Myofibrils

Each myofibril is surrounded by a membranous network called the *sarcoplasmic reticulum* (SR).

Table 3.1 The hierarchy of muscle organisation

Whole muscle	Bundles of fascicles	Surrounded by connective tissue sheath (epimysium)
Muscle fascicles	Groups of muscle fibres	Surrounded by connective tissue sheath (perimysium)
Muscle fibres	Bundles of myofibrils (about 2000) arranged in parallel	Surrounded by connective tissue sheath (sarcolemma) 10–100 μm diameter, multinucleated
Myofibrils	String of sarcomeres arranged in series	Surrounded by sarcoplasmic reticulum and T tubules about 1 μm diameter
Sarcomeres	Functional unit of muscle contraction	Composed of myofilaments and structural, non-contractile proteins
Myofilaments	Actin (thin) and myosin (thick) filaments	

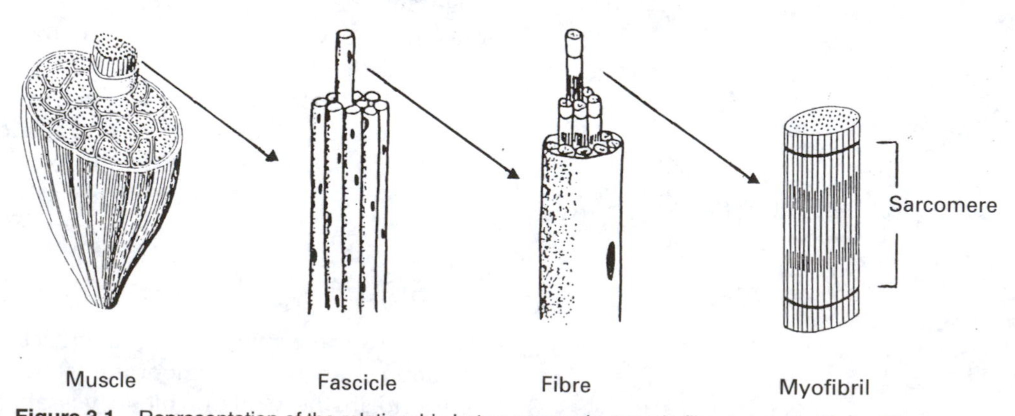

Figure 3.1 Representation of the relationship between muscle, muscle fibres and myofibrils. The striations in the myofibrils are the result of the arrangement of the actin and myosin filaments. Each myofibril consists of adjacent sarcomeres connected to each other in series. These run the length of the fibre.

The interior of the SR is quite separate from the contents of the fibre and contains the calcium which is necessary for the interaction of actin and myosin and the generation of force. The signal for calcium release from the SR is the arrival of the action potential from the neuromuscular junction. This travels over the surface of the fibres and is transmitted into the interior by a series of invaginations in the surface membrane called the T (tubular) system. The SR is in close proximity with the T tubules and ensures an effective calcium release, as shown in Figure 3.2. Once the action potential has passed, calcium is pumped back into the SR and the muscle relaxes.

The sarcomere

These are the structural units for muscle which repeat along the length of each fibril. They are bound at each end by a Z line which connects adjacent sarcomeres, shown in Figure 3.3 above. The actin (thin) myofilaments are a structural part of the Z line and the myosin (thick filaments) interdigitate with the actin filaments, but are not connected directly to the Z line. The alignment of the actin and myosin filaments give the striations to skeletal muscle.

However, the myosin filaments do not lie free in the sarcomere. They, and the architecture of the entire sarcomere, are maintained by a number of structural proteins. They include some for which the purpose is unclear and some, such as dystrophin, appear to be absent or reduced in muscular dystrophies.

The actin filament is a globular protein which appears to be double helical strands with

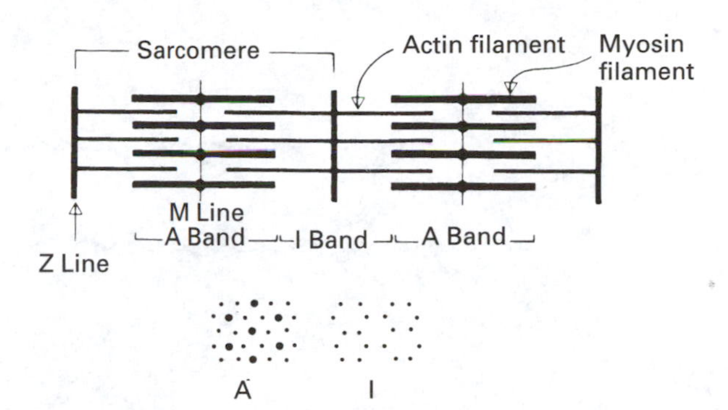

Figure 3.3 Two adjacent sarcomeres. This pattern is repeated along the length of the fibre. The actin (thin) and myosin (thick) filaments are arranged precisely and give the striations to skeletal muscle. The A band contains both types of filament, while the I band contains only actin filaments. Cross sections through the A and I bands are shown below.

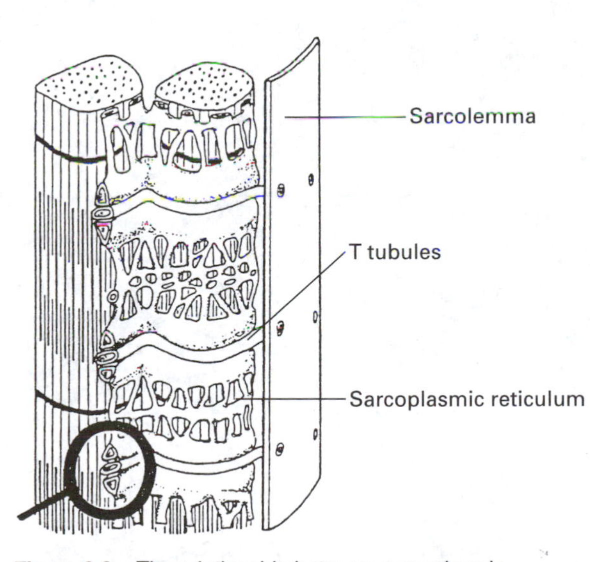

Figure 3.2 The relationship between sarcoplasmic reticulum (SR), T tubules and muscle fibres. The T tubules are invaginations of the sarcolemma and allow transmission of the action potential into the interior of the fibre. The membranous bags forming the SR are wrapped around sections of each myofibril and have a T tubule on either side. They contain calcium which is released on the arrival of an action potential and cause the interaction between actin and myosin which results in force generation. The area circled is called a triad and is a T tubule with a portion of SR on either side (reproduced with permission from Jones & Round 1990).

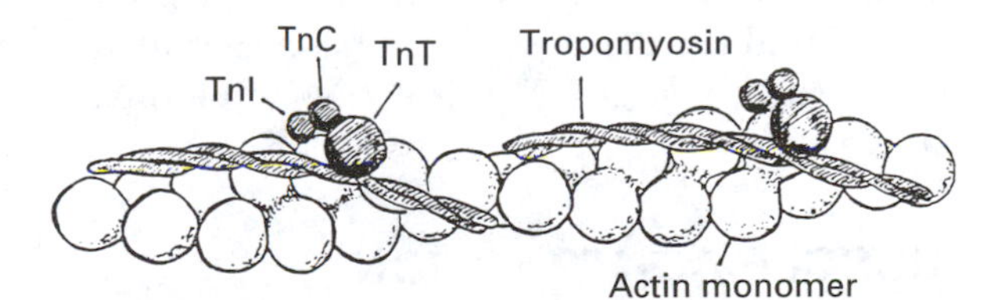

Figure 3.4 A section of an actin filament. The double strands of actin monomers form a groove into which the regulatory protein tropomyosin fits. The protein troponin is made of three subunits and is located at intervals along the tropomyosin. Tropomyosin blocks the binding sites for myosin until caused to move by the binding of calcium to troponin C. Troponin T binds troponin to tropomyosin and troponin I inhibits tropomyosin when there is no calcium release (reproduced with permission from Jones & Round 1990).

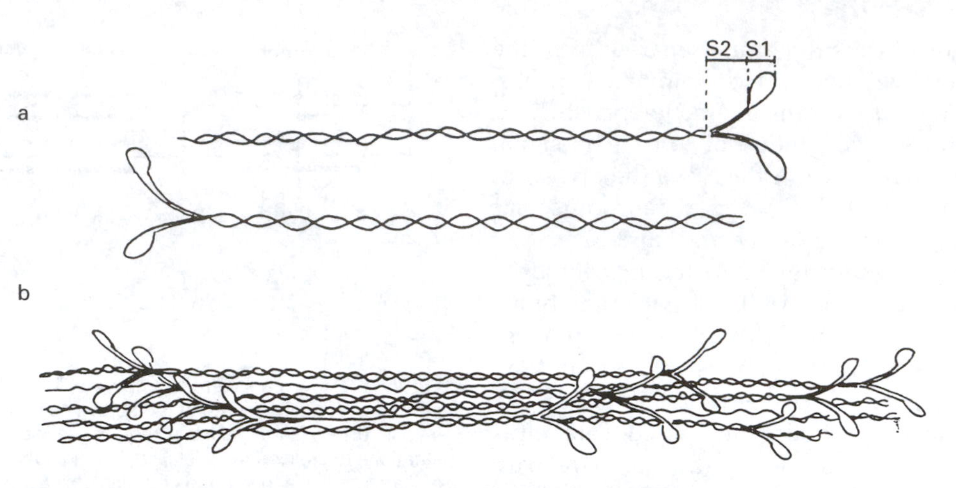

Figure 3.5 a) Myosin molecules consist of two identical chains. The globular head (S1 fragment), which combines with actin, and the extensible neck region (S2 fragment) are connected to a long tail or backbone. b) The molecules are packed together to form the thick filaments. They are rotated so that the heads are arranged around the filament.

tropomyosin and three troponin (Tn) subunits TnC, TnT and TnI (Figure 3.4). The regulatory protein tropomyosin blocks the myosin binding sites until it is caused to move and uncover them, when calcium binds to TnC. TnT binds troponin and tropomyosin and TnI inhibits tropomyosin in the absence of calcium.

Myosin filaments are two identical chains arranged in an antiparallel fashion. The globular head of the S1 fragment is attached to the S2 fragment. This is a flexible neck region which connects with the long tail of the molecule. The S1 fragment forms the cross bridge with the actin filament and causes force generation.

Individual myosin molecules pack together to form the thick filament with the heads projecting out around the filament which has a central area free of cross bridges. This is shown in Figure 3.5 above.

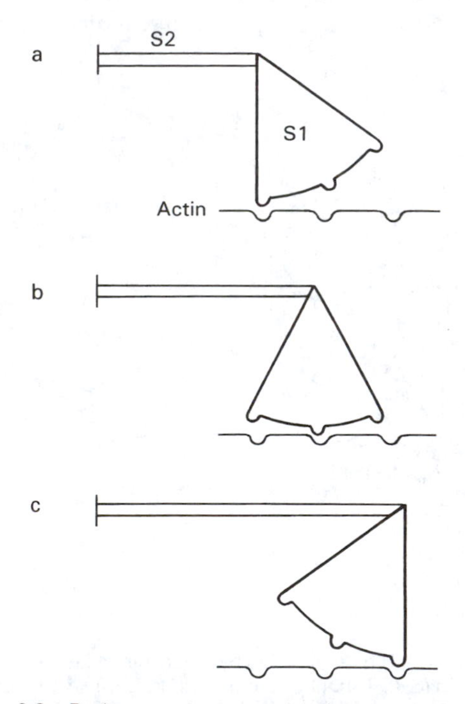

Figure 3.6 During muscle contraction the cross bridge or S1 fragment of the myosin molecule attaches to a binding site on the actin filament (a). During an isometric contraction the myosin head rotates (b and c) and pulls the actin filament (to the left), but the relative position of the actin and myosin filaments remains constant and the S2 portion of the myosin molecule is stretched and force is generated. If the muscle shortens during contraction, phases a, b and c occur as above. Then the actin filament is pulled towards the myosin molecule and the stretch on the S2 fragment decreases.

MUSCLE CONTRACTION

Figure 3.6 is a diagrammatic representation of a muscle contraction. When a muscle contracts, the filaments themselves do not change length. However, the sarcomeres, and therefore fibres, may change length because of changes in the amount of overlap between actin and myosin filaments (Huxley & Simmons 1971). The myosin head attaches to a binding site on the actin filament and rotates, thus pulling the actin filament toward the centre of the sarcomere and exerting a passive extension force on the S2 fragment. The myosin heads are released and attempt to attach to another binding site and the cross bridge cycle continues as long as the muscle is activated and the energy requirements are met.

Factors affecting force generation

Studies on isolated muscle preparations have shown that the force which can be generated by a fully activated muscle fibre, or even single sarcomere, is affected by a number of factors. These are intrinsic properties of skeletal muscle and also apply to intact human muscles.

Length

There are two components which affect the force that is generated at different muscle lengths. These components are active and passive tension.

Active tension is the force generated by the cross bridges and is illustrated in Figure 3.7. The muscle isolated has been moved through the full range of length and stimulated electrically at numerous different lengths while tension is measured as the length remains constant. The main point is that there is an optimum length for force generation. Force declines when the muscle is activated at either longer or shorter lengths (Gordon et al 1966) and at the extremes of length no force is generated.

The explanation for this lies in the amount of overlap between the actin and myosin filaments and underlies the sliding filament theory of contraction (Huxley & Simmons 1971). At optimal length there is sufficient overlap for numerous cross bridges to form and force generation is not impaired by the length of the sarcomere. At

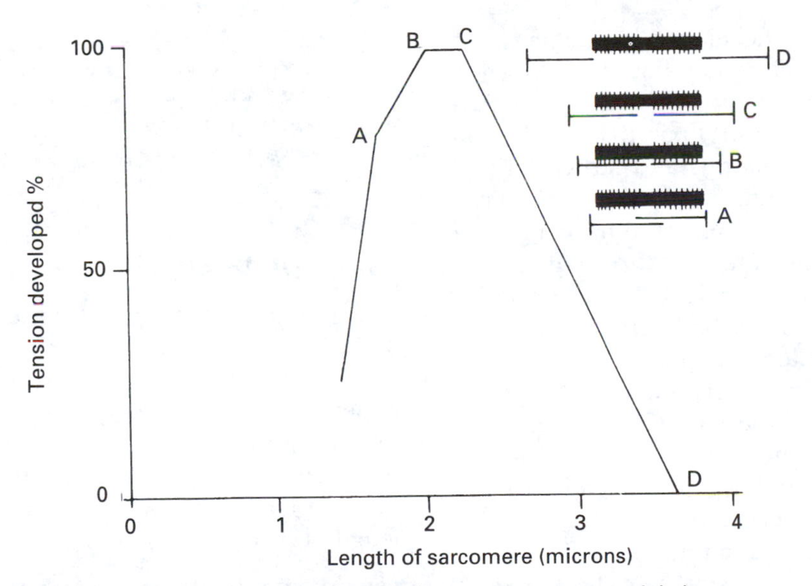

Figure 3.7 The length:tension for a single sarcomere, measured during a series of isometric contractions at different sarcomere lengths. On the ascending limb, force increases with length. At sarcomere lengths shorter than 2 μm (B), the thin filaments begin to overlap and at still shorter lengths (A) they come into contact with the Z lines. These factors both reduce the force that can be generated. On the descending limb (C–D), force declines as length is increased. As length increases the amount of overlap between thick and thin filaments decreases and so fewer connections can be made between them. The optimal conditions for force generation occur on the plateau (B–C) (after Gordon et al 1966 — reproduced with permission from Guyton 1976).

lengths shorter than optimal, the actin filaments from the two ends of the sarcomere come closer together and progressively interfere with force generation. At lengths longer than optimal, the amount of overlap between actin and myosin filaments decreases as the number of cross bridges which can form decreases.

Muscles in intact animals and humans do not reach the extremes of length in Figure 3.7 because of anatomical constraints. However, a number of intact human muscles demonstrate a length:tension relationship which clearly shows an optimal length and decreasing force as length changes in either direction. Other muscles show flatter curves and this is probably due to biomechanical changes and also orientation of fibres within the muscle (described later).

Some muscles in the intact human body develop maximal active tension at approximately the midpoint of joint range, but relatively few muscles have been studied systematically in this respect.

Passive tension is developed if a resting muscle which is not being stimulated is stretched. The origin of passive tension is in the connective tissue in series and parallel with the myofilaments. The tendons and structural proteins, particularly titin, form the series elastic component, while the surrounding connective tissue represents the parallel component.

Total tension can be calculated when active and passive tension are both measured; this is graphically illustrated in Figure 3.8. This initially minimises the effects of active tension lost on the descending limb of the length:tension relationship. However, further increases in length cause a rapid increase in total tension and can be sufficient to cause muscle rupture.

In the living body most muscles which cross a single joint cannot be stretched to the point where passive tension contributes significantly to total tension. However, muscles acting over more than one joint may be stretched sufficiently for passive tension to limit joint range.

It is important to remember that the length: tension relationship describes the effect of length on isometric (static) tension. In order to describe movement it is necessary to understand the effect on force of contraction type and also velocity.

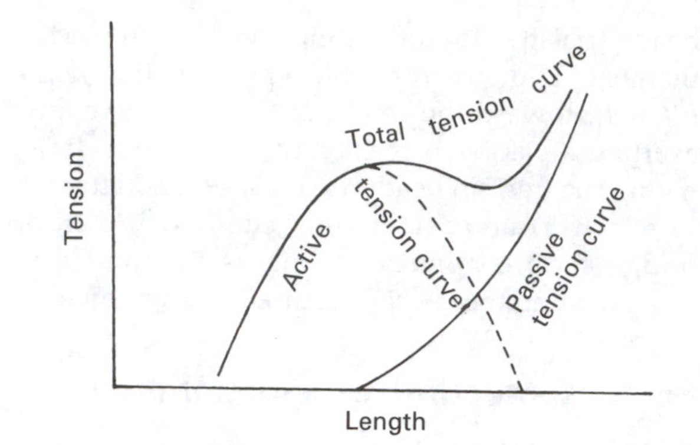

Figure 3.8 The effect of length on total tension. The active length:tension relationship of a single sarcomere is shown here and in Fig. 3.7. In intact preparations the non-contractile components start to exert passive tension, at approximately mid-length, which increases with length.

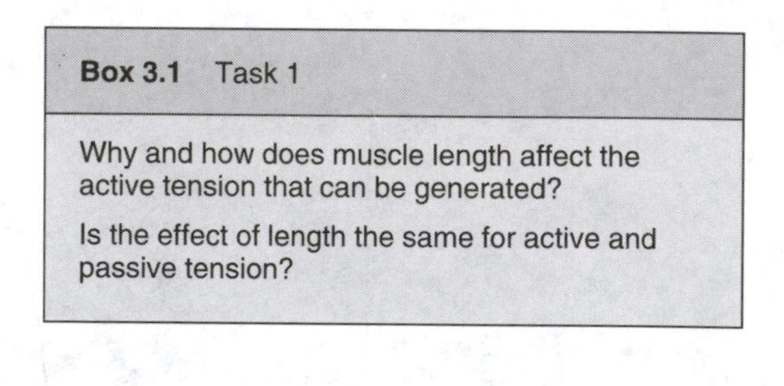

Box 3.1 Task 1

Why and how does muscle length affect the active tension that can be generated?

Is the effect of length the same for active and passive tension?

Types of muscle contraction

The sliding filament theory of muscle contraction dictates that an active muscle will attempt to shorten. Whether or not length changes occur depends on the external resistance offered. It is important to realise that the term 'contraction' is used to describe an active muscle and relays no information about whether or not it changes length during activity:

- An *isometric* (static) contraction occurs when there is no external movement, because the internal tension generated by the muscle is equal to the external force.
- When there is movement a *dynamic* contraction takes place and the muscle may become longer or shorter.
- A muscle will shorten and perform a *concen-*

tric contraction if the force generated is greater than the external force.

- An *eccentric* contraction occurs when a muscle generates a lower force than the external load and is lengthened by it.

The characteristics of isometric, concentric and eccentric contractions and the forces generated by each type of activity are very different, as shown in Table 3.2 (Jones & Round 1990, Lieber 1992).

Isometric contractions occur when there is no movement and no external work is done by the muscle. The maximal force that can be generated and the energy cost required for a given amount of force are intermediate between concentric and eccentric contractions.

Although the muscle does not change length in gross terms, there is some internal shortening as the sarcomeres shorten slightly due to filaments sliding past each other. In the intact body, the elastic tendons are slightly stretched.

Concentric contractions occur when the muscle shortens, movement occurs and external work is done. A muscle contracting concentrically will always generate less force than when contracting isometrically. The actin binding sites are moving past the myosin cross bridges and it takes a certain amount of time for cross bridges to attach to, and detach from, the actin binding sites. Therefore, the number of attached cross bridges, and thus force generation, is less than during an isometric contraction. This rapid cycling of cross bridges has a high energy cost and, therefore, the highest oxygen demand and heat production occurs during concentric contractions.

Eccentric contractions occur when active muscle is lengthened by an external resistance. Work is done *on* the muscle, rather than *by* it and so negative work is said to be done. During an eccentric contraction a muscle generates a force higher than it is capable of under isometric or concentric conditions. This is thought to be because the tension generated by cross bridge formation is increased by the additional component of elastic force caused by the stretch of the neck of the myosin molecule (S2 fragment). The cross bridges remain attached to the actin binding site unless ripped away by the stretching force.

Curiously, in view of their high force generation, eccentric contractions are performed at a very low energy cost. This is thought to be because when the myosin is pulled away from the actin filament, it is in the correct position for subsequent reattachment to another binding site and does not require energy to move into the attachment position.

The high force generation and low energy cost of eccentric exercise is still not fully explained.

The performance of unaccustomed, high force eccentric activity will cause muscle fatigue, pain and damage in excess of that caused by isometric or concentric contractions and this is thought to be due to mechanical damage (Clarkson & Newham 1995).

Functional examples. The rather complicated concept of these types of muscle contraction can be illustrated by considering the right quadriceps muscle. If a person is standing still on their right lower limb, their quadriceps will be working isometrically to prevent the knee flexing. If they are sitting down and wish to stand up, the muscle will work concentrically. If they then decide to sit down slowly, the muscle will be working eccentrically as it controls movement velocity by opposing the force of gravity.

Table 3.2 The characteristics of the three types of muscle contraction. The determining factor for movement is whether the internal force generated by a muscle is the same, lower or higher than the external forces

Type of contraction	Function	External force (relative to internal)	External work by muscle	Force generated	Energy cost (O_2 demand)
Concentric	Acceleration	Less	Positive	Lowest	Highest
Isometric (static)	Fixation	Same	None	Intermediate	Intermediate
Eccentric	Deceleration	Greater	Negative	Highest	Lowest

Thus, isometric contractions are generally used for fixation, concentric ones for acceleration and eccentric for deceleration.

Further classification of contraction type

Muscle contractions may also be isotonic or isokinetic in nature. An *isotonic* contraction is one in which the force remains constant throughout. An *isokinetic* contraction is dynamic and is performed at a constant velocity throughout. These are considered in more detail in Chapter 13.

Functional activity

It is relatively unusual for a muscle to perform only one type of contraction in any particular activity. Most involve a mixture of varying proportions of the different types and all have important functional roles.

An eccentric contraction often precedes a concentric one and it is thought that this utilises the energy stored during eccentric activity and increases mechanical efficiency (Komi 1986). A common example of this 'stretch-shortening' cycle is in the calf muscles where the active muscles are stretched prior to shortening during the push off phase of walking, running and jumping, as illustrated in Figure 3.9.

Velocity of contraction

The amount of tension that a fully activated muscle can generate also varies with the velocity of contraction. If an active muscle is either allowed to shorten, or is stretched at a range of different velocities and the force generated is measured at each velocity, then the force:velocity relationship can be determined (Fig. 3.10). This also shows the effect of contraction type on force as discussed above.

It can clearly be seen that, while eccentric force remains above and concentric force below isometric (zero velocity), velocity has a marked and different effect on the two types of dynamic contraction. Concentric force decreases with velocity while eccentric force increases.

Concentric contractions

The higher the velocity of shortening, the shorter is the time available for the myosin cross bridges

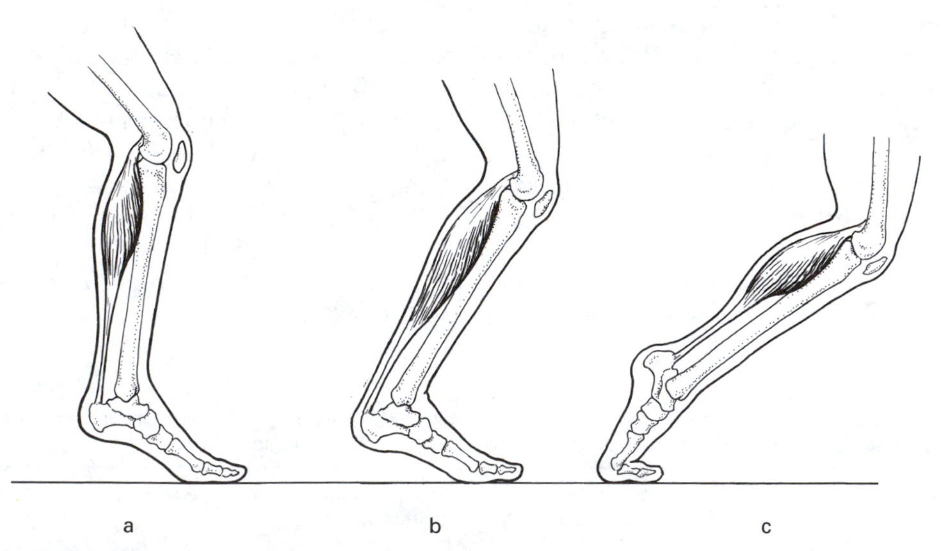

Figure 3.9 The stretch-shortening cycle of the calf muscles during walking, jumping and running. Just before contact the muscles are activated (a) in order to resist the forces of impact during which they are stretched and perform an eccentric contraction (b). This is immediately followed by a shortening (concentric) contraction during push off (c).

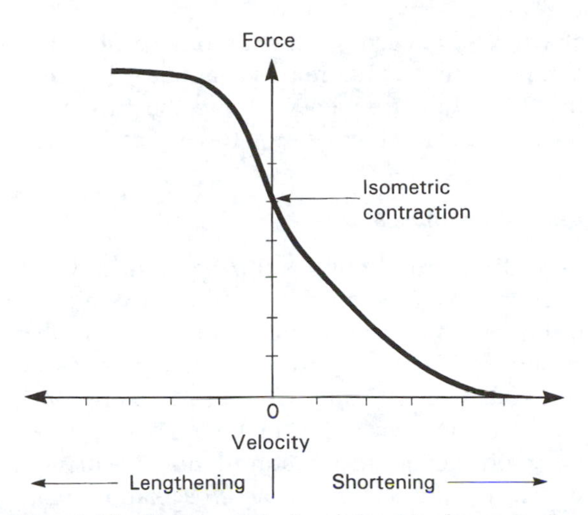

Figure 3.10 The force:velocity relationship of a maximally activated muscle. Concentric force is always greater than isometric and decreases as velocity increases. Eccentric force is always greater than isometric, it initially increases with velocity, then remains relatively constant.

to attach to the actin binding sites moving past them. The proportion of cross bridges that manage to attach to the actin filament in the region in which the cross bridges can exert a useful force decreases as velocity increases. Therefore, the number of attached cross bridges, and the force they can exert, decreases with velocity.

Maximal velocity of shortening (V_{max})

Eventually a velocity is reached at which no force can be sustained; this is the maximal velocity of shortening (V_{max}). The V_{max} differs between individuals and also in different muscles in the same person or animal. It is largely determined genetically by the fibre types within muscle and is little affected by physical training.

Eccentric contractions

In contrast, eccentric force increases with velocity and plateaus at about 1.8 times the isometric force. As the velocity of stretch increases, the extensible S2 portions of the cross bridges produce more passive, elastic force. The plateau of the eccentric force:velocity relationship suggests that skeletal muscles are relatively resistant to stretch and this is useful in many normal movement patterns.

Functional implications

Activities of daily life frequently demonstrate the practical consequences of the effects of velocity on force generation. The heavier a weight is, the slower we are able to move or lift it using concentric contractions. However, eccentric contractions such as those involved in lowering heavy weights, whether an inanimate object or the weight of the body, become faster as the weight increases.

Box 3.2 Task 2

Why are different forces generated by a fully activated muscle depending on whether it stays the same length (isometric), gets shorter (concentric) or longer (eccentric)?

Is the effect of velocity similar for concentric and eccentric force?

Power

In many activities the functional requirement is for power, i.e. the rate of doing work, rather than force. Power is the product of force and velocity, therefore, during isometric contractions (zero velocity) and at maximal velocity (zero force) the power output is zero. Figure 3.11 shows the power output at different velocities of concentric contraction.

Maximal power output usually occurs at about two-thirds of the maximal velocity.

Frequency of stimulation

When a muscle fibre is stimulated at an intensity above the threshold for motor activation it will generate force. The force generated is strongly influenced by the frequency of stimulation. A single impulse will result in a mechanical response called a twitch. If stimulation frequency is increased, the force initially increases and then remains relatively constant despite substantial increases in force, as shown in Figure 3.12.

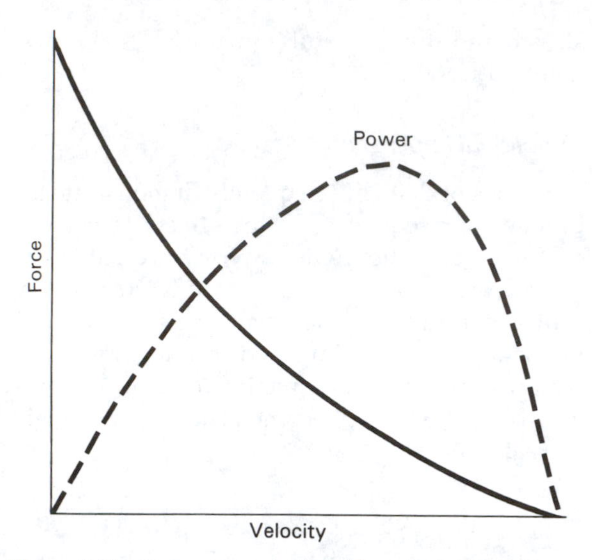

Figure 3.11 The power output at different velocities of concentric contraction. This is derived from the force:velocity relationship (solid line) as shown in Fig. 3.10.

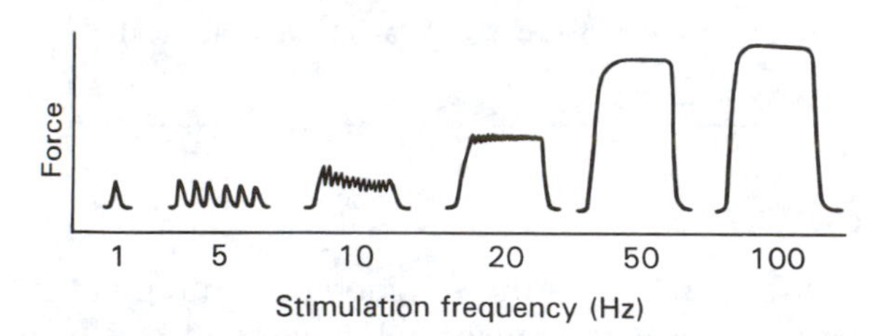

Figure 3.12 The relationship between force and the frequency of external electrical stimulation. Note that increased frequency initially increases force and reduces oscillations. However, increasing the frequency above that which produces a fused tetanic contraction does not increase force.

As the stimulation frequency initially increases, two observations can be made. The first is that the contraction becomes smoother because the muscle has less time to relax between consecutive stimuli. The second observation is that the force increases with stimulation frequency because the next impulse arrives before the muscle has completely relaxed and the impulse is superimposed on the remaining tension. This is known as the summation of force. When stimulated at a sufficiently high frequency (termed the *fusion frequency*), the muscle will produce a smooth, tetanic force in which there is no relaxation between individual stimuli. The frequency at which this occurs depends on the fibre type and is discussed later in this chapter. An increase in stimulation frequency above the fusion frequency does not increase the force of contraction.

Functional implications

Recordings from muscles during voluntary contractions have shown that the physiological firing rates are usually at a frequency lower than the fusion frequency found when muscles are stimulated electrically (Binder-Macleod 1992). However, we can easily make a smooth voluntary contraction using stimulation frequencies which would show marked oscillation using external electrical stimulation. This is because during external stimulation the active fibres are being stimulated synchronously, but in voluntary activity they are all firing asynchronously, smoothing out force oscillations.

THE MOTOR UNIT

Each individual muscle fibre generates a force so small as to be impractical for even the most delicate movements. Therefore, the system is designed such that a group of muscle fibres share common innervation from a single alpha motor neurone (gamma motor neurones innervate the fibres within the muscle spindle). This functional grouping is called a motor unit and is composed of the cell body of the alpha motor neurone (the anterior horn cell in the spinal cord), the motor neurone itself and the muscle fibres innervated by it. Figure 3.13 illustrates the motor unit. If a motor neurone fires, all the muscles in that unit will contract at the same time, producing a synchronised electrical discharge (action potential) which can be measured by electromyography and also the generation of force. The size of both action potential and force are proportional to the number of muscle fibres within the motor unit. There is a range of motor unit size within a single muscle and also between muscles. Muscles requiring the precise regulation of small forces, such as the small hand muscles, have smaller motor units which may contain only 10 fibres. Large postural muscles, such as the quadriceps,

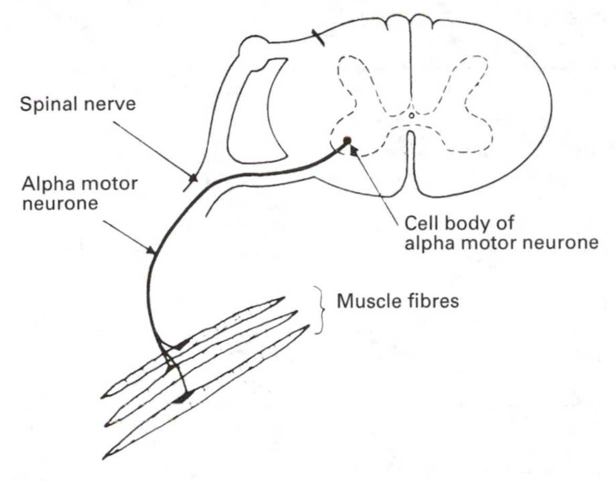

Figure 3.13 A single motor unit is composed of the motor neurone and the muscle fibres which it innervates. The cell body of the motor neurone is in the anterior horn cell in the spinal cord and its axon is one of the motor nerves in a mixed peripheral nerve.

have much larger motor units which may have several thousand fibres.

The fibres belonging to an individual motor unit are scattered throughout a muscle, thus adjacent fibres are unlikely to belong to the same motor unit. All the fibres within a single unit are the same fibre type.

Gradation of muscle force

It is clear that both voluntary and reflex muscle force can be precisely controlled. There are two ways in which force can be varied, motor unit recruitment and rate coding. Both have been shown to be used in voluntary contractions of human muscle, but it is unclear to what extent they are employed. It is possible that large postural muscles, which do not require fine control of force, predominantly use motor unit recruitment. Those needing fine control, such as the hand muscles, may rely more on rate coding.

Motor unit recruitment

The force generated can be increased by activating more motor units and this is termed motor unit recruitment. The smaller motor units have the most excitable motor neurones and therefore are recruited first. As more force is required, the larger, and progressively less excitable, motor neurones are recruited in an orderly fashion. This has become known as the *size principle* (Hennemann et al 1974).

Rate coding

The force of active motor units can also be varied by the frequency of stimulation of the motor neurone and by utilising the force:frequency characteristics. Recordings from single motor units have shown that the firing rate varies considerably even within a constant low force contraction. Initially, a short burst of firing may be used to generate relatively high forces by the motor unit, but rapidly decrease to maintain force.

Box 3.3 Task 3

What is a motor unit? What is their role in the generation of different levels of force?

FIBRE TYPES

The observation that some muscles are dark and others light, as in the leg and breast muscle of a chicken, is an indication that not all muscle fibres are the same. It was thought previously that discrete fibre types existed, but it seems that there is a spectrum of fibre types. Considerable variations in histochemistry, contractile properties and the type of metabolic fuel used have been identified. This is shown in Table 3.3.

The colour differences exist because of the different levels of myoglobin which is red. It can be seen from Table 3.3 that Type I fibres are specialised to use oxidative metabolism and are resistant to fatigue. They contain a lot of myoglobin and are relatively slow to contract and relax. They are recruited early in low force muscle activity due to their small axon size and therefore fatigue resistance is important. The number of muscle fibres in slow motor units is small and so motor unit recruitment can result in fine grada-

Table 3.3 Examples of differences between fibre types. Note that different terminology exists for types of muscle fibre and motor unit

Property	Type I	Type IIa	Type IIb
Muscle fibre type	Slow oxidative (SO)	Fast oxidative glycolytic (FOG)	Fast glycolytic (FG)
Motor unit type	Slow (S)	Fast fatigue resistant (FR)	Fast fatigable (FF)
Motor unit size	Small	Medium	Large
Twitch tension	Low	Moderate	High
Mechanical speed	Slow	Fast	Fast
Fatigability	Low	Low	High
Mitochondrial enzyme activity	High	Medium	Low
Glycogenolytic enzyme activity	Low	Medium	High
Myoglobin content	High	Medium	Low
Capillary density	High	Medium	Low

tions of force. They rely on oxidative (aerobic) metabolism and are developed so that the delivery and utilisation of oxygen is maximised.

Type II fibres are subdivided into Types IIa, b and c. The Type IIb fibres contrast sharply to Type I fibres in almost all respects. They are fast to contract and relax and rely on glycolytic metabolism and intramuscular stores of fuel. Due to the large axon diameter they are recruited only during high force contractions and fatigue rapidly. This is illustrated in Figure 3.14. The motor units contain relatively large numbers of fibres and so recruitment of additional units causes relatively large force increases.

Type IIa fibres and their motor units are intermediate between Types I and IIb and span a broad range of the characteristics of both. The subgroup IIc is found mainly in regenerating fibres and in the embryo.

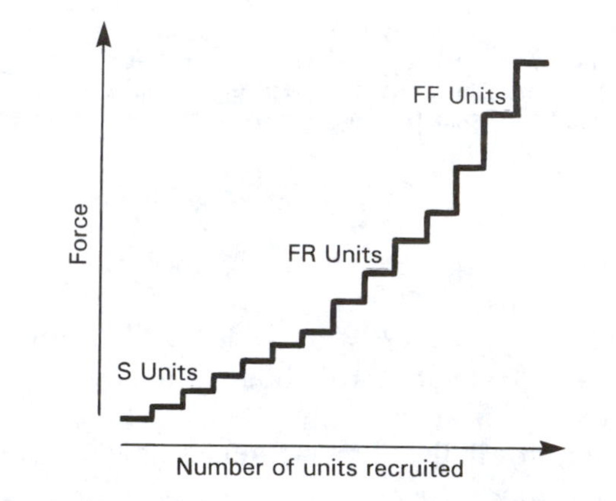

Figure 3.14 The regulation of force by recruitment of motor units. At low forces only the small, slow (S) units are recruited. As force increases, the larger fast, fatigue-resistant (FR) and then the largest fast, fatigable (FF) units are recruited. The S units have relatively few muscle fibres and the FF units have the most. Therefore the increment in force as a new unit is recruited also varies.

Functional aspects

During a low force contraction only the Type I fibres are recruited, therefore, they are the ones used mainly for normal activity which does not require maximal or high force contractions. They are well suited to this role by their fatigue resistance. As the force of contraction increases, the Type IIa and then IIb fibres are progressively recruited and during maximal contractions all the motor units are active. However, maximal force rapidly declines due to the high rate of fatigue in the fast fibres.

Fibre type is largely determined genetically and is governed by the activity in the motor neurone. Training with voluntary activity may shift the characteristics of motor units and their muscle fibres, but does not actually transform motor units and muscle fibres into different types.

GROSS MUSCLE STRUCTURE

Within a single muscle, groups of fibres are organised into fascicles. Figure 3.15 shows that the arrangement of the fascicles varies considerably in different muscles, and this has an impact on muscle function. Muscles arranged with the fibres in parallel with the line of pull (strap and

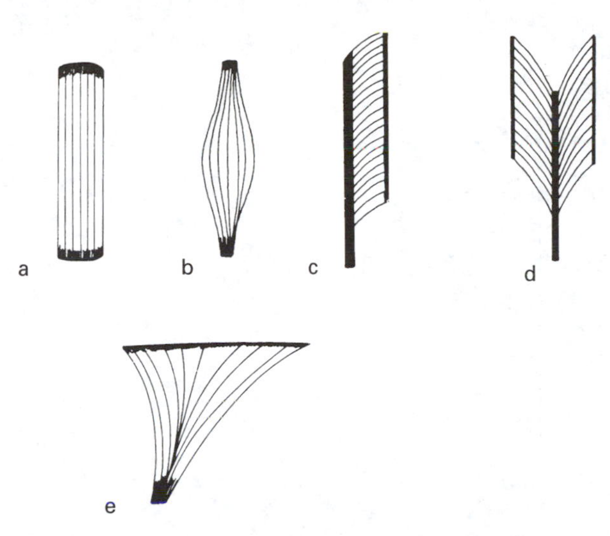

Figure 3.15 Some of the common arrangements of muscles. Strap (a) and fusiform (b) muscles have a tendon at either end and relatively long muscle fibres which run between the tendons. The muscle fibres are arranged in parallel, or are very similar to, the angle of pull of the tendon. Unipennate (c) and bipennate (d) muscles have a central tendon into which short muscle fibres are inserted. The fibre direction is different to the angle of pull of the tendon. Triangular muscles (e) display mixed characteristics.

also fusiform structure to a large extent) usually have a tendon at each end which inserts into bone. In these muscles the fibre length is long and similar to the muscle length. In muscles with a pennate structure, the fibres are inserted into a longitudinal tendon and lie at an angle to the line of pull of the muscle. In this case the muscle fibres are much shorter than the whole muscle and also shorter than the fibres in a strap or fusiform alignment.

Effect of fibre alignment on muscle performance

An important concept is that we basically require muscles for two purposes: one is to produce power and cause movement, the other is to generate relatively static force for the maintenance of posture. Therefore, force and power are the key requirements and power is the product of force and velocity. It can be seen that force is necessary for all muscle activity, but power, and thus velocity, are essential for rapid movement. Parallel fibres are usually seen in muscles which are fast acting, such as sartorius, whereas pennate arrangements are seen more in muscles required for strength, such as gluteus maximus.

Determinants of force

The force a muscle can generate in any given situation is proportional to its cross-sectional area, i.e. the number of sarcomeres in parallel. Along the length of a fibre the tension generated by adjacent sarcomeres is equal and opposite at the central Z line and therefore they cancel each other. The only forces transmitted through the muscle attachments are those generated by the sarcomeres at either end of the muscle. Therefore, force is independent of fibre length (Fig. 3.16).

Muscles which are mainly required for static activity tend to have a large cross-sectional area. A pennate structure has the advantage that more fibres can be packed into the same cross-sectional area (compare Fig. 3.15a and c). The disadvantage is that the force transmitted to the tendon is the cosine of the angle between the angle of pull of the tendon and the fibres and therefore some of the generated force is lost.

Muscle cross-sectional area can be increased by strength training (see Ch. 13) and this increases force generation. However, in pennate muscles with a large angle between the line of pull and fibre alignment, an increased cross-sectional area may result in an increased angle of pennation and therefore a relatively small increase in force.

Determinants of velocity

The maximal velocity at which a muscle can contract is determined by its length, i.e. the number of sarcomeres arranged in series, and is independent of the cross-sectional area. At the start of muscle activity all the sarcomeres begin to contract at about the same time and velocity. If a muscle contained only one sarcomere which shortened by 1 μm in 0.1 seconds (s), the shortening velocity would be 10 μm/s. If the muscle contained 100 sarcomeres in series, each would shorten at the same time and velocity, so the total

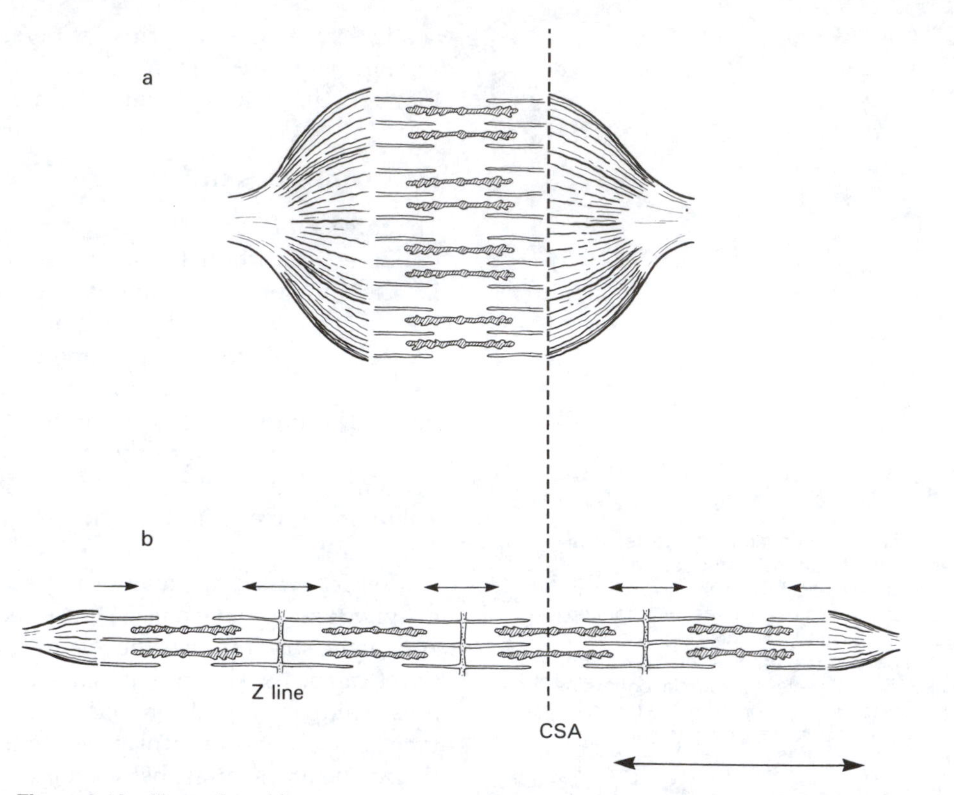

Figure 3.16 Illustration of four sarcomeres arranged in parallel (a) and in series (b). When they are arranged in series, the forces from adjacent sarcomeres at each Z line are equal and opposite. Therefore the only force transmitted by the muscle is that from the outside half of the two sarcomeres at the end of the fibre. The same force is generated by each fibre, irrespective of its length. With a parallel arrangement the force of each sarcomere is transmitted to the tendon. Therefore the same four sarcomeres will generate four times as much force as when they are arranged in series. Note that the cross-sectional area (CSA) is proportional to the force that is produced.

velocity of shortening of the whole muscle would be 1 mm/s.

The maximal velocity of shortening of a single cross bridge and sarcomere is governed by the fibre type and activity of the enzyme myosin ATPase and appears to be relatively unaffected by training.

Muscle length, strength and power

We have already seen that strength is proportional to the cross-sectional area and velocity to length of a muscle. Therefore, a short, fat muscle will generate more force and have a lower velocity of shortening than a long, thin one. However, as power is the product of force (cross-sectional area) and velocity (length), it is proportional to volume.

Muscle attachments

Most muscles are connected at each end to at least two bones through a tendinous attachment. One of the bones remains relatively steady and anchors one end of the muscle. Therefore, the force of the contracting muscle moves the bony lever attached to the other end of the muscle.

The traditional anatomical description of muscle attachments are for the more proximal one to be called the *origin* and the distal one the *insertion*. The proximal attachment is usually the more stationary, but there are instances where this is not the case. To avoid confusion in movement analysis, the attachments are often referred to as either *fixed* or *moving*.

ROLES OF MUSCLES

An individual muscle rarely works alone. Usually several different muscles are active simultaneously in even the simplest movement. The forces generated by the active muscles may vary considerably throughout a movement. Each muscle can fulfil several roles in different movements or patterns of movement. These roles are discussed below.

The prime mover or agonist is the muscle which plays the major role in initiating, carrying out and maintaining a particular movement. An example is the biceps brachii during elbow flexion.

Assistant movers are those muscles which perform a movement similar to the prime mover, but which play a less significant role in a particular movement. This is the role played by the brachialis in elbow flexion.

Stabilisers or fixators are muscles which contract to control the position of a bone so that it may act as a steady base from which the agonist can act. They thus provide a fixed attachment for another muscle. During elbow flexion the shoulder girdle muscles act as fixators to control the position of the arm.

A synergist is a muscle which acts simultaneously with one or more muscles to produce a movement which neither could produce alone. All muscles in the team are called synergists. There are two types of synergists: true and helping.

True synergists act, for example, when using the long finger flexors as agonists to grip an object in the hand. The unwanted action of these muscles in flexing the wrist needs to be controlled or opposed. This is done by the simultaneous contraction of the wrist extensors acting as true synergists.

Helping synergists act simultaneously. Flexor carpi radialis and extensor carpi radialis longus are usually antagonistic to each other in producing wrist flexion and extension, respectively. However, they can act simultaneously as helping synergists to produce radial deviation at the wrist.

Antagonists are muscles which act in a direction which is opposite to the agonist. They are often inhibited by a reflex mechanism originating from the agonist and known as *reciprocal inhibition*. However, during reciprocal and rapid movements they are often activated at the end of joint range and perform an eccentric contraction to decelerate the movement. This aids movement control and can offer protection against musculoskeletal damage. During elbow flexion the triceps are antagonists and may be either relaxed or active.

Box 3.4 Task 4

Consider an individual muscle and all the activities in which it can be involved. Identify the role it is playing and what type of contraction it is performing. Also consider the role and type of activity for other muscles which are working simultaneously.

RANGE OF MOVEMENT

During a dynamic contraction muscles move through a range of movement. This is called full range when the muscles move throughout the full anatomical position from the shortest to longest possible length. The extremes of both short and long length in an intact body are much less than those which would be possible if the muscle were freed from its bony attachments. This is because of the limitations imposed by the anatomical arrangement of joints and soft tissue such as joint capsules and ligaments.

Movement through the range will be from long to short length for the agonist during a concentric contraction. Conversely, during an eccentric contraction the agonist will move from short to long length.

The full range of muscle excursion can be subdivided, as shown diagrammatically in Figure 3.17. In the *outer range* the muscle length moves between its longest length and the midpoint of the range. The *inner range* is between the

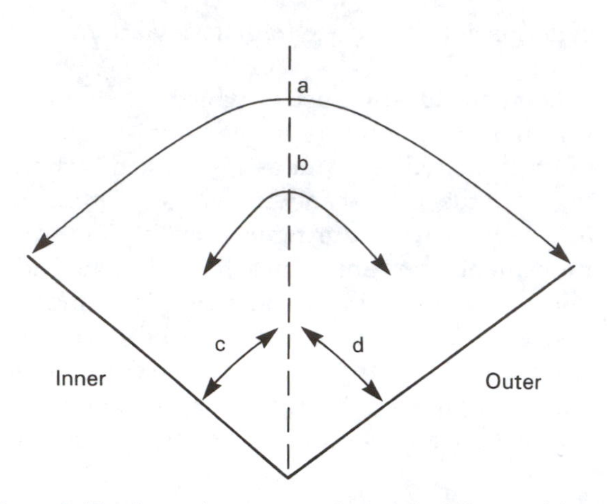

Figure 3.17 A muscle can move through its full range (a) or parts of it. The dotted line indicates mid-position. The muscle may move in mid-range (b), inner (c) or outer (d) range.

shortest length and the midpoint of range. In the *middle range* the muscle changes its length from the middle positions of the inner and outer ranges.

It is important not to confuse this terminology with that of the range of joint motion since muscle and joint range may not be the same, particularly in muscles acting over more than one joint.

Active and passive insufficiency

Muscles which cross only one joint are usually capable of shortening and lengthening sufficiently to allow full range of anatomical movement, but this is not necessarily the case for muscles crossing one or more joints. Multiple joint muscles causing simultaneous movement at all the joints crossed may reach a length at which they can no longer generate a useful active tension. At this point the muscle is said to be actively insufficient. This occurs when the hamstrings are used to simultaneously flex the knee whilst extending the hip.

If a multiple joint muscle is unable to stretch across the joints enough to allow their full anatomical range, they are said to be passively insufficient. An example of this is when the length of the hamstrings prevents touching the toes or floor with extended knees, as they are stretched over both hip and knee (Gowitzke & Milner 1980, Lehmkuhl & Smith 1983). Increasing the passive tension in a multiple joint muscle may cause joint movement called *tenodesis*. This occurs when moving from flexion to extension of the wrist when the fingers are relaxed. The fingers automatically extend during wrist flexion and flex during wrist extension as shown in Figure 3.18.

HUMAN MOVEMENT

The reader should have a knowledge of human anatomy as found in numerous basic textbooks. Further information on movement analysis and biomechanics can be found in Palastanga et al (1990), Broer & Zernicke (1979), Kapandji (1978), Basmajian & Deluca (1979). In this section the movements possible at each body segment, and how they are brought about, are considered in biomechanical terms.

The body is a series of long and short bones connected at junctions or joints. For movement to occur, the junctions must allow for free movement in the directions that their design allows. Movement is produced by the internal forces generated by muscle contraction or by external forces such as gravity and manual or mechanical forces. It occurs at joints and is contained by ligaments.

In order to understand the way in which the

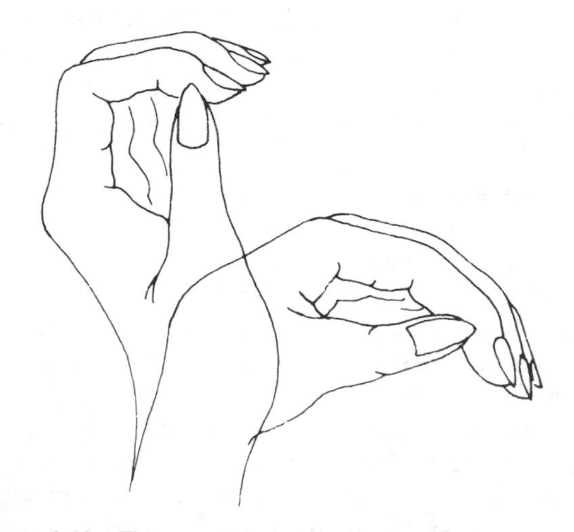

Figure 3.18 The tenodesis action of the long finger flexors causes finger flexion to increase as the wrist is extended.

body moves it is necessary to understand how it is constructed and formally described (Fig. 3.19).

The skeleton is designed to absorb and dissipate stress and comprises the long bones of the limbs and short bones of the vertebral column. The hand and foot have short irregular bones proximally and long bones distally. The scapula and pelvis are irregular flat bones. Their purpose is to provide shape to the body, to provide attachments for muscles, ligaments and joint capsules and also to dissipate stress generated by movement.

Forces and stresses

Each time we take a step and the heel strikes the floor, a ground reaction force occurs that has to be transmitted through the body. Stress forces travel in straight lines. Every time they meet an interface some are absorbed and others are reflected. Those absorbed will be transmitted to the soft tissues outside the bone. Therefore, at heel strike some of the stress force is transmitted into the foot bones and dissipated by them, the rest is transmitted through the tibia. As the tibia is not straight, the stress force will meet at least one interface before arriving at the knee and therefore lose some of its strength. This continues through the body segments so that only a relatively small stress force remains to cause a jarring at the junction of the skull and the vertebral column. All long bones are slightly curved and this enhances the dissipation of stress within the body (Leveau & Bernhart 1984).

Movement occurs in the limbs at synovial joints. The joint surfaces are lined with hyaline cartilage and the joint is surrounded by a strong fibrous capsule lined with a synovial membrane that produces synovial fluid. The combination of fluid and the shiny cartilaginous surfaces produces an almost friction-free environment which enables the muscles to concentrate their energies on moving the body levers, rather than on overcoming friction. However, when the joint surfaces become worn with age or disease this is no longer the case. Patients with arthritis find it much more difficult to move and more exhausting; one of the reasons is the friction that they have to overcome when moving an affected joint.

All synovial joints have only one position where the surfaces fit precisely together and there is maximal contact between the opposing surfaces (MacConaill & Basmajian 1977). This is called the close packed position and it permits no movement. When not in this position, the joint is said to be in the loose packed position and movements of spin, roll and glide may occur. Each joint has a least packed position in which the capsule is at its most lax. Joints tend to assume this position when there is inflammation in order to accommodate the increased volume of synovial fluid.

The hip joint

The hip is a ball and socket joint with the rounded head of the femur fitting into the deep socket of the acetabulum on the pelvis.

Movements

The anatomical movements that can occur at this joint are flexion, extension, abduction, adduction, external (lateral) rotation and internal (medial) rotation. When these are combined a circular arc of movement, known as circumduction, occurs. These movements are performed by large muscle groups which work in a coordinated way to produce the anatomical movements. They are large and strong as, for much of the time, they are having to counter the effects of gravity (Fig. 3.19).

Biomechanics

In order to flex the hip when standing, the flexor muscles have to counter the effects of gravity acting on the lower limb. The centre of gravity of a straight lower limb is approximately half-way along its length, i.e. at the knee joint.

With the hip acting as a fulcrum, the force that the body needs to overcome is the product of the lower limb weight and the perpendicular distance between the muscle attachment and the

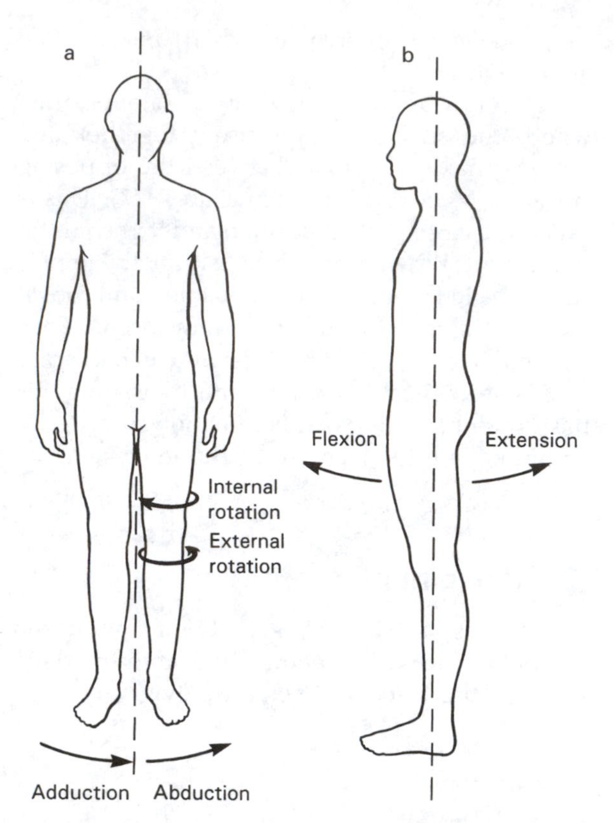

Figure 3.19 a) The sagittal plane (dotted line) is the midline of the body and divides it into right and left halves. Abduction and adduction are movements away from and towards the sagittal plane respectively. Rotation may be toward (internal rotation) or away from (external rotation) this plane. b) The frontal plane (dotted line) divides the body into front and back halves. Flexion is a movement about this plane in which the angle between the surfaces of two adjacent surfaces decreases as the joint is bent. Extension is the opposite movement and causes the angle to increase (except in the thumb where these movements occur in a frontal plane). The arrows show the direction of movement for the hip joint. They are not the same for all joints and, for example, are reversed for the knee.

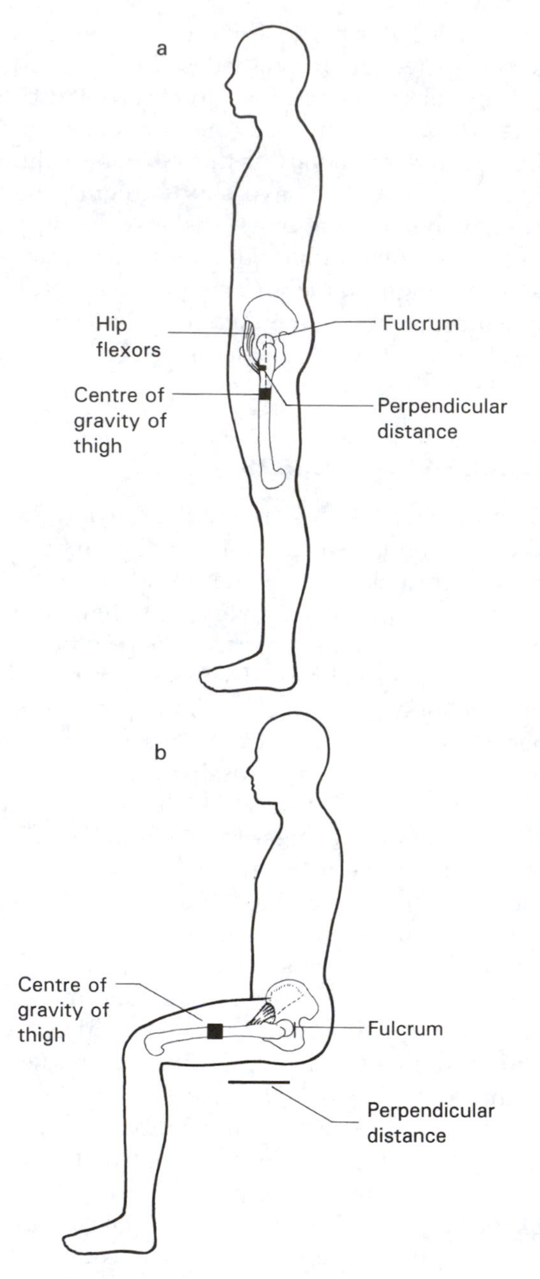

Figure 3.20 In standing (a) the perpendicular distance between the fulcrum and muscle attachment on the thigh is much smaller than when sitting (b). Therefore, the hip flexors need to generate greater forces in sitting.

fulcrum centre (Fig. 3.20). The forces exerted are referred to as *turning moments*. Consider the performance of the same limb action as above when lying supine (straight leg raise).

In the first example the muscle had to generate considerable force at approximately its middle length, which is close to the optimal length for force generation as shown by the active length:tension curve. In the second example the muscle was being asked to work very hard at a longer length, where its ability to generate force is reduced.

During a straight leg raise, the lumbar spine tends to develop a lordosis due to the vertebral attachments of the hip flexors. This can be prevented if the contralateral knee is flexed and the foot rested on the supporting surface, allowing the muscles to work at a more effective length.

Patients with weak hip flexors may have a well preserved gait on level ground as they can use momentum to aid forward propulsion. However, climbing up stairs may present a problem due to the increased antigravity forces required as the lever arm moves further from the fulcrum. The abductors and adductors are most commonly used to stabilise the pelvis, an action which requires considerable strength. A part or all of the body weight is transmitted through each hip, depending on whether one or two limbs are being used for weight bearing. This is shown in Figure 3.21.

The main role of the hip extensors is to propel the body upwards against gravity as when walking upstairs. The hip flexors raise the leg so that the foot can be placed and then the hip extensors raise the body over the foot. In this situation both muscle groups are working concentrically and need to be strong.

The muscles around the hip are shaped like a fan. Depending on the position of the hip in relation to the torso, different muscles are mechanically better suited to produce rotation; the abductors and gluteus maximus produce a great deal of rotation (Fig. 3.22).

Box 3.5 Task 5

Consider the differences between the activity of the agonists during hip abduction when (a) standing on one leg and (b) side lying.

What are the actions of the hip extensors during ascending and descending a flight of stairs?

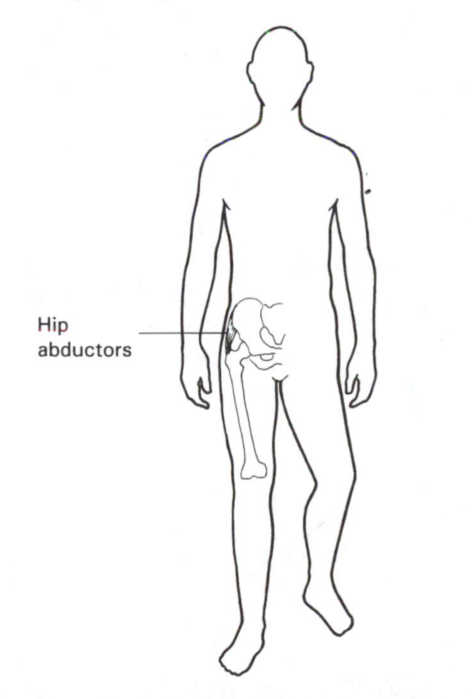

Figure 3.21 The pelvis is stabilised during single leg weight bearing by the action of the hip abductors. When standing on the right leg, the pelvis tends to drop on the left. This is prevented by an isometric contraction of the hip flexors on the right.

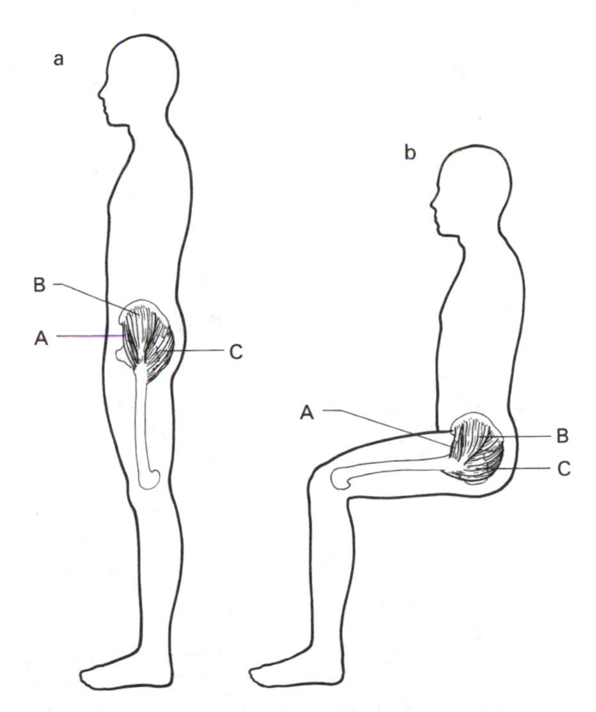

Figure 3.22 The glutei are fan shaped and their action depends on the relative position of the hip and thigh due to the angle of pull. a) In standing gluteus minimus (A) causes medial rotation, gluteus medius (B) causes abduction and the horizontal fibres of gluteus maximus (C) produce lateral rotation of the femur. b) In sitting the angle of pull is different and gluteus medius (B) produces medial rotation.

The knee joint

In sharp contrast to the hip, the knee joint is inherently unstable due to the nature of its bony construction. Two fibrocartilaginous menisci greatly aid stability between the rounded femoral condyles, which sit on an almost flat tibia. Much of the stability of the knee comes from the ligaments. The back of the patella articulates with the femur and is covered in hyaline cartilage. This allows it to move freely over the femur during movements at the knee. The anterior capsule is attached to the sides of the patella.

Movements

The main movements that occur at the knee are extension and flexion. As flexion increases it is possible to gain rotation. The amount of rotation possible slowly increases as the knee flexes, reaches a peak at 90° and thereafter declines.

While the hamstring muscles produce flexion and the quadriceps produce extension of the knee, many of the daily knee movements are caused by the quadriceps. They do this by working concentrically to produce extension and then eccentrically to control flexion speed which is usually caused by gravity.

Biomechanics

The presence of the patella improves the mechanical advantage of the quadriceps by increasing the distance from the patella tendon to the centre of axis of the knee joint.

In daily life the quadriceps muscles nearly always work in their inner to mid-range. During climbing the foot is placed on the step and the quadriceps work in order to propel the body upwards onto an upright knee (Fig. 3.23).

The hamstring muscles often work in con-

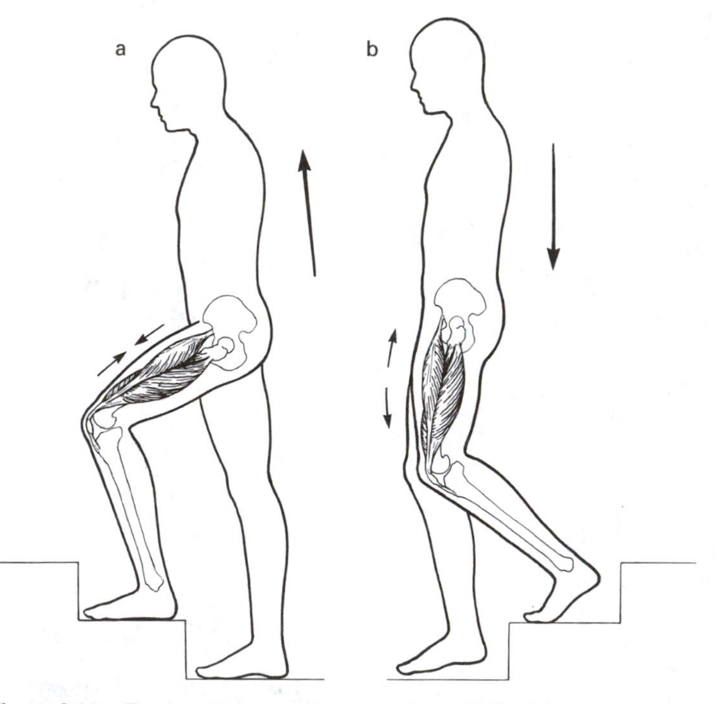

Figure 3.23 The quadriceps are important for activities such as ascending and descending stairs, but their action is different in the two situations. When going up stairs (a) they work concentrically to extend the knee of the leading leg. They work eccentrically when going down stairs (b) as they control the knee of the trailing leg. Without this eccentric contraction, the body weight would cause the knee to flex in an uncontrolled manner.

junction with the quadriceps. The hamstrings flex the knee but extend the hip. When the quadriceps work to extend the knee, the hamstrings may be working simultaneously to bring about hip extension. The hamstring tension aids knee stability by helping to prevent the tibial shift that would otherwise stress the cruciate ligaments. The opposing forces of the two muscle groups at the knee allows the rolling action of the femoral condyles on the flat tibial condyles. This muscular stability is particularly important to those with permanent damage or loss of knee ligaments, particularly the cruciates.

Medial and lateral rotation of the knee are produced by the medial and lateral hamstrings, respectively. This action of the hamstrings, and their ability to rotate the knee, is particularly important as the knee is frequently used in a semiflexed position which permits rotation. Rotation cannot occur when the knee is extended or locked. If we step forward onto a flexed left leg and turn to the left, the foot and tibia are fixed and the lateral hamstrings cause lateral rotation.

Often a twist or a turn at the knee occurs very rapidly. If not controlled, the sudden forced rotation can injure menisci or ligaments.

The popliteus muscle plays a part in knee rotation. The tension it generates in a fully extended knee allows the medial femoral condyle to slide against the tibia. This moves the knee out of the locked position and allows flexion to occur.

Functionally, the knee is very rarely held in the fully extended, locked or close packed position, where no movement can occur and there is maximum congruence between the bony surfaces. To achieve this position, rotation occurs in the final few degrees of extension by the action of the horizontal fibres of vastus medialis.

During weight bearing the lateral condyles of the tibia and femur reach full congruence with the lateral meniscus a few degrees before the medial edge. The action of vastus medialis pulls the patella medially and allows the medial femoral condyle to slide posterolaterally on the medial tibial condyle, bringing the medial structures into total congruence. This action is reversed by the popliteus muscle.

Weakness in the action of vastus medialis in particular gives rise to what is termed *quadriceps lag*. In this situation the passive range of movement is full, but the last few degrees of extension cannot be achieved actively. When the knee is locked, the bony congruence prevents the soft tissues being stressed and, therefore, clinical knee examinations are usually carried out in slight flexion.

The ankle joint

The ankle joint is the articulation between the lower end of the tibia and fibula with the superior surface of the talus. The tibia, the more medial of the two bones, is wider across the talus, the fibula being simply a strut on the lateral side. The two bones are strongly held together at the lower tibiofibular joint.

The capsule has two ligament components to reinforce it. On the medial side it has a very strong ligament, triangular in shape, and on the lateral aspect a weaker ligament, originating from the fibula, consisting of three bands.

Immediately distal to the talus is the calcaneus and the articulation of these two bones, the subtalar joint. It is inappropriate to consider the ankle without this joint, which is strongly reinforced by a capsule and an interosseous ligament which permit very little movement.

Movements

At the ankle joint there is active dorsiflexion, plantar flexion and some inversion and eversion when the foot is plantar flexed. At the subtalar joint, very slight inversion and eversion are regarded as accessory, rather than active, movements. It is important that this tolerance to movement exists to allow for the dissipation of stress and also for accommodation in the fore foot during weight bearing. An excess of movement at this joint would lead to an unstable platform for the ankle joint.

The muscles that operate the ankle joint are in four groups. The dorsiflexors, or anterior tibial muscles, cause pure dorsiflexion only when acting simultaneously. Acting separately the tibialis anterior has a secondary action of foot inversion. The two lateral peronei are primarily evertors of the foot while dorsiflexion and plantar flexion are assisted by peroneus brevis and longus respectively. The posterior deep flexors primarily assist plantar flexion. Tibialis anterior and posterior assist each other to produce inversion. The power of the plantar flexion comes from the gastrocnemius and soleus muscles, whereas flexion of the fore foot on the hind foot is from the deep flexors.

Biomechanics

Most daily activity requires that our bodies are propelled forwards. The plantar flexors work strongly, and often simultaneously, as an anti-gravitational force. The gastrocnemius and soleus muscles, therefore, work powerfully together concentrically to produce plantar flexion at the ankle joint. Gastrocnemius is used more when power and speed are required in activities such as running, jumping or climbing stairs. During the push off phase in walking and running these two muscles can be seen to work strongly. Gentle walking may use only the soleus muscle, but as speed and power requirements increase, the gastrocnemius becomes more active. In dorsiflexion, achieved by the combined action of the anterior tibial muscles, power is usually less important than in the calf muscles. Frequently the foot is simultaneously dorsiflexed and inverted, as when clearing the foot off the ground during walking. Tibialis anterior plays a very important role in achieving this activity. Reduced strength and power from these muscles causes a drop foot. Often these muscles work eccentrically to control plantar flexion against gravity.

In routine activity the foot rarely operates in full dorsiflexion, the close pack position. Therefore, there is some surface incongruity at the ankle joint. Stability is created by the muscles acting like guy ropes around a ship's mast. Each muscle has its counterpart, e.g. tibialis anterior and posterior. This enables the fast reactions required on uneven ground and the lack of congruence is an important stabilising influence.

The reaction speed of the peronei is thought to be important in preventing lateral ankle sprains. Following a ligament sprain at the ankle, the muscular response time is decreased (Lentell et al 1990). For this reason, balance and proprioception should be assessed during rehabilitation.

The foot

The tarsal and metatarsal bones are held together by strong ligaments and capsules. Numerous synovial junctions exist between the various bones.

Movements

Between each junction some movement occurs. It is normally thought of as an accessory movement where one bone can move against its neighbour, rather than an active, volitional movement.

Biomechanics

These subtle movements are essential to allow the foot to mould and operate effectively on uneven ground.

Box 3.6 Task 6

Compare the actions of moving from sitting to standing and walking. Consider the range of movement of the lower limb joints and which muscles produce both movement and stability.

The vertebral column

The adult lumbar and cervical spine are characterised by a lordosis which is due to a slight wedge shape in both vertebral bodies and intravertebral discs.

The vertebral column is designed to accept the vertical stresses and strains of everyday life and also to protect the very delicate spinal cord and neural structures that run down its centre. Strands of bone fibres run across the width of the vertebral body and give strength and enhance the ability to resist compressive forces. The ability to dissipate stress is important in preventing the skull from jarring down onto the top of the neck. The vertical stresses are passed posteriorly throughout the lamina to the pedicles and finally to the spinous processes and into the soft tissues. The pedicles, acting almost vertically compared with the horizontal lamina, comprise hard cortical bone around a hollow centre. The design of the pedicles allows for stresses and strains to be transmitted along their edges and, therefore, distortion may occur without causing fractures (Bogduck & Twomey 1991). The discs are equally designed to accept stresses and allow distortion to accommodate for movement of the column.

Strong ligaments run between the bones throughout the entire length of the vertical column. The ligamentum flavum is particularly important as it contains a high number of elastic fibres that are never fully relaxed. In natural standing the ligament exerts a vertical compressive force at each level.

The major muscles producing extension of the vertical column are the posterior erector spinae muscles. These vary enormously in length and are arranged in layers. Spinal flexion is produced by the muscles on the anterior abdominal wall, collectively known as the abdominals.

Movement

Movement is dependent on the shape of the interlocking facets of the synovial joints of the articulating vertebral pillar which occur at every level. In the lumbar spine these joints allow the movements of flexion, extension and lateral flexion and little or no rotation occurs. The thoracic region is much more rigid, particularly the upper thoracic cage which protects the heart and the lungs. The movement that does occur is mostly that of rotation. In the neck, with the exceptions of C1 and C2, the articulating pillar allows for flexion, extension, rotation and lateral flexion. The movement between C1 and C2 is mostly rotation, while movement between the skull and C1 is mostly flexion and extension as in nodding.

Biomechanics

Once again gravity is important in this area. The muscles which produce or control the movement depend on the starting position. If a person standing upright bends forward at a rate slower than that dictated by gravity, the erector spinae work eccentrically to control the movement and prevent the torso from falling forwards. When returning to the upright position, the same muscles act concentrically as extensors.

However, when moving from supine lying to sitting, the abdominals, particularly rectus abdominis, work concentrically to bring the thoracic cage closer to the pelvis and cause spinal flexion against gravity. When returning to supine lying, these muscles work eccentrically to control the movement against gravity.

The internal and external oblique muscles work to produce trunk rotation. Transversus abdominis controls and restrains the abdominal contents. The thoracic and abdominal cavities are separated by the diaphragm which is under semiautomatic control. As it contracts and relaxes, it alters the pressures within the thoracic and abdominal cages. When contracting, it descends towards the abdomen, increasing the size of the thoracic cage and allowing the lungs to expand, so increasing the intake of air. When relaxing, the muscle rises in the thorax, forcing air out of the lungs. This accounts for the majority of the work of breathing at rest. The rib cage moves in conjunction with the movement of the diaphragm.

Box 3.7 Question 1

Consider the actions of quiet breathing and coughing. Which muscles are more involved in coughing than in natural breathing, and why do you think that this occurs?

In coughing, an increase in expulsion of the thoracic content is required. Therefore, the abdominal pressure is maximised by the simultaneous activity of the diaphragm, abdominal and pelvic muscles.

The weak link in the lumbar spine appears to be the fibrocartilaginous disc. In midlife the nucleus dries, the annulus becomes fairly brittle and is subjected to greater stresses thus becoming prone to microfractures (Bogduck & Twomey 1991). Studies of the stresses likely to cause damage in the lumbar spine indicate that disc pressure is greatest during sitting and flexion and least during lying and intermediate during standing and walking (Nachemson & Morris 1964).

Many activities require movement of the hands towards the floor, thus increasing the lever, or moment arm, across the lower back. The weight of the torso, acting approximately halfway along its length, increases its perpendicular distance from the fulcrum at L5 and also increases the pressure in the lumbar spine. Therefore, during lifting or hip flexion with extended knees, the pressure on the lumbar spine is high. If a load is added at the hands, the perpendicular distance from the centre of gravity of the torso to the fulcrum increases.

Keeping the movement arm as short as possible, i.e. keeping the hands close to the trunk, helps to decrease the affective stresses in the lumbar spine. Studies using electromyography and biomechanical modelling techniques using the pressure data of Nachemson & Morris (1964) have led to the recommendation of crouch lifting to reach the floor and lift objects. This literature is well reviewed by Chaffin & Anderson (1990), Basmajian & Deluca (1979), and Singleton (1986). In practice, however, the recommended posture is often impractical and creates an increased leverage over the knee joint which can make the lifting of certain shaped objects difficult. It is often suggested that increasing the power and strength of the muscles surrounding the abdominal cavity will assist and protect the lumbar spine. By increasing the pressure and strength of these muscles, it should be possible to increase the pressure exerted on the abdominal cavity. Since the pressure on contained fluids increases in all directions, this would assist in decompression of the loaded lumbar spine and decrease the stresses upon it. However, it is generally recommended that the muscle strengthening exercises should not involve spinal movement. Static muscle contractions help increase strength without distorting the disc.

Box 3.8 Question 2

Imagine lying flat on the floor with the knees bent and feet resting on the floor. Gently tuck in the chin and raise the shoulders only off the floor. Hold this position and then return slowly to the start position. Which muscles are contracting and how are they working?

Rectus abdominis works concentrically when raising the shoulders, isometrically to hold them and eccentrically when returning to the start position.

Box 3.9 Question 3

Consider the same starting position, lying flat on the floor with the knees bent, but this time take your right arm over to your left knee. Which muscles are working and how?

The right external oblique and left internal oblique work concentrically when rising and eccentrically when returning to the start position.

The shoulder complex

The arm comprises a collection of complex joints which are often thought to start with the glenohumeral joint. However, the sternoclavicular and acromioclavicular joints need to be considered along with the shoulder.

Movements

The clavicle is able to elevate, depress and rotate. The greatest movement is elevation and depression at the distal end. At the medial (sternal) end, small gliding movements occur against the sternum. If the sternal end depresses, the acromial end will elevate. The clavicle is also capable of protraction and retraction; if the sternal end protracts the acromial end will retract and vice versa. The scapula, strongly associated with the clavicle, is also capable of elevation and depression, protraction, retraction and rotation and any movement that occurs at the acromial end of the clavicle is mimicked by the scapula.

Movements of the scapula and clavicle are closely associated with those of the glenohumeral joint. A strong fibrous capsule is loose inferiorly. The joint is fundamentally designed to supply mobility, rather than stability, to the upper limb. The joint is capable of the same six anatomical movements available at the hip. Circumduction allows for a huge excursion and combination of movements.

In conjunction with movement of the elbow and hand, the glenohumeral joint, along with scapula and clavicular movements, enables the hand to be placed practically anywhere on the body.

The movements of the scapula and clavicle are performed mostly by muscles attached to the scapula. The rhomboids retract and serratus anterior protracts the scapula. The large, triangular trapezius operates the scapula by working in several ways. If it acts as a single unit, the scapula retracts. If the upper fibres alone contract, the scapula will rise at the acromial end and external rotation occurs. If the lower fibres act independently, the reverse happens, the scapula is depressed at the acromial end and the inferior angle is internally rotated. The middle fibres acting alone cause retraction. Levator scapula assists the upper fibres of trapezius in scapula elevation and the fibres of latissimus dorsi assist the lower fibres of trapezius in scapula depression.

There are two types of muscle that control movement of the glenohumeral joint. Small muscles travelling from the scapula to the humerus control the position of the head of the humerus in the glenoid cavity. There are also large and powerful muscles which run from the torso to the humerus.

Biomechanics

The small muscles are known collectively as the rotator cuff and consist of supraspinatus, infraspinatus, subscapularis, teres major and minor and are assisted by the long heads of triceps and biceps. Their role is vital in controlling and stabilising the head of the humerus.

The alignment of supraspinatus enables it to initiate abduction, while that of infraspinatus is at a good angle for initiating lateral rotation and assisting adduction. It is also able to assist in depressing the head of the humerus within the glenoid during abduction. Subscapularis is at the correct angle for medial humeral rotation and assisting infraspinatus in controlling the descent of the humeral head during movement and keeping it in contact with the glenoid cavity. These actions are reinforced by teres major and minor, along with the long heads of biceps and triceps. In stabilisation of the shoulder joint, they are assisted by the long heads of biceps and triceps. These muscles also initiate flexion and extension, respectively, and help to keep the humeral head in contact with the glenoid cavity. Together they are vital in maintaining stability of the humeral head and provide a fixed point of attachment for the prime movers.

The head of the humerus glides down within the cavity during arm abduction, backwards during flexion and forwards during extension. The rotator cuff muscles are responsible for this fine control. They act in a coordinated way like guy ropes to control the humeral head and alter its position in relation to the glenoid to keep maximum congruity. They can be seen in many instances almost as elastic ligaments because of their subtle control which provides stability for the anatomically unstable joint. As the tendons of supraspinatus, infraspinatus and teres minor travel underneath the acromium, they are protected by one of the largest fluid bursa in the body. This bursa prevents the tendons becoming trapped as the humeral head rises in abduction.

The main power muscle across the glenohumeral joint is the deltoid which forms the shape of the shoulder and causes abduction, flexion and extension. This muscle can act in unison as an abductor, while the anterior fibres alone cause flexion. The middle fibres abduct and the posterior fibres extend the glenohumeral joint. This fan-shaped muscle behaves like the glutei at the hip joint. The fibres recruited depend on the position of the arm and torso, those at the best mechanical advantage are recruited first. For the majority of daily activities the arm is abducted. This is initiated by supraspinatus to give deltoid the correct angle of pull to perform the power action required. Without this action of supraspinatus, contraction of the deltoid would result in the humerus rising in the glenoid cavity. Consider the action of abducting a straight upper limb and taking it to full elevation when standing. The centre of gravity of the upper limb is approximately at the elbow and the perpendicular distance to the fulcrum is at its greatest as the upper limb goes through 90°. At this point deltoid has to work hardest and it is also at its middle length range and at the strongest point on the active length:tension curve. Either side of this point the perpendicular distance from the elbow to the fulcrum decreases, as does the length of the movement arm forces that have to be overcome. Therefore, less strength is required from the muscle which is working in less favourable areas of the active length:tension curve and so is capable of producing less power.

Box 3.10 Task 7

Look at a colleague and see how the scapula, clavicle and glenohumeral joint move together in order to produce elevation of the arm.

The movement is complex; elevation through abduction is performed by abduction of the glenohumeral joint, but simultaneously there is a rotation of the scapula and clavicle. To gain full elevation there is a lateral rotation of the glenohumeral joint for the last 30°.

Other powerful shoulder muscles are the pectoralis major and latissimus dorsi. These muscles together produce adduction. They are essential in activities that require stabilisation of the shoulder when the arm and the body are held close together. They also bring the arm and trunk forcibly towards each other in activities such as climbing. Strength in these muscles is necessary for activities such as crutch walking, where the body weight is taken through the upper limbs and the torso moves forwards and backwards.

When performing 'press ups', the scapula needs to be anchored and prevented from moving away from the chest wall in order to provide for the glenohumeral movement. The proximal fibres of serratus anterior work with the rhomboids as stabilisers while trapezius generates both power and stability. This allows subtle glenohumeral movement and the deltoid to generate powerful contractions.

The elbow joint

By comparison with the glenohumeral, the elbow joint is anatomically extremely stable. The convex ulna tightly hugs the lower end of the humerus and glides around the humerus, allowing flexion and extension. The flat radial head simply glides around the lower end of the humerus. The radius and the ulna unite at the superior radioulnar joint. The elbow joint is held together by two very strong collateral ligaments. From the ulna a very strong annular ligament hugs the rounded head of the radius, enabling it to rotate against the ulna. The capsule encompasses the elbow joint and the superior radioulnar joint.

Movements

The muscles acting across this joint are mainly the triceps posteriorly and the biceps and brachialis anteriorly which produce extension and flexion, respectively. The radial attachment of biceps also enables it to cause supination at the wrist.

Biomechanics

In everyday life much of the activity at the elbow joint requires flexion or extension to be controlled against gravity. For example, biceps will work concentrically when flexing the elbow to raise a weight against gravity, isometrically to maintain a given elbow position and eccentrically when controlling the rate of extension due to gravity.

The inferior radioulnar joint, just above the wrist, is very different to the superior radioulnar joint. Here the radius is allowed to rotate across the ulna by a very loose capsule. In the anatomical position, the forearm is held in supination; rotation or pronation is allowed through approximately 180°.

Wrist and hand

The wrist joint is normally defined as the radiocarpal joint since the radius alone articulates with the carpal bones. The ulna articulates with the carpal bones through a disc.

Movements

This joint allows for flexion, extension, radial and ulnar deviation. The flexors and extensors originate from the common flexor and extensor tendons on the medial and lateral humeral epicondyles, respectively. The muscles operating the wrist attach distally to the carpus and metacarpal bones. Running centrally across the wrist are the flexors and extensors of the fingers, flexor digitorum and extensor digitorum. Flexor carpi ulnaris and extensor carpi ulnaris have dual functions: working together they cause ulnar deviation at the wrist, independently they either flex or extend. Extensor carpi radialis longus and brevis, along with flexor carpi radialis, also have dual function: working synergistically they radially deviate the wrist, while independently they assist either extension or flexion of the wrist. Extensor digitorum with flexor digitorum profundus and superficialis are responsible for extension and flexion, respectively, of the wrist and the fingers.

The muscles of the thenar eminence are extremely important in assisting and generating force, as well as the position of the thumb. The hypothenar eminence is much less developed but it is important in forming an opposition to the thumb. The interossei and lumbricales also work in assisting actions for the fingers.

Biomechanics

The importance of the hand is that its anatomical and muscular construction enables it to be extremely flexible and yet produce a fairly rigid structure when required. The carpal bones can be made fairly rigid by muscular activity, as when pushing against a heavy door. Small and subtle movements also enable delicate actions such as cupping the hand and the delicate movements vital for everyday activity. The versatility of the hand, particularly the thumb, gives humans their unique ability to undertake a broad spectrum of activities ranging from precision activities to making a fist.

Box 3.11 Task 8

Consider the coordinated activity of opening a door, turning the handle and pulling the door towards you. What movements are occurring in the various joints; which muscles are active and how are they working? Start with the hand, work up the forearm to the elbow and finally the shoulder.

Consider the activity of eating an apple. Pick it up from the table and go through the motion of trying to eat, thinking of the movements of the arm and hand and the muscle activity necessary to produce them.

REFERENCES

Basmajian J V, Deluca C J 1979 Muscles alive, 5th edn. Williams and Wilkins, Baltimore

Bogduck N, Twomey L T 1991 Clinical anatomy of the lumbar spine, 2nd edn. Churchill Livingstone, Edinburgh

Broer M, Zernicke R F 1979 Efficiency of human movement, 4th edn. W B Saunders, Philadelphia

Binder-Macleod S A 1992 Force–frequency relation in skeletal muscle. In: Currier D P, Nelson R M (eds) Dynamics of human biologic tissues. F A Davis, Philadelphia

Chaffin D B, Anderson G B J 1990 Occupational biomechanics. John Wiley, New York

Clarkson P M, Newham D J 1995 Associations between muscle soreness, damage and fatigue. In: Gandevia S C, Enolta R M, McGomas A J, Stuart D G, Thomas C K (eds) Advances in experimental medicine and biology. Fatigue—neural and muscular mechanisms. Plenum Press, New York, pp 457–470

Gordon A M, Huxley A F, Julian F J 1966 The variation in isometric tension with sarcomere length in vertebrate muscle fibres. Journal of Physiology 184: 170–192

Gowitzke B A, Milner M 1980 Understanding the scientific bases of human movement, 2nd edn. Williams and Wilkins, Baltimore

Guyton A C 1976 Textbook of medical physiology, 5th edn. W B Saunders, Philadelphia
Hennemann E, Clamann H P, Gillies J D, Skinner R D 1974 Rank order of motoneurons within a pool, law of combination. Journal of Neurophysiology 37: 1338–1349
Huxley A F, Simmons R M 1971 Proposed mechanism of force generation in striated muscle. Nature 233: 533–558
Jones D A, Round J M 1990 Skeletal muscle in health and disease. Manchester University Press, Manchester
Kapandji I A 1978 The physiology of the joints. Vol 2 Lower limb. Churchill Livingstone, Edinburgh
Komi P V 1986 The stretch shortening cycle and human power output. In: Jones N L, McCartney N, McComas A J (eds) Human muscle power. Human Kinetics, Champaign, Illinois
Lehmkuhl L D, Smith L K 1983 Brunnstrom's Clinical Kinesiology, 4th edn. F A Davies, Philadelphia
Lentell G L, Katzman L L, Walters M R 1990 The relationship between muscle function and ankle stability. Journal of Orthopaedic and Sports Physical Therapy 11: 605–610
Leveau B F, Bernhardt D B 1984 Developmental biomechanics. Effect of forces on the growth, development and maintenance of the human body. Physical Therapy 64: 1874–1881
Lieber R L 1992 Skeletal muscle structure and function. Williams and Wilkins, Baltimore
MacConaill M A, Basmajian J V 1977 Muscles and movements – a basis for human kinesiology, 2nd edn. R E Kreiger, New York
Nachemson A L F, Morris J M 1964 In vivo measurements of intradiscal pressure. Journal of Bone and Joint Surgery 46A: 1077–1092
Palstanga N, Field D, Soames R 1990 Anatomy and human movement. Butterworth Heinemann, Oxford
Singleton W T 1986 The body at work, 2nd edn. Cambridge University Press, New York

CHAPTER CONTENTS

4

The neural control of human movement

J. L. Crow

OBJECTIVES

When you have completed this chapter you should be able to:

1. **Describe the roles of the different levels of the central nervous system in the neural control of movement**
2. **Explain the various interconnections between the different levels of the central nervous system and the periphery**
3. **Understand how information from the periphery can alter or shape a movement response**
4. **Demonstrate a knowledge of the overall control of the body with specific reference to rapid movements**
5. **Consider the close relationship between postural control and active movements.**

INTRODUCTION

Human movement is a complex affair. Normal movements are often automatic in nature and only come under volitional control when circumstances change or as a consequence of new experiences. The unique functional ability of humans is possible through the quality of movement produced from normal coordinated patterns of movement. Such patterns are produced on a background of normal sensory information and feedback, normal tone, reciprocal innervation, normal balance and postural reactions (Todd & Davies 1986).

The control of movement has traditionally been considered to be hierarchical with the highest, cortical level of control organising voluntary, skilled movement. However, in reality there is no separation between voluntary movements and the background of postural control that maintains the body in an upright position with the aid of automatic reflexes and responses. Therefore, parallel systems of control, with integration of all levels rather than just a serial hierarchy, may be a more appropriate description. All levels of control, from the spinal cord up to the cerebral cortex, are necessary and integrated to provide the base of axial stability for more normal distal mobility and skilled or refined coordinated limb movements (Kidd et al 1992).

An understanding of the neural mechanisms involved in normal movement control is essential for the assessment and treatment of disorders of movement resulting from musculoskeletal injury and damage to the nervous system. The neural control of posture and movement, whether voluntary or automatic, will be considered briefly as an overall system and then subdivided into centres (levels) of control within the central nervous system (CNS).

A basic knowledge of neuroanatomy and neurophysiology is assumed when reading this chapter and will not be covered directly. Rather, the applied neurophysiology of the control of movement will be discussed. Therefore, the reader may find it useful to recap their knowl-

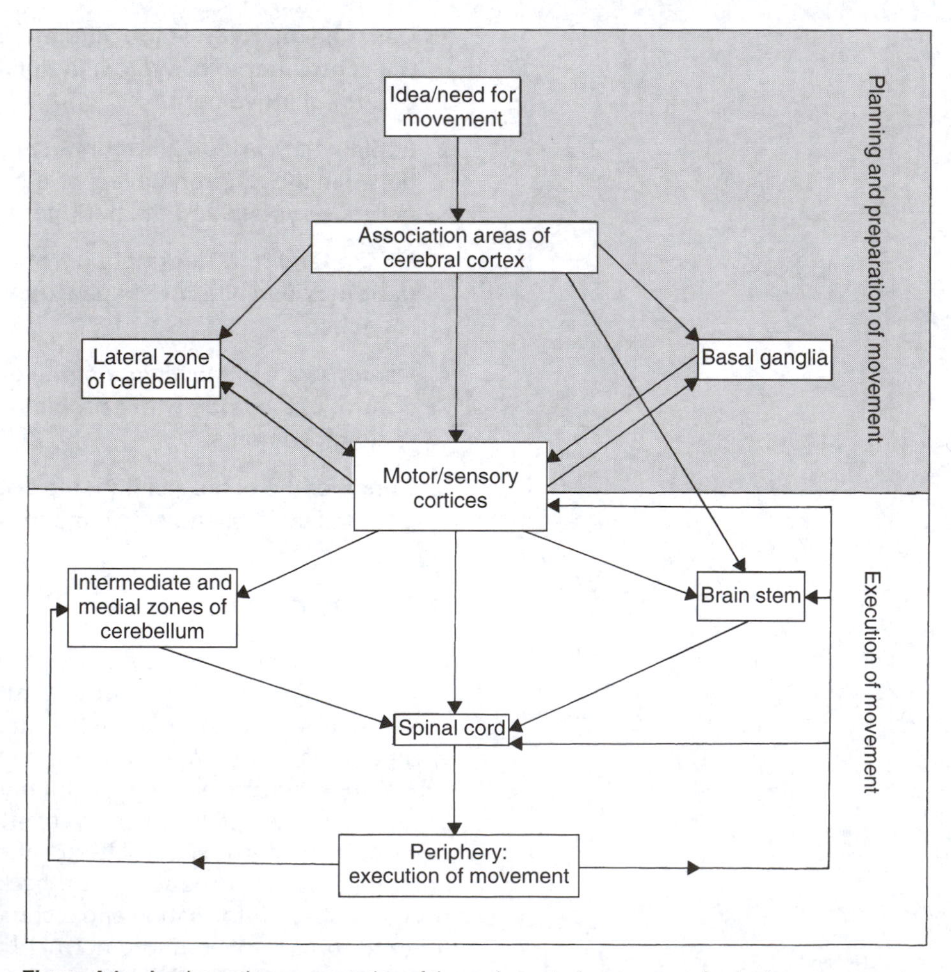

Figure 4.1 A schematic representation of the main control processes of voluntary movement.

edge of the basic neurophysiology with particular reference to the role of the muscle spindle and the stretch reflex prior to studying this chapter.

The development of motor control is well described elsewhere and provides a useful background to normal child growth and development. This section will, therefore, restrict itself to the control of posture and movement of the normal adult.

OVERVIEW OF MOVEMENT AND POSTURAL CONTROL

For any motor function to occur an input of sensory information is required. Sensory information is integrated at all levels of the nervous system and causes appropriate motor responses, i.e.

- Spinal cord: simple reflexes (automatic, stereotyped reflex movement); the peripheral execution level of movement.
- Brain stem and basal ganglia: more complicated responses (postural and balance reactions) able to affect the spinal cord to produce or change automatic movement.
- Cerebrum: most complicated responses controlled (variable and adaptable skilled voluntary movement based on stored programmes of learnt movements). The motor, sensory and associated areas begin the chain of commands for most movements, though this may be initiated or modified at lower levels.
- Cerebellum: overall planning, timing and predictive function to produce coordinated skilled and rapid movements.

Figure 4.1 illustrates this movement control system with the interconnections between the main levels of the central nervous system suggesting a concurrent parallel organisation (Kidd et al 1992).

ORGANISATION OF THE SPINAL CORD FOR MOVEMENT CONTROL

The grey matter of the spinal cord is the integrative area for the cord reflexes and other automatic motor functions. As the region for the peripheral execution of movements, it also contains the circuitry necessary for more sophisticated movements and postural adjustments.

Sensory signals enter the cord through the sensory nerve roots and then travel to two separate destinations:

- Same or nearby segments of the cord where they terminate in the grey matter and elicit local segmental responses (excitatory; inhibitory; reflexes, etc.).
- Higher centres of the CNS, i.e. higher in the cord, and brain stem cortices where they provide conscious (and unconscious, i.e. cerebellum) sensory information and experiences.

Each segment of the cord has several million neurones in the grey matter which include sensory relay neurones, anterior motor neurones and interneurones.

Interneurones are small and highly excitable with many interconnections, either with each other or with the anterior motor neurones. They have an integrative/processing function within the spinal cord as few incoming sensory signals to the spinal cord or signals from the brain terminate directly on an anterior motor neurone. This is essential for the control of motor function. One specific type of interneurone is called the Renshaw cell, located in the anterior horn of the spinal cord. Collaterals from one motor neurone can pass to adjacent Renshaw cells which then transmit inhibitory signals to nearby motor neurones. So stimulation of one motor neurone can also inhibit the surrounding motor neurones. This is termed **recurrent** or **lateral inhibition**. This allows the motor system to focus or sharpen its signal by allowing good transmission of the primary signal and suppressing the tendency for the signal to spread to other neurones (Eccles 1977).

In addition to interneurones there are also propriospinal fibres that run from one segment of the cord to another, so providing pathways for multisegmental reflexes, i.e. those reflexes that coordinate movement in different limbs simultaneously.

MUSCLE RECEPTORS — THEIR ROLE IN MUSCLE CONTROL

For control of motor function, muscles need to be

activated by excitation of the motor neurones. There is also a need for continuous feedback from each muscle to the central nervous system giving the status of that muscle, i.e. length, instantaneous tension, and rate of change of length and tension. Muscle spindles detect rate and changes in the length of a muscle whereas golgi tendon organs detect degree and rate of change of tension. Signals from these sensory receptors operate at an almost subconscious level, transmitting information into the spinal cord, cerebellum and cerebral cortex where they assist in the control of muscle contraction.

The muscle spindle has both a static and dynamic response. The primary and secondary endings respond to the length of the receptor, so impulses transmitted are proportional to the degree of stretch and continue to be transmitted as long as the receptor remains stretched. If the spindle receptors shorten, the firing rate decreases. Only the primary endings respond to sudden changes of length by increasing their firing rate and only whilst the length is actually increasing. Once the length stops increasing, the discharge returns to its original level, though the static response may still be active. If the spindle receptors shorten, then the firing rate decreases.

Control of the static and dynamic response is by the gamma motor neurone. Normally, the muscle spindle emits sensory nerve impulses continuously with the rate increasing as the spindle is stretched (lengthened) or decreasing as the spindle shortens.

The stretch or myotactic reflex can therefore be considered as:

Dynamic: that is an instantaneous, strong reflex elicited from primary endings only, with the function to oppose sudden changes in length of muscle as the muscle contraction opposes the stretch.

Static: that is a weaker reflex elicited from primary and secondary endings which continues for a prolonged period with the function to continue to cause muscle contraction as long as the muscle is maintained at excessive length.

The stretch reflex also has the ability to prevent some types of oscillation and jerkiness of body movements even if the input is jerky, i.e. a damping function (Palastanga et al 1990).

Role of the muscle spindle in voluntary motor activity

When the motor cortex or other areas of the brain transmit signals to the alpha motor neurones, the gamma motor neurones are nearly always stimulated simultaneously, i.e. a coactivation of the alpha and gamma systems so that intra- and extrafusal muscle fibres (usually) contract at the same time. This stops the muscle spindle opposing the muscle contraction and maintains a proper damping and load responsiveness of the spindle regardless of change in muscle length. If the alpha and gamma systems are stimulated simultaneously and the intra- and extrafusal fibres contract equally, then the degree of stimulation of the muscle spindle will not change. If the extrafusal fibres contract less because they are working against a great load, the mismatch will cause a stretch on the spindle and the resultant stretch reflex will provide extra excitation of the extrafusal fibres to overcome the load (Fig. 4.2).

The gamma efferent system is excited/controlled by the bulboreticular facilitatory region of the brain stem with influences from impulses transmitted to that region from the cerebellum, basal ganglia and cerebral cortex. The bulboreticular facilitatory area is concerned with antigravity muscle contractions (antigravity muscles have a higher density of muscle spindles). Therefore, the gamma system has an important role in controlling muscle contraction in positioning different parts of the body and for damping the movement of the different parts (Guyton 1991).

The golgi tendon organ

The golgi tendon organ, as a sensory receptor in the muscle tendon, detects relative muscle tension. Therefore, it is able to provide the CNS with instantaneous information of the degree of tension on each small segment of each muscle. The

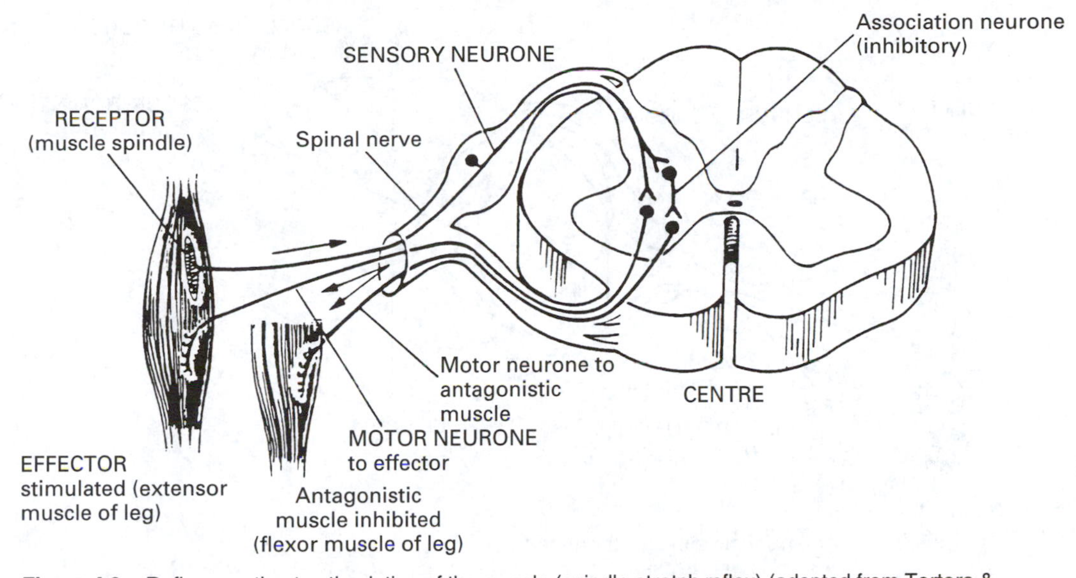

Figure 4.2 Reflex reaction to stimulation of the muscle (spindle stretch reflex) (adapted from Tortora & Anagnostakos 1987; reprinted by permission of HarperCollins Publishers, Inc.).

golgi tendon organ is stimulated by increased tension. When the increase in tension is too great, the tendon reflex response is evoked in the same muscle and this response is entirely inhibitory. The brain dictates a set point of tension beyond which automatic inhibition of muscle contraction prevents additional tension. Alternatively, if the tension decrease is too low, then the golgi tendon organ reacts to return the tension to a more normal level. This leads to a loss of inhibition so allowing the A-alpha motor neurone to be more active and increase the muscle tension.

Information from the golgi tendon organ has both a direct and indirect effect. Signals travel to local areas of the spinal cord and excite a single inhibitory interneurone which then directly inhibits the A-alpha motor neurone to that individual muscle only. Information is then also spread to all the higher centres through indirect means (Guyton 1991).

The golgi tendon organ is similar to the muscle spindle in that it has a static and a dynamic response. The static response provides a low level of steady state firing proportional to the muscle tension. The dynamic response allows rapid/intense firing when muscle tension suddenly increases.

If the golgi tendon organ is stimulated by an increased tension, the signals transmitted cause an inverse stretch reflex effect in the respective muscle, which is entirely inhibitory (Fig. 4.3). This serves as a negative feedback to prevent development of too much tension. If the tension is extreme, the golgi tendon organ tendon reflex effect is to cause sudden relaxation of the entire muscle, a lengthening reaction. This protective mechanism prevents muscle tears or tendon avulsions.

With the golgi tendon organ being in series with the muscle fibres it is stimulated by both passive and active stretch. A passive stretch has a slow response as the more elastic fibres take up most of the stretch first. It actually takes a strong active contraction of the muscle to produce the relaxation of autogenic inhibition (Ganong 1991). Table 4.1 summarises and compares the roles of the golgi tendon organ and the muscle spindle.

With the stretch reflex, as the supplied muscle is excited, the antagonistic muscle is simultaneously inhibited to allow movement to occur (Figs 4.2 and 4.4). This phenomenon is called reciprocal inhibition (postsynaptic) with the neural mechanism termed reciprocal innervation

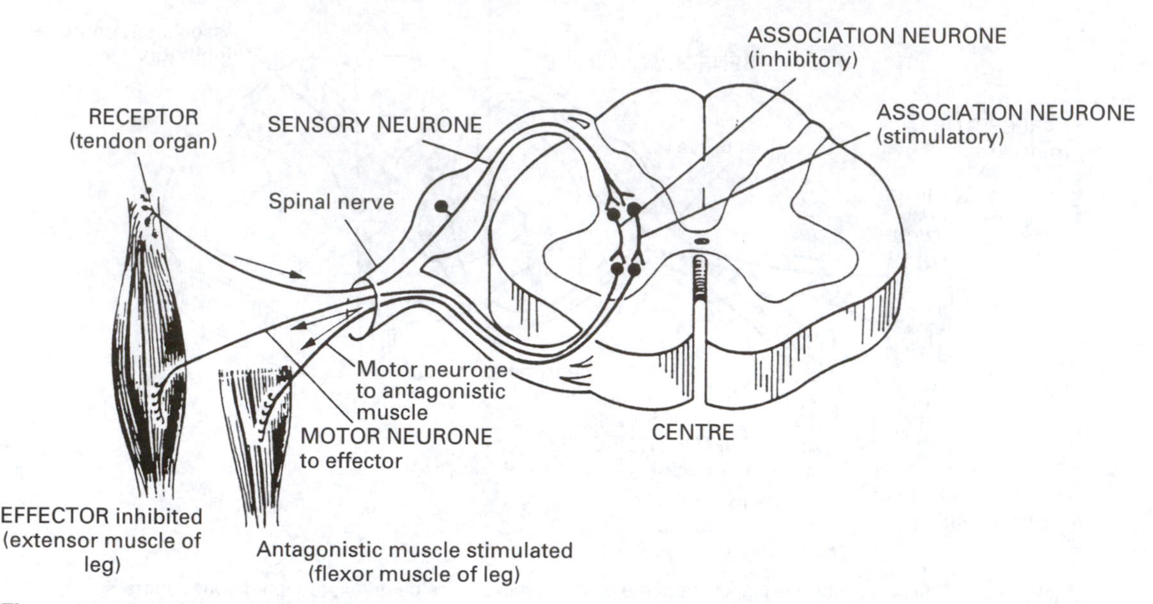

Figure 4.3 Reflex response to stimulation of the golgi tendon organ (tendon reflex) (adapted from Tortora & Anagnostakos 1987; reprinted by permission of HarperCollins Publishers, Inc.).

Table 4.1 Comparison of the roles of golgi tendon organs and muscle spindles

Golgi tendon organ	Muscle spindle
Detects relative muscle length	Detects relative muscle tension
In series with the muscle fibres	In parallel with the muscle fibres
Tendon reflex is a feedback mechanism to control muscle tension	Stretch reflex is a feedback mechanism to control muscle length
Has dynamic and static responses	Has dynamic and static responses
During active concentric movement	
Increase in tension in the tendons	Decrease in muscle length
During passive movement	
Increase in tension in the tendons	Increase in muscle length

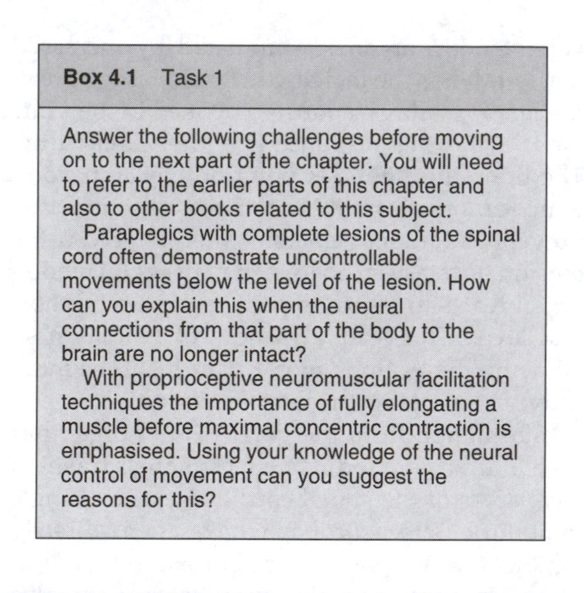
Box 4.1 Task 1

Answer the following challenges before moving on to the next part of the chapter. You will need to refer to the earlier parts of this chapter and also to other books related to this subject.

Paraplegics with complete lesions of the spinal cord often demonstrate uncontrollable movements below the level of the lesion. How can you explain this when the neural connections from that part of the body to the brain are no longer intact?

With proprioceptive neuromuscular facilitation techniques the importance of fully elongating a muscle before maximal concentric contraction is emphasised. Using your knowledge of the neural control of movement can you suggest the reasons for this?

or a flexor reflex. With the inverse stretch reflex, the direct effect to the muscle supplied is one of inhibition and relaxation with additional information being sent to the motor neurones supplying the antagonist muscle causing excitation in that muscle group (Fig. 4.3).

BRAIN STEM AND BASAL GANGLIA LEVEL OF CONTROL OF POSTURE AND MOVEMENT

The principal role of the brain stem in control of motor function is to provide background contractions of the trunk, neck and proximal parts of limb musculature so providing support for the body against gravity. The relative degree of contraction of these individual antigravity muscles

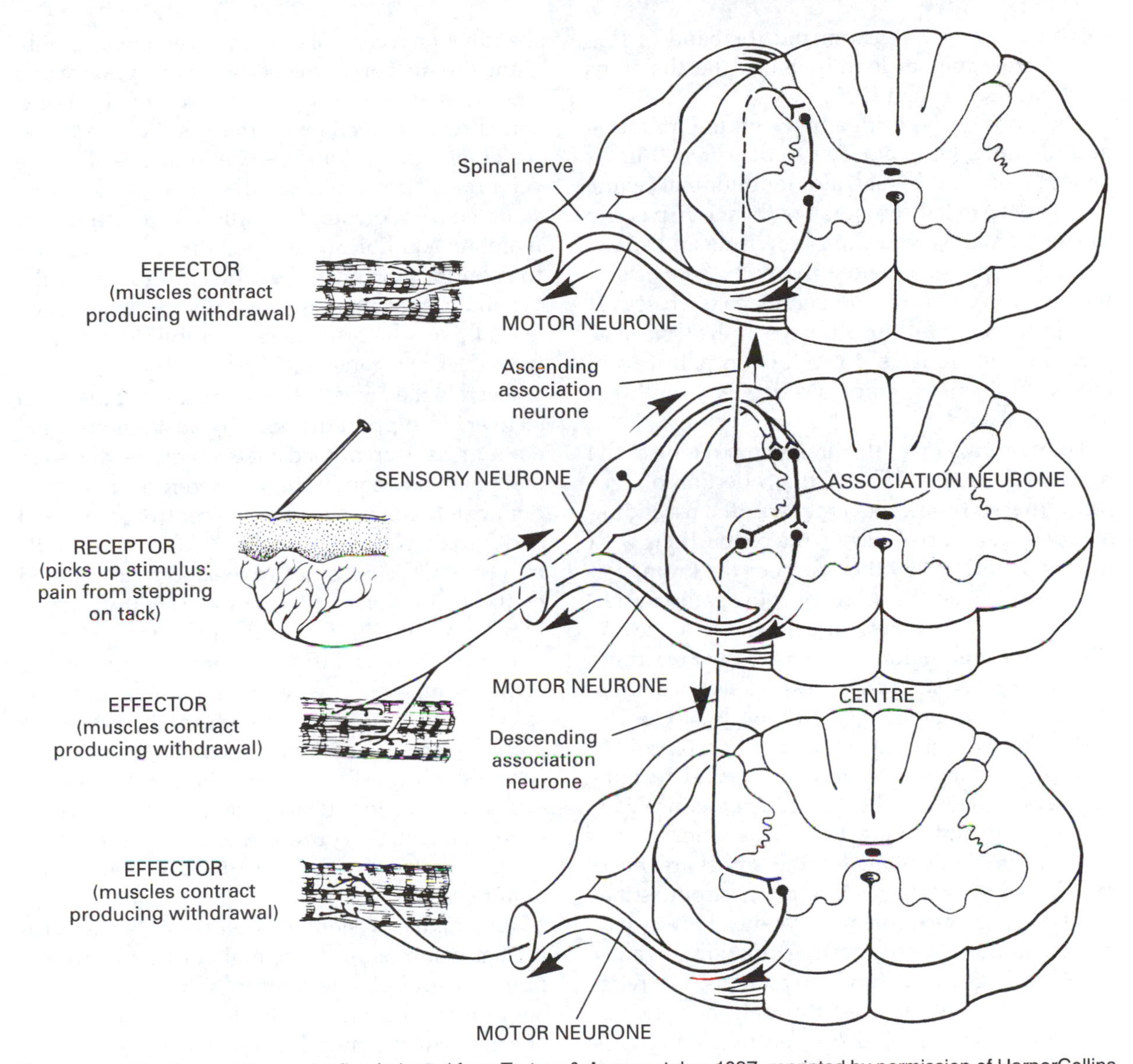

Figure 4.4 The flexor withdrawal reflex (adapted from Tortora & Anagnostakos 1987; reprinted by permission of HarperCollins Publishers, Inc.).

is determined by equilibrium mechanisms, with reactions being controlled by the vestibular apparatus which is directly related to the brain stem region.

The brain stem connects the spinal cord to the cerebral cortex. It is comprised of the midbrain (mesencephalon), pons and medulla oblongata. The central core of this region is often referred to as the recticular formation. This region of the central nervous system comprises all the major pathways connecting the brain to the spinal cord in a very compact, restricted space. It is also the exit point of the cranial nerves from the central nervous system.

The midbrain connects the brain with the pons. It consists of several elevations which contain many ascending and descending fibre pathways, specifically the auditory, visual and vestibular connections, plus the substantia nigra and red nucleus which are relay stations from the cerebellum and cerebrum, and the band of sensory fibres (medial lemniscus) to the thalamus (Hubbard & Mechan 1987).

The pons is the bridge between the midbrain and medulla, lying anterior to the cerebellum. It consists of transverse and longitudinal white fibres scattered throughout with nuclei passing to and from the cerebellum, cerebrum and spinal cord. The medulla, below the pons, connects to the spinal cord and is the main area of crossover (the pyramids) of the pathways in the CNS. The medulla functions as a conduction pathway of motor and sensory impulses between the brain and spinal cord.

The central core of the brain stem region is the reticular formation. This is the collection of neurones (nerve fibres, etc. that form the ascending and descending pathways), the connections with the cerebellum and other parts of the brain, collateral fibres and the origin of some pathways. In addition, most of the cranial nerves leave the CNS from this region. Therefore, the area functions as a relay station processing sensory information and organising motor output (Gordon 1990). The cells are clustered together to provide centres, such as the cardiovascular and respiratory centres, with vital yet discrete functions generally at an automatic level. Examples of the types of functions related to these cells are: heart rate, blood pressure, respiration, gastrointestinal function, eye movements (eye–head coordination), equilibrium, supporting the body against gravity and gross postural movements. The reticular formation influenced through the reticular activating system also controls alertness (Ganong 1991).

It is through the integration of the information reaching the reticular formation that axial postural control and gross movements are controlled. Input to the reticular formation is from many sources, including the spinoreticular pathways, collaterals from spinothalamic pathways, vestibular nuclei, cerebellum, basal ganglia, cerebral cortex and hypothalamus. The smaller neurones make multiple connections within the area whereas the larger neurones are passing through, being mainly motor in function.

The vestibular nuclei are very important for the functional control of eye movements, equilibrium, the support of the body against gravity and the gross stereotyped movements of the body. The direct connections to the vestibular apparatus of the inner ear and cerebellum, as well as the cerebral cortex, enable the use of preprogrammed, background attitudinal reactions to maintain equilibrium and posture. Working with the pontine portion of the reticular formation, the vestibular nuclei are intrinsically excitable; however, this is held in check by inhibitory signals from the basal ganglia (Guyton 1991).

Overall, the motor related functions of the brain stem are to support the body against gravity, generate gross, stereotyped movements of the body and maintain equilibrium. Therefore, it is predominately concerned with the control of the axial and proximal limb muscles (trunk and girdle movements). This is not in isolation but assisted by the integration of information from the cerebellum, basal ganglia and cortical regions.

The brain stem influences motor control directly through the descending pathways of the spinal cord and indirectly through ascending pathways to higher centres where their role is controlling overall activity of the brain and so of alertness (Gordon 1990). The descending pathways, which help control axial and girdle movements, can be divided into medial and lateral motor systems.

The medial system descends in the anteromedial columns of the spinal cord and projects directly (through interneurones) to medial motor neurones of the anterior horn influencing groups of proximal limb muscles and axial body regions, often bilaterally. It is important for organising and controlling whole body movements which require groups of muscles working together. The vestibulospinal pathways are specifically related to the position and movement of the head, making them important in the organisation of postural movements in balance control. The reticulospinal pathways influence postural movements and locomotion through pontine portions that are inhibitory (Guyton 1991), so facilitating the support of the body against gravity by exciting the antigravity muscles. The tectospinal pathways link the midbrain with the cervical and upper thoracic spinal cord and are important for organising and orientating movements of the head and neck.

The lateral system descends in the dorsolateral columns of the spinal cord and directly or indirectly innervate the motor neurones to distal muscles of the limbs, i.e. for more discrete muscle actions rather than muscle groups that are concerned with the individual, agile and skilled movements of the extremities (Kidd et al 1992).

The basal ganglia region forms part of the internal structure of the cerebral hemispheres. It comprises various poorly understood structures such as the caudate nucleus and lentiform nucleus which seem to serve as side loops for feedback of information between the cerebral cortex and thalamus above and the brain stem and spinal cord region below. This region is involved in the modulation of all types of movement by feedback control (Gordon 1990).

It is believed that the basal ganglia play an essential role in the initiation of most activities of the body in association with the sensory cortices. Information is then returned to the motor cortex to refine the activity. This negative feedback loop provides the inhibition of unwanted movements which allows the fine control or regulation of gross intentional movements which are normally performed subconsciously.

Working in association with the brain stem region, the globus pallidus operates as a motor relay station. It helps to control the axial and limb girdle movements, so providing the background attitudinal movements and stability of the body and proximal parts of the limbs which then allows more discrete motor functions to be performed (Guyton 1991).

CEREBRAL CORTEX (CORTICAL) CONTROL OF MOVEMENT

Posture and equilibrium are controlled subconsciously by the brain stem and spinal cord; however, the cerebral cortex is the main centre for the control of voluntary movement. It works with the information it receives from the cerebellum, basal ganglia and other centres in the CNS to bring movements under voluntary control.

The cerebral cortex provides the advanced intellectual functions of humans, having a memory store and recall abilities along with other higher cognitive functions. The cerebral cortex is, therefore, able to perceive, understand and integrate all the various sensations. However, its primary function is in the planning and execution of many complex motor activities, especially the highly skilled manipulative movements of the hand (Gilman & Newman 1987).

The motor cortex occupies the posterior half of the frontal lobes. It is a broad area of the cerebral cortex concerned with integrating the sensations from the association areas with the control of movements and posture. It is closely related to other motor areas including the primary motor area and the premotor or motor association area. The primary motor area contains very large pyramidal cells which send fibres directly to the spinal cord and anterior horn cells via the corticospinal pathways. In contrast, the premotor area has a few fibres connecting directly with the spinal cord but it mainly sends signals into the primary motor cortex to elicit multiple groups of muscles, i.e. signals generated here cause more complex muscle actions usually involving groups of muscles which perform specific tasks, rather than individual muscles. This area connects to the cerebellum and basal ganglia which both transmit signals back, via the thalamus, to the motor cortex. Projection fibres from the visual and auditory areas of the brain allow visual and auditory information to be integrated at cortical level to influence the activity of the primary motor area. The premotor area is activated when a new motor programme is established or when the motor programme is changed on the basis of sensory information received, i.e. exploring a new environment. The supplementary motor area is thought to be the site where external inputs and commands are matched with internal needs and drives to facilitate formulation (programming) of a strategy of voluntary movement. So altogether the motor cortex and related areas of basal ganglia, thalamus and cerebellum constitute a complex overall system for voluntary control of muscle activity (Afifi & Bergman 1986).

The primary motor area has a topographical

representation of the body on the surface of the cerebral cortex. This demonstrates the connections of the cerebral cortex to the different areas of the body with the size of the representation being proportional to the degree of innervation in that area, so the hands are exaggerated in size as the refined skilled movements produced by the hand require more innervation to achieve that level of control (Palastanga et al 1990).

For voluntary movement to occur, information has to reach the motor cortex, mainly from the somatic sensory systems plus auditory and visual pathways. The sensory information is processed with information from the basal ganglia and cerebellum to determine an appropriate course of action. As can be seen there are many two-way pathways between the various centres of the CNS.

The organisation of the motor cortex is in vertical columnar arrangements which act as functional units. Information enters a column from many sources and is amplified as necessary to produce appropriate muscle contraction. Each functional unit is responsible for directing a group of muscles acting on a single joint. So movements, not individual muscles, are represented in the motor cortex. Individual muscles are represented repeatedly, in different combinations, amongst the columns. Neurones of the motor cortex, having axons in the corticospinal pathways, function chiefly in the control of the distal muscles of the limb. They function to:

- change their firing rate in advance of limb movements
- fire at a frequency that is proportional to the force to be exerted in a movement and not in relation to the direction of the movement (Gilman 1990).

Each time the corticospinal pathway transmits information to the spinal cord the same information is received by the basal ganglia, brain stem and cerebellum. Nerve signals from the motor cortex cause a muscle group to contract. The signal then returns from the activated region of the body to the same neurones that caused the contraction, providing a general positive feedback enhancement if the movement was successful and recording it for future use.

At any segment of the spinal cord multiple motor pathways enter/terminate in the cord from the brain or higher centres. Generally, the corticospinal and rubrospinal pathways lie in the dorsal portions of the lateral columns of the spinal cord and terminate on the interneurones when concerned with the trunk, leg and arm areas of the cord. However, at the cervical enlargement where the hands and fingers are represented, the motor neurone supplying the hands and fingers lies almost entirely in the lateral portions of the anterior horns. In this region a large number of corticospinal and rubrospinal fibres terminate directly onto the anterior horn cells, i.e. a direct route from the brain in keeping with a high representation for fine control of the hand, fingers and thumb in the primary motor cortex (Ganong 1991).

The spinal cord can provide specific reflex patterns of movement in response to sensory nerve stimulation which are also important when the anterior horn cells are excited by signals from the brain, i.e. the stretch reflex is functional all the time helping to dampen the motor movements initiated from the brain. For example, when brain signals excite agonist muscles it is not necessary to inhibit the antagonistic muscles at the same time because reciprocal innervation will occur through the flexor reflex (Fig. 4.4).

CEREBELLAR CONTROL OF MOTOR FUNCTIONS

The cerebellum is vital for the control of very rapid muscular activities such as running, talking, playing sport or playing a musical instrument. Loss of the cerebellum leads to incoordination of these movements such that the actions are still available but no longer rapid or coordinated. This is due to the loss of the planning function.

The cerebellum monitors and makes corrective adjustments in the motor activities elicited by other parts of the brain. It is able to do this as it continuously receives information from the motor control areas of the cerebrum on the desired motor programme and from the periphery to determine the status of the body parts.

With these feedback systems the cerebellum compares the actual instantaneous status of each part of the body, as depicted by peripheral information, with the status that is intended by the motor system. Corrective signals can then be transmitted if necessary to alter levels of activation (Gordon 1990).

Although the cerebellum is anatomically divided into three lobes (anterior, posterior and flocculonodular lobes), functionally, the anterior and posterior lobes which control movement are divided longitudinally into a medial zone or the vermis and the intermediate and lateral zone from the cerebellar hemispheres (Fig. 4.5). The medial zone controls motor function of the axial body (neck, trunk and limb girdles), whereas the intermediate zone is concerned with control of the distal portions of the upper and lower limbs. The lateral zone operates more remotely to provide overall planning of sequential motor movement such as timing and coordination (Ghez & Fahn 1985).

Extensive input and output systems operate to and from the cerebellum. Input pathways to the cerebellum from the cerebral cortex, carrying both motor and sensory information, pass through various brain stem nuclei before reaching the deep nuclei of the cerebellum. Likewise,

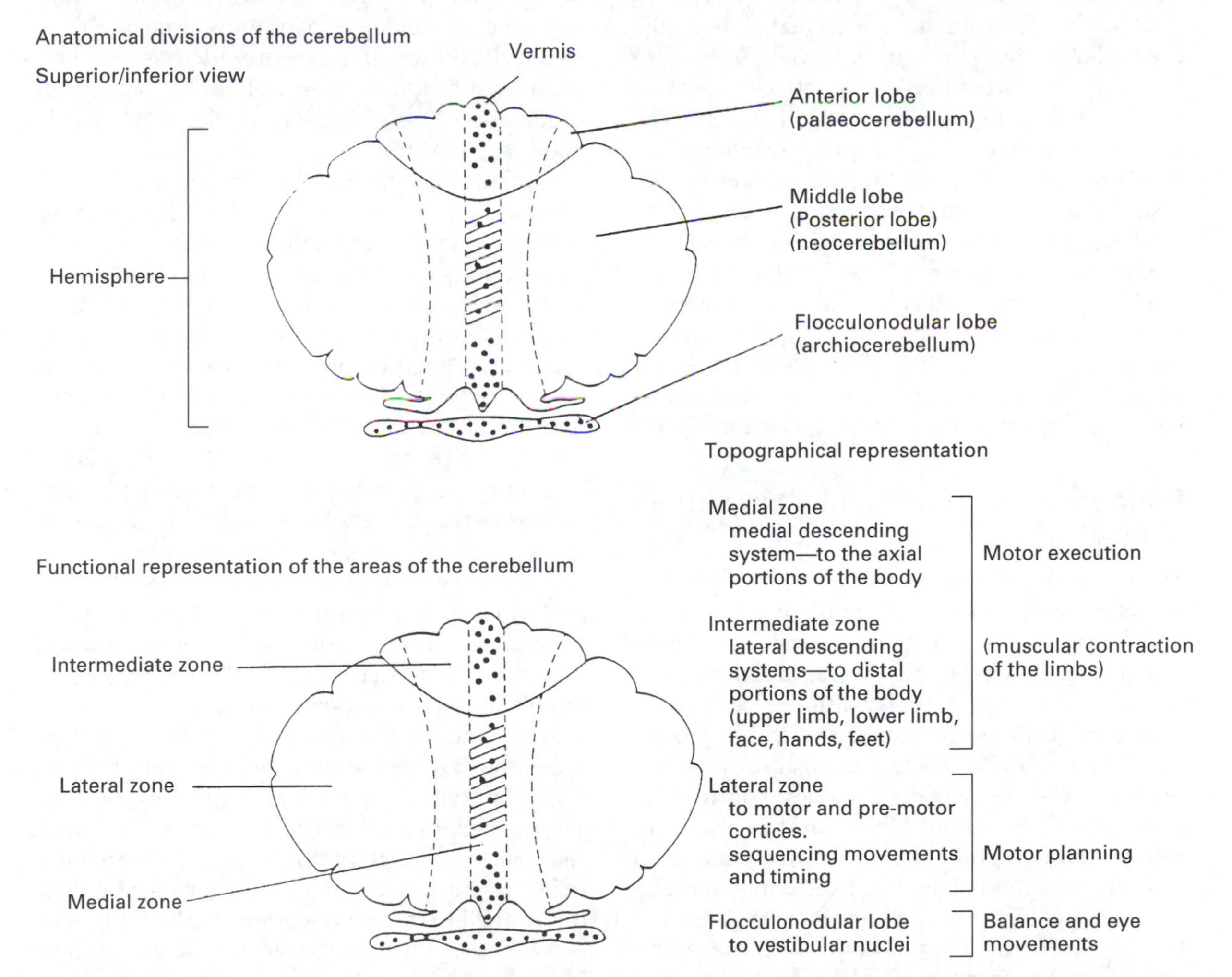

Figure 4.5 Anatomical divisions, functional divisions and representation areas of the cerebellum (adapted from Ghez & Fahn 1985).

output from the three zones of the cerebellum exit via the deep nuclei with the fibres from the lateral zone passing to the cerebral cortex to help coordinate voluntary motor activity initiated there. Fibres from the intermediate zone usually pass to the thalamus then onto the basal ganglia, cerebral cortex or brain stem regions to coordinate reciprocal movements, whilst those from the medial zone pass to the brain stem region to function in close association with the equilibrium apparatus to control postural attitudes of the body. Output from the deep nuclei of the cerebellum is continually under the influence of both excitatory and inhibitory influences, excitatory from the afferent fibres that enter the cerebellum and inhibitory from the functional unit of the cerebellar cortex, the Purkinje cell. A balance between the two effects is relatively constant with a continuous level of stimulation from both sources. However, when rapid movements are required, the timing of the two effects on the deep nuclei is such that excitation appears before inhibition. A rapid excitatory signal is initiated to modify the motor programme followed by an inhibitory signal, which is negative feedback having the effect of damping the movement and preventing overshooting. The cerebellum has a purely input and output system with no reverberating pathways (Guyton 1991, Gordon 1990).

Functions of the cerebellum in motor control

The cerebellum does not initiate motor activities but plays an important role in planning, mediating, correcting, coordinating and predicting motor activities, especially for rapid movements. The medial zone of the cerebellum represents the axial and girdle parts of the body with a predictive function for the control of posture and equilibrium. The intermediate zones control the distal parts of the limbs which produce the more refined and highly skilled movements in a smooth, coordinated and controlled manner. The lateral zones, with no topographical representation, oversee the overall control of the entire body with the function of planning and correct timing of sequential movements.

The medial zone of the cerebellum with the flocculonodular lobe works with the spinal cord and brain stem regions to control postural and equilibrium movements and provides a predictive function for rapid movement. Information from the muscle spindles is used to enhance and prolong the spinal cord stretch reflex. Thus, postural attitudes and adjustments of axial and limb girdle parts of the body are controlled with the help of the vestibular apparatus. During rapid movements the continuous feedback of the body's status by the cerebellum provides up to date and predictive information on body positions to other parts of the motor system. It is thought that the rapidly conducted information from the vestibular apparatus is used within a feedback control circuit to provide this instantaneous correction of postural motor signals in preparation or anticipation of the body's needs (Ghez & Fahn 1986).

Working with the basal ganglia and thalamus, the intermediate zones of the cerebellum help to control voluntary movement by utilising feedback circuits from the periphery and the brain. The distal parts of the limbs are controlled by information from the motor cortex and from the periphery and this information is integrated in the cerebellum. This provides smooth, coordinated movements of agonists and antagonistic muscle groups, allowing the performance of accurate, purposeful intricate movements which are especially required in the distal part of the limbs. This is achieved by comparing the intentions of the higher centres of the motor cortex with the performance of respective parts of the body. In general, the motor cortex sends more information than is needed so the cerebellum has to inhibit the motor cortex at an appropriate time **after** the muscles have begun to move. This provides both a damping function and control of ballistic (rapid) movements with a specific intention. The cerebellum assesses the rate of movement and calculates the length of time required to reach the point of intention. Inhibitory signals are then transmitted to the motor cortex to inhibit the agonist and excite the antagonist muscle groups and apply the brakes so that the movement stops on target. There is no time for feedback control of such highly skilled movements, therefore they have to be previously planned or learnt. The damping function prevents overshooting and pendular movements. This point of reversal of excitation depends on the rate of movement plus previously learnt knowledge of the movement. Consequently, voluntary movements, despite their name, are not purely voluntary but are also controlled at a subconscious level. Likewise, ballistic movements need the excitation from the cerebellum to get the speed of a movement and then inhibition to stop the movement. The lateral zone of the cerebellum is concerned with the overall control of movement of the entire body by having a planning and timing function working with the premotor, sensory and association areas of the cerebral cortex. Planning the sequence of movements is necessary for a smooth and orderly transition of one movement to the next. The lateral zones draw on the cerebellum's predictive ability to recognise the pattern of action, how far different parts of the body will move in a given time and, so, the next movement. A timing function is then needed to predict and control the transition from one movement sequence to the next in a coordinated and accurate manner. This is possible because a two-way traffic between the sensory and premotor areas of the cerebral cortex and the lateral zones of the cerebellum exists. Thus, the deep nuclei of the cerebellum, through the lateral zones, are already involved in the activity pattern of the next movement at the same time as the present movement is occurring (Guyton 1991).

In addition, extra motor predictive functions from other senses (auditory and visual) are alerted to environmental factors so providing additional situational detail. Such information is important in interpreting spatiotemporal relationships and recognising the specificity of the movement, i.e. how rapidly the body is approaching an object or an object is approaching the body. This is essential in avoiding collisions.

Overall, the cerebellum serves as an error correcting device for goal directed movements. It receives information on body position and movements in progress then computes and delivers appropriate signals to the brain stem effector centres to correct posture and smooth out movements. The cerebellum is important in the process of learning and acquisition of motor skills.

Box 4.2 Task 2

Ataxia is a disorder of movement control resulting in slow, uncoordinated motor actions. Some patients with this problem only show these clinical features when their eyes are closed. Can you explain why?

SENSORY FEEDBACK CONTROL OF MOTOR FUNCTIONS

Once a movement is learnt, a sensory engram is established in the sensory cortex and used as a guide for the motor system of the brain to reproduce the same pattern of movement. If the engram and pattern of movement do not match, then the additional motor signals automatically activate appropriate muscles for the correct performance of the task. The sensory engram is projected according to a time sequence and the motor control system automatically follows from one point to the next, so it is not a controlling system but rather a following process. If the motor system fails to follow the pattern, this is fed back to the sensory cortex and corrective signals are transmitted to the appropriate muscles.

The establishment of rapid motor patterns which are too quick for sensory feedback are based on motor engrams. Rapid movements such as typing are initially learnt slowly so there is time for sensory feedback to guide a movement through each step. Successive performance of a skilled activity results in an engram for the activity being laid down in the motor control areas as well as the sensory system. The motor engram causes a precise set of muscles to go through a specific sequence of movements to perform a task. The pattern can then be performed without sensory feedback, but the sensory system still determines if the act was performed correctly. In addition, a variety of patterns are stored to allow for the variety of movement necessary in different situations (Guyton 1991).

In summary, the control of movement is a complex process involving more and higher levels of the CNS as the movement becomes more complicated. Learning a new movement can be a slow process and takes time and practice, but once established, the learnt movement and variations on that theme become more automatic, adapting to the minor alterations necessary for a potentially changing environment. As a totally integrated system, with many direct and indirect feedback circuits, information is constantly being gathered on the status of the body parts for the present and for the future. This is frequently compared with past movement activities to see if a stored posture and movement pattern is already available. An efficient system can then utilise previous experiences with less effort demanded of the CNS enabling attention to be given to other aspects of a task.

Box 4.3 Task 3

Here are two final challenges. As before, you should use the information in this chapter plus other sources of information on neurology to work out the answers. If you are able to complete these final two tasks, then you will have a reasonable grasp of the basic relationship between neurological control and human movement. If you find the tasks difficult, it would be advisable to undertake some more reading.

Draw a flow chart that details the hierarchical levels of control of movement in the central nervous system. Add to this the alternative parallel routes connecting the different levels of control. Now consider in turn the result of damage to each of the levels of control. Would normal movement still be possible?

Why do postural attitudes adopted following musculoskeletal injury continue long after the injury is resolved?

REFERENCES

Afifi A K, Bergman R A 1986 Basic neuroscience: a structural and functional approach, 2nd edn. Urban & Schwarzenberg, Baltimore

Brooks V B 1986 The neural basis of motor control. Oxford University Press, Oxford

Carpenter R H S 1990 Neurophysiology, 2nd edn. Edward Arnold, London

Eccles J C 1977 The understanding of the brain, 2nd edn. McGraw-Hill, New York

Ganong W F 1991 Review of medical physiology, 15th international edn. Lange Medical, New Jersey

Ghez C, Fahn S 1985 The cerebellum. In: Kandel E R, Schwartz J H (eds) Principles of neural science, 2nd edn. Elsevier North Holland, New York

Gilman S, Newman S W 1987 Manter and Gatz's essentials of clinical neuroanatomy and neurophysiology, 7th edn. F A Davis, Philadelphia

Gordon J 1990 Disorders of motor control. In: Ada L, Canning C (eds) Key issues in neurological physiotherapy. Butterworth Heinemann, London

Guyton A C 1991 Basic neuroscience — anatomy and physiology. W B Saunders, Philadelphia

Hubbard J L, Mechan D J 1987 Physiology for health care students. Churchill Livingstone, Edinburgh

Kandel E R, Schwartz J H, Jessell T M 1991 Principles of neural science, 3rd edn. Elsevier North Holland, New York

Kidd G, Lawes N, Musa I 1992 Understanding neuromuscular plasticity. Edward Arnold, London

Palastanga N, Field D, Soames R 1990 Anatomy and human movement, student edn. Butterworth Heinemann, London

Shepherd G M 1988 Neurobiology, 2nd edn. Oxford University Press, New York

Todd J M, Davies P M 1986 Hemiplegia — assessment and approach. In: Downie P A (ed) Cash's textbook of neurology for physiotherapists, 4th edn. Faber, London

Tortora G J, Anagnostakos N P 1987 Principles of anatomy and physiology, 6th edn. Harper Collins, New York

Tortora G J, Grabowski S R 1992 Principles of anatomy and physiology, 7th edn. Harper Collins, New York

5

Motor learning

R. A. Charman

CHAPTER CONTENTS

OBJECTIVES

On completion of this chapter you should be able to:

1. **Demonstrate an understanding of cybernetics by analysing simple movements using the concepts of cybernetics**
2. **Define what is meant by skill, and discuss skill in terms of its main components, the factors involved in skill learning, the effects of different repetition magnitudes upon skill memory and performance, and the main skill learning strategies**
3. **Describe the general sequence of skill acquisition and changes in skill performance from the initial stage of peripheral programming, through the transfer stage to the end goal of central programming**
4. **Distinguish the main characteristics of movements classified as discrete, continuous and serial, using examples of each type of movement to illustrate the differences between them**
5. **Relate the concept of body image and body image projection to the performance of common tasks using tools and the importance of body image projection for the competent use of prostheses and other forms of aids used in rehabilitation**
6. **Discuss in broad outline the neurological mechanisms involved in skill acquisition and the relevance of mental practice of a skill in**

improving the rate of skill acquisition and level of performance in clinical rehabilitation.

INTRODUCTION

Previous chapters have outlined the biomechanical principles involved in bodily movement and the musculoskeletal and neuromuscular basis of movement and posture. Studies of the mechanics underlying everyday activities, from writing to walking, running, jumping, ascending and descending stairs, show that the magnitude of physical forces involved in moving the required body masses range from the very small and localised to the very large and multiple. High speed athletic and sports activities involve rapid generation and control of huge physical forces without error.

The computational brain power that is required to plan, initiate, control and coordinate these interacting mechanical forces within the necessary constraints of balance and equilibrium is correspondingly enormous. Tens of thousands of motor units in hundreds of muscles may need to be activated and deactivated to exert precise contractile (tensile) forces upon dozens of skeletal levers to move them through the correct arcs of movement, in the correct sequences, at the correct time, in the correct trajectories, through the correct segments of space, at the correct rates of acceleration, deceleration or constant velocity, to ensure the precise execution of a required movement to the exact completion of a predetermined goal.

Such feats of movement are often accomplished within an external world of fast changing physical and spatial variables, as in gymnastics, athletics, sports and many manually skilled occupations. The possible implications that any such changing variables may have upon the successful completion of a particular task must be instantly evaluated, and any necessary change in movement strategy reprogrammed and executed. The sensorimotor memory capacity necessary to hold a vast range of compensatory movement strategies for any given skill, together with the correct selection and execution of a required motor strategy, places the control systems of the central nervous system (CNS) into a 'state of the art' category of performance that is far beyond the most advanced computer programme yet available.

When it is remembered that the circuitry of the nervous system is constrained to using impulse trains that travel at speeds of between 1 m to 120 m per second, and requires a timescale of several milliseconds to do its computations, whereas computers use impulses travelling close to the speed of light (3×10^6 ms) and compute on a timescale of nanoseconds (10^{-9} s), then the performance of the nervous system is truly awesome.

These points have been emphasised because, when control of movement is considered in these terms, it can be seen that learning a new motor skill cannot be an easy task because the programming that will be required to perform that skill in all of its variations can only be 'written' and stored in the CNS through endless sessions of trial and error practice. The CNS is not supplied in advance with software programmes for voluntarily learned skills such as tennis, for example: it is a *self-learning* and *self-programming* system that must train its own circuitry to learn to discriminate between error and accuracy, and modify its own synaptic mechanisms into circuit programmes that accurately represent the bodily expression of the desired skill (Rosenbaum 1991, Singer 1980).

As yet the means by which the CNS achieves this are not known, despite our present knowledge of neural pathways, synaptic transmitters, synaptic modulators and inhibitory/excitatory electrical biasing of neurone membrane polarity. At present we can only use explanatory analogies taken from related sciences, such as computer programming and cybernetics, and use them as models of probable brain functioning and problem solving when considering how a new motor skill is acquired. Until the mystery of how objective brain mechanisms, functioning at some unknown level of complexity, somehow manage to create the subjective reality of consciousness which, with all of its thoughts, experiences, emotions and desires, provides the driving force for all voluntary activities is solved, then modelling by helpful analogies from other sciences is the only recourse available.

Two main working hypotheses have been adopted for this chapter on motor learning. One is that the science of **cybernetics** can supply some valuable models concerning the factors involved in successful voluntary movement, and the other is that the two related concepts from the **computer sciences** of *peripheral programming* and *central programming* help to explain why the early stages of learning a new skill are so slow and frustrating (peripheral programming) compared to the competent ease of proficiency (central programming).

To save endless distinction between conscious mind and physical brain the term *brain* will be used to mean mind and brain working together in voluntary activities unless a particular distinction is made to emphasise a point.

VOLUNTARY MOVEMENT — A CYBERNETIC MODEL

Voluntary movement is movement with purpose. It is the motor outcome of a conscious decision to do something; for example, commencing a tennis serve, or practising partial weight-bearing walking with elbow crutches. Voluntary movement is therefore *intentional,* and when something is done with intent it means that it is directed towards a predetermined goal or goals. In other words, intentional movement has objectives. A mental picture of the desired outcome is first constructed in the 'mind's eye' and the sequence of intentional movement that is expected to achieve the chosen objective is commenced as appropriate. The outcome of voluntary movement is judged as successful, or unsuccessful, by mentally comparing what actually happens with what was intended to happen.

Because the body is a physical object that moves in three-dimensional space to achieve physical objectives, its operation is subject to the same physical laws and forces as any other object that has mass. As the brain is the control centre for movement it has to know what is occurring in the stationary or changing personal space of the body itself, how this relates to what is happening in external space, and what bodily forces and force directions must be employed to ensure that the interaction between the body and the environment actually achieves the movement goals of voluntary intention.

The problems that the brain has to solve to achieve its movement goals are, in principle, the same as any other system that is trying to reach, or maintain, predetermined objectives. It therefore seems reasonable to assume that the brain must employ similar strategies of problem solving. For example, the brain of a soldier who is attempting to shoot and hit a moving target is trying to solve the same problems of anticipatory spatial tracking and range finding as the control system of an anti-aircraft gun that is programmed to track and destroy an aircraft.

The science that deals with such problems of self-correction and control, whether they concern, for example, temperature control, a controlled rate of chemical reaction, radio frequency detection, or a given rate of output, is known as **cybernetics,** and it is this science that provides many useful models for explaining the successful execution of voluntary movement and the balance and postural reflexes upon which such movement occurs.

Cybernetics — an introduction

Cybernetics is the theory and science of communication and control mechanisms that operate in living and non-living systems. The term *cybernetics* was coined by the brilliant American mathematician and philosopher Norbert Wiener (1894–1964) and used as the title for his 1948 epoch-making book in which he presented a formal exposition of the science and its application across a wide range of scientific disciplines. He adapted the term 'cybernetics' from the Greek *kybernetes,* or *cybernetica,* which means *steersman,* as in steering a boat using a rudder. Plato (c428–c348 BC) used the example of a steersman in a series of books, known as the *Republic,* that examined the role of society and government. Wiener considered that the job of steersman, which was used by Plato as an analogy of government policies directed towards political objectives, provided an excellent illustration of the basic concept involved in communication and control systems, and it will be used here to introduce the science and show how it can be applied to intentional movement (Wiener 1961).

The job of a steersman is to keep a ship on a given, or predetermined, course, say due east. As the contrary forces of wind and current flow continually push the forward direction of the ship away from its correct course, the steersman has to continually compensate for this unwanted deviation by swinging the tiller bar to alter the angle of the rudder blade in the water to bring the ship back into line. In effect, he compensates for any unwanted drift away from the correct course by turning the ship in the opposite direction by use of the rudder until the error is cancelled. If he, through inexperience and misjudgment, attempts to compensate by steering further *into* the unwanted deviation, the ship will diverge further and further off course. If he consistently overcompensates or undercompensates for each deviation, the course of the ship will oscillate widely about the mean of the correct course which will add extra miles to the voyage and be an inefficient use of the motive power of the ship.

In cybernetic terms, the steersman is acting as a **self-regulating steering device** that has been attached to the ship for the duration of its journey, and his function in this role will now be considered with the relevant cybernetic terms given in italics.

The forward velocity (speed and direction) of the ship can be considered as its **output**, as it is the outcome **product** of the motive power, in this case the sail power, of the ship. The predetermined direction of the ship's course, which has been decided upon by the captain, is taken by the steersman as his directional **reference value**, against which any unwanted change in course must be assessed, and the given destination is his **reference target**, towards which he must aim the ship. Any variation of course away from the correct course will be directional **output error**, and any inaccuracy on arrival of the ship will be **target error**.

All self-regulating devices need to be continually *updated* as to what is happening and must therefore possess *sensors* that are selectively sensitive to the appropriate stimuli. In this case, the steersman will use his visual, auditory, balance, touch and kinaesthetic sensory systems as his sensors, and they will continuously send **data**, in the form of travelling nerve impulse trains, to his brain as a response to external and internal bodily stimuli. These streams of data are **coded** for each sensory system and his brain **decodes** them upon reception into **information** concerning the effects of wind and currents upon the course and movement of the ship, and the effects of any corrections that he makes. His brain, in effect, is **monitoring** the **messages** that his senses are sending and acting as a **comparitor** to compare what is happening, in terms of any course **error**, to the given **norm** of the directional **reference value**. When all of this information has been **processed**, his brain will then act as an **evaluator** to assess the implications of any reported error and determine the appropriate magnitude of **error correction response**. The principle upon which this correction is based is called **negative feedback**.

Negative feedback

The essential principle of error correction in self-regulating systems is that the magnitude of output error away from the reference value, target, or signal, determines the magnitude of error correction response. This is the principle of negative feedback in which the term 'negative' means to negate, or cancel out, the error, and 'feedback' means a return effect, or response, of the self-regulating system back onto the output. In effect, if an output that has been set to a given reference level is connected to a self-regulating mechanism, it can control the input in an opposite way to the error so that the resulting change of throughput (input arriving at output) **cancels** the output error. The system is self-governing.

How does the self-governing steersman effect the required error correction? The steersman is manually connected to the rudder, which is the physical effector mechanism that alters the course of the ship by holding the tiller. The error correction signals are sent from his brain to selectively activate his muscles to provide the power that will adjust the angle of the tiller until the new angle of the rudder cancels out the unwanted course error and brings the ship back onto its correct course. The steersman, the tiller, and the rudder act as a coordinated effector device that converts error correction instructions into actual course correction reactions by altering the input of directional steering to affect the output of course direction.

A novice steersman will be relatively ineffective as a self-regulating device compared to an experienced steersman as he has little experience in anticipating the possible effects of a given wind and/or current flow upon the ship, and can only react to course error after it has occurred, as described above. If, through inexperience, he attempts to correct course error by steering *into* the deviating course of the ship he will have responded by providing positive feedback, which can only make matters worse. The term 'positive' in this sense means to positively reinforce whatever is influencing the output away from the norm (a typical example of **positive feedback** from everyday life is when a loudspeaker is too close to a microphone and its amplified sound is fed back into the microphone to the point where the amplifier cannot cope and howling distortion occurs).

An experienced steersman can **anticipate** what is likely to happen and can initiate predictive strategies of anticipatory feedback that will minimise unwanted changes (errors) in course direction (output).

Both normal intentional movement and abnormal intentional movement can be analysed in terms of cybernetic principles and mechanisms, as can the principles and methods of movement rehabilitation, so this science, in effect, provides a unifying concept that underlines this chapter. For a full discussion of cybernetic principles applied to human movement see Rosenbaum (1991) and Schmidt (1982).

Skill — definitions, components, levels, learning strategies

Everyone knows what they mean when they say that someone is skilled at some particular task, especially if they cannot do it themselves. What they mean is that the other person can perform that activity without apparent difficulty and with little, or no, error. The underlying implication is that skilled performance is not a natural inheritance of movement ability, or we would all possess the same range of specialised skills by natural development, but is something that must be learned by effort.

Box 5.1 Task 5.1

Task analysis in cybernetic terms
First list the cybernetic terms that have been employed in this introduction for use as reference, then divide into small groups and work through Task A, analysing performance in cybernetic terms.

Task A. Place three or four obstacles, such as upright skittles or gym stools, into a straight line at a distance of about 1.5 bench lengths apart and place a separate pair a further 1.5 bench lengths away at the far end to act as a narrow entrance just wide enough to pass through. Two students lift a bench to thigh height, facing each other at opposite ends, and line up with the obstacles some 1.5 lengths away at the near end. The task is to pass as closely as possible on alternate sides of the obstacles and finish by passing through the narrow entrance. The backward walking student (or forward walking if blindfolded) at the 'prow end' is controlled by the forward walking steersperson. At no time must the 'ship' touch, or pass over, the obstacles or it will be 'sunk'. The observers analyse the performance.

Task B. Devise, practise, and analyse in cybernetic terms exercises involving: a) counteracting varying and unexpected sideways or vertical forces when attempting to follow a predetermined path, either with, or without, carrying a load; b) counteracting predictable sideways or vertical forces; and c) throwing and catching a light ball and a heavy ball alternately.

To explore this issue, the following definitions of skill are offered to formally identify some key aspects of skilled performance:

- Perfect skilled movement is the accomplishment of a motor task *without error*.
- Movement is considered to be skilled when the performance *exactly matches* the requirements of the task.
- Skilled movement achieves its objectives with the *least* effort.

- Skilled movement comprises the *most effective form of physical interaction* that a person can undertake with the environment.

The last definition summarises the previous definitions. There is no error, so no wasted muscular effort in recovering the body from the wrong position or direction of movement. If the movement has exactly matched the requirements of the task, then it has been completed with the least effort because no muscle has been contracted unnecessarily. Motor skill, in sum, is the selective use of one's own bodily forces to so effectively interact with the environment, whether it involves picking up a box and putting it on a shelf, tying a shoelace, driving a car or jumping on the spot, that the environment has been used, or manipulated, to meet the exact requirements of the task, neither more, nor less. O'Brien & Hayes (1995) neatly define skill as 'characterised by spatial and temporal accuracy in performing a movement task'.

Three changing components of skill

Skill in the performance of a given task can only be obtained by the corrected repetition of the desired activity until error has been minimised or eliminated. The level of skill competence can be roughly assessed at any point on the learning curve by the changes that are occurring in the following three components that have been chosen as representing some major aspects of skill development. The term 'inversely proportional' is used in a purely descriptive sense, as it would be difficult to quantify in practice.

Level of motor skill competence. Any increase in skill competence can be considered as being inversely proportional to the amount of unnecessary muscle activity that occurs during its performance. In effect, increasing competence in performing the correct skill movements is correlated with a decrease in the performance of incorrect skill movements, or incompetence of movement, to the point where, hopefully, no unwanted muscle contraction occurs.

Level of subjective skill performance. Any increase in skill competence is inversely proportional to the degree of consciously directed control over individual components during performance. In effect, as skill competence increases, the effortful concentration on each part of the movement decreases as it becomes more automatic. When all that is required of conscious effort is the decision to commence the performance of the skill, to determine how long it will continue, and to end its performance, then the skill is well learned.

Level of neurological skill competence. Any increase in actual skill competence, and any reduction in consciously directed control, can be considered as physical and mental outcome measures of successful reprogramming changes that have occurred in the nervous system. These outcome measures are expressions of improving neurological skill competence at all levels of the nervous system, including cognitive (as an ability to mentally understand the skill as a whole), sensory (as an ability to recognise and interpret bodily and environmental cues related to the skill) and motor (as an ability to combine the subroutines of the skill into seamless combinations).

At motor neurone level this increase in skill competence is expressed as an increasing ability of the descending motor control systems to selectively inhibit motor neurone activity that would otherwise cause unnecessary and disruptive muscle activity, and selectively enhance the recruitment of those motor neurones controlling wanted muscle activity.

In practical terms there is as observed, and subjectively experienced, reduction in clumsiness and error during practise of the skill towards the goal of error-free performance.

Learning a new skill

What must be done to acquire a new skill? The answer has been summed up by Kottke et al (1978) as the five P's of *perception, precision, perpetual practice, peak performance* and *progression.*

Perception. The term *perception* is used to mean the conscious perception and analysis of skill performance and planning future practice strategies to correct for known error. This is the slow and uncertain stage of learning.

Precision. This is the repetition of the corrected subroutine of the skill at high levels of precision following conscious analysis of the causes of error and formulation of strategies of correction, usually from the simple to the complex.

Perpetual practice. The desired internal programming of the skill can only be achieved by developing the correct synaptic linkages and establishing them in long-term motor memory, known as a motor engram. This requires perpetual practice, as Kottke puts it, until this occurs. The implications of various amounts of repetition practice on the establishment of a motor engram will be considered under the heading of 'How much practice is required?'.

Peak performance. Peak performance means performance at the level of near maximal, or maximal, efficiency for the given stage of learning with the minimum percentage of error, and this requires frequent repetition to be retained.

Progression. There should be planned progression in the level and complexity of performance to include mastery of more and more variables to increase the boundaries of achievement.

How much practice is required?

When Kottke (1980) reviewed the literature on attaining skill proficiency his discussion on this subject came under the heading of 'Millions of repetitions are needed for peak skill', which is quite a sobering thought! He summarised the effects of different numbers of repetitions in establishing a motor engram of the desired skill along the following lines (additional comment in brackets):

- Tens of repetitions: create a conscious testing and awareness of the skill but little, if any, motor memory retention.
- Hundreds of repetitions: create a fragile motor engram that fades quickly.
- Tens of thousands of repetitions: create a fair engram in which any necessary speed and force of performance can increase (this could be classed as the late novice stage of performance).
- One hundred thousand repetitions: create a reasonably capable motor engram with significantly increased levels of sustained skill competence (the good amateur stage of performance).
- Millions of repetitions: create the near perfect motor engram of skill performance. This level applies to the competent performance of everyday skills, such as dressing, washing, eating at table, walking, running, climbing stairs, gardening, cooking, driving a car and so on, when they have become the automatic activities that are done 'without thought' as completely routine. Specialised vocational skills, such as playing a violin, touch typing, basketball playing and snooker are specialised adaptations of the everyday skills that are taken for granted, and require millions of repetitions in their own right to attain the levels of recognised, top class, professionalism (this is the ultimate stage of specialised skill performance that is dependent upon endless practice).

Some examples to illustrate this point (Kottke et al 1978) are:

- learning to walk (to age 6) — 3 million steps
- parade ground marching (end of army basic training) — 0.8 million steps
- hand knitting — 1.5 million stitches
- violin playing to a professional level — 2.5 million notes (4500 hours of practice)
- baseball throwing (pitcher) — 1.6 million throws
- basketball into net from any angle — 1 million throws
- gymnast — repetitions unknown, but at least 8 years of daily practice.

The implications for movement rehabilitation therapy are considerable, especially where the therapeutic intention is to replace disorganised patterns of movement, whether *congenital* (as in cerebral palsy) or *acquired* (as in head injury syndromes or hemiplegia following cerebrovascular accident), with normal patterns of movement. The mechanisms of neuroplasticity by which new patterns of synaptic activity are formed as movements are learned are not biased in favour of establishing normal movement over the already present abnormal movement. In fact, in congenital motor disorders the ineffective 'abnormal' *has been*, and *is*, the 'normal'. Therefore, whatever system of movement therapy is employed, it can only replace 'abnormal' with the desired normal if it can establish motor engram dominance by the necessary number of repetitions. To establish the guidelines for effective repetition of therapeutic movement correction would seem to be a very important area for research.

Forming a motor engram

If a new skill is to be learned and retained, then the CNS must be able to form the appropriate engram pattern(s), store the engram in long-term memory, and release the motor patterns as necessary. An engram is an encoded imprint, and its formation is dependent upon the efficiency of the neuroplasticity mechanisms in forming the engram. This will be different for every individual, and factors that may affect the neuroplasticity processes could include age (either too old or too young for the complexity involved), natural aptitude, physique, motivation, existing competence in related skills, frequency and duration of practice sessions, and ability to form an imaginative picture of performance.

Anyone suffering from neurological disorders that impair understanding, concentration, attention span and motivation will have impaired abilities to activate the neuroplasticity processes that are necessary to encode long-term engrams. Where the problem is one of dementia, such as Alzheimer's disease, the rate of short-term forgetting may equal, or exceed, the rate of short-term skill remembering, so that no transfer to long-term memory is possible.

Learning — unstructured and structured

Unstructured learning. Unstructured learning occurs when the learner repeatedly attempts to mimic the behaviour and movement of others. This is the normal way in which the young of most species, including human beings, learn the basic skills of everyday life. Such learning either occurs without any particular encouragement or, as is more likely, with general encouragement by older siblings and adults but no specific instruction or demonstration. In neurological terms, as tract myelination and maturation of the growing brain and spinal cord occurs, the basic survival reflexes that are present at birth start to fade and become replaced by the motor patterns of intentional movement. This is modelled upon what the infant, toddler and young child sees others doing and wishes to join in and do for themselves. Repetition leads to the developing engrams of voluntary movement and an increasing repertoire of motor skills (Whiting 1975).

Box 5.2 Task 5.2

Unstructured learning
Divide into threes. One of a pair secretly devises a task or exercise that follows an internally planned sequence and can be repeated in sets of repetitions without error. The learner attempts to mimic the performance with no help from the skilled performer. The third person observes and notes any changes in the level of unskilled performance over several set repetitions. What is the learner mentally doing during each performance? What skill variables change during learner practice?

Structured learning. Structured learning supersedes unstructured learning and can be undertaken at any age. Structured learning often takes the form of a social contract in which the person with the desired skills takes on the role of teacher, coach, or trainer and devises a programme of instruction which the learner agrees to follow until a certain level of agreed proficiency is attained.

Everyone has undergone the familiar experience of learning a new skill through structured learning, whether it be the therapeutic skill of joint mobilisations, of safe transfers following the misfortune of spinal trauma and resulting paraplegia, or driving a car. The purpose of the following section is to analyse the experiential sequence of skill learning under a series of headings to highlight the processes involved.

The decision to learn — from peripheral programming to central programming

Before any intention to learn a particular skill can

be formed, the existence of the skill must be known to the intending learner. At first reading this seems to be such a statement of the totally obvious that it hardly merits space, but it is an absolutely crucial prerequisite for the following reason. The knowledge that a particular skill exists is represented in the mind of the intending learner as an imagined *ghost model* of its performance. This ghost model includes an imaging of what it must be like to perform the skill, and 'what it must be like' is a sensorimotor image that is likely to include the context, such as a golf course, or gymnasium, together with the imagined satisfaction of achievement. The more the intending learner knows about the skill, the more vivid the imagery, and the implications of imagining a skill on the rate of motor engram formation will be discussed in a later section.

The extremely important point here is that if someone, through sheer lack of motivation, limited understanding, disabling mental disorder or loss of imaging ability through brain damage, will not, or cannot, form a ghost model of the skill and commit it to memory, they cannot proceed to the next stage which is the decision to acquire the skill. This situation has considerable implications for physical therapy as it may severely limit the level of possible success unless the patient develops the will to succeed.

The stage of peripheral programming

From the standpoint of cybernetics, the problem during the early stages of learning a new skill is that the reference values that guide performance have not been established in practice experience nor has the sense of error. In other words, there is no central engram of the skill. So knowledge of what is happening as the movement is being performed is based upon the information that is fed back by the peripheral nervous system, including kinaesthetic, vision, hearing and balance, as the skill is being practised. There is, therefore, a timelag in milliseconds before the brain knows whether or not a particular segment of movement has been performed correctly.

The system that controls this stage of learning is *conscious attention* to each part of the movement. This requires the full concentration of focused attention throughout performance and is always very slow, cautious and uncertain compared to competent practice. There are several important reasons for this stage of movement inefficiency which need discussion.

As mentioned above, no sensorimotor skill engram yet exists upon which to base the performance. It is literally a stage of trial effort and analysis of error. Therefore, there is little, or no, *anticipatory* preprogramming of alpha and gamma motor neurone firing thresholds in continuous advance of the actual movement. This lack of anticipatory preparation means that the changing sequence of body postures that provide the basis upon which a particular skilled movement depends is uncertain, and may even be inappropriate, leading to error and possible injury if the intended movement is forceful, as in lifting.

To compensate for the absence of central knowledge of the skill, conscious practice is characterised by excessive co-contraction of postural and limb musculature in an inappropriate attempt to prevent error and correct it as it occurs. A striking example of this co-contraction rigidity during this stage of peripheral programming practice is the postural rigidity adopted when attempting to skate or ski, or slide glide along a slippery surface. The performance is marked by a total absence of appropriate, *anticipatory* changes in posture to compensate for the lack of normal, stabilising, frictional contact of footwear on non-slippery surfaces. The central engram for the latter remains dominant, despite being completely inappropriate, and automatically attempts to adjust to frictionless uncertainty of balance by anxious co-contraction.

The lack of a skill engram means that conscious programming of successive movements are, in effect, based upon guesswork as to the requirements of the task and the selection of movement in terms of posture, range, velocity and strength. This is why this stage of practice requires ceaseless concentration.

Excess co-contraction of muscle is very energy intensive, leading to rapid muscle fatigue and lack of endurance. It also generates a barrage of uncomfortable sensory input into consciousness which reduces conscious endurance of effort. Intense concentration is usually accompanied by prolonged breath-holding during the period of effort, and breath release upon cessation. This leads to a feeling of uncomfortable respiratory distress that further reduces voluntary endurance.

Conscious attention is unable to shift motor attention at more than three times per second, as the consciously controlled error reaction time, based upon sensory feedback and information processing, is about 300 milliseconds per item of performance. Conscious attention is a narrow focus activity of single sequential processing known as *single channelling*. During the performance of most skills many events, both internal and external, are occurring at the same time, either in expected step with each other in simultaneous and serial continuity, or are unexpected and need to be evaluated. Competent skill performance therefore requires a control system that has the necessary capacity for *parallel processing* of all the relevant variables as they occur, as well as sequential processing of each one. Such control systems are known as *multichannelled*.

The subjective discomfort of attempting to maintain prolonged periods of unusual concentration, together with the inevitable poor performance at this stage, is highly stressful and requires considerable emotional commitment to sustain perseverance. The gap between reality and the ideal is wide and discouraging.

In summary, early attempts at performing the whole sequence of a new skill, or relearning an old skill after illness or injury, are subjectively tiring and objectively clumsy. Occasional success is not easily repeated. This is the stage of least skill and most error. It usually leads to periods of extreme self-doubt and discouragement and, in rehabilitation, therapists need all their skills of encouragement and motivation to counteract this negative reaction.

Skill subroutines

Once the whole skill has been attempted, the most practical method of learning is to break it down into easily understood sections, or subroutines, and practise each of them in their own right with repeated emphasis upon correct postural positioning and correct sequencing of movement. Conscious experience of the body image during practice is correlated with actual achievement, and there is a combined therapist/learner emphasis upon sensation associated with success. This strengthens sensorimotor memory of good performance and inhibits associations with failure.

With time and practice the motor improvement becomes correlated with a reduction of subjective tension and intensity of sustained concentration during performance. The learner finds that an increasing number of subroutines are becoming 'automatic' on performance.

Transitional transfer stage

The probable reason for this improvement in subroutine performance is that repetition is effecting a transfer from short-term to long-term memory. The patterns of neural activity are thought to move from cortically based conscious control, dependent upon peripheral feedback programming, to subcortically based central programming which is shared between the cerebellum, basal ganglia and associated brain stem nuclei.

During this interim stage, any practice of the whole skill is usually experienced as an exasperating mix of accurately performed subroutines, which proceed at the appropriate rate, and slower subroutines that are still under conscious control. This uneasy mix leads to unpredictably erratic performance, sometimes almost error free, sometimes with considerable error. Every learner driver knows this experience all too well.

Central programming

At some point all of the subroutines are transferred into long-term memory and a comprehensive motor engram of the skill comes into existence. The subroutines are merged into a seamless performance and are available for intentional use when required.

What is consistently present in the centrally programmed motor engram, but conspicuously absent in the peripherally programmed conscious control stage, is non-conscious anticipatory preprogramming of musculoskeletal activity which eliminates error and matches the performance with the task. Lifting and transferring patients is a good example. An experienced physiotherapist, nurse or porter whose performance is based upon central programming, sees the patient, estimates their probable weight, plans the transfer manoeuvre and accurately matches their preparatory posture and lifting power against anticipated load almost without conscious calculation. At non-conscious level their central programming runs as a *feedforward* system that is not immediately dependent upon peripheral feedback. The inexperienced cannot do this because they have no internalised motor engram which is available to run the movement programme.

Box 5.3 Task 5.3

Assessing the difference between peripheral and central programming
Working in pairs, decide upon two different skills, whether balance, striking or throwing, one of which should be easy and familiar to you and the other difficult and unfamiliar. If each partner has a skill that the other does not possess, the differences will be more sharply defined. Practise the chosen two skills, noting the reasons that make, for you, one skill easy and one skill difficult to perform.

Because the computation of central programming is run at non-conscious levels, it is no longer immediately dependent upon the return of sensory information. Movement sequencing in specialised skills, such as playing the violin or piano, or responding to a fast game of skill such as squash or table tennis, can proceed at a much higher frequency of accurate serial movements per second (up to 20) than it could when dependent upon sensory return to conscious awareness (up to 3 per second).

This does not mean that sensory return is no longer important. Both McGuigan (1994) and Rothwell (1994) have reviewed the literature on the effects of experimental deafferentation in animals, and comparable traumatic loss of sensation in humans, on control of movement. They found that while central programming is surprisingly good in the short term, the loss of retrospective peripheral return eventually causes a breakup of movement.

The fundamental difference between central programming and peripheral programming is that the former 'unrolls' in a *proactive*, anticipatory sequence, whereas the latter waits for return information and is *reactive* in operation. Another major difference between the two is the speed of sensory processing. The 300 millisecond response time of conscious processing is replaced by the 10–30 millisecond response times of non-conscious central programming (McGuigan 1994).

Box 5.4 Task 5.4

Skill subroutines
Using whichever was the difficult skill practised by you in completing the task for Box 5.3, analyse the whole skill into a series of well defined components or subroutines. With the other member acting as coach, practise one subroutine, noting any change in performance ability, and then reintegrate it into the whole skill, noting the effect upon overall performance.

Classification of motor skills

For purposes of description and analysis, motor skills can be classified as *discrete, continuous* and *serial* (Singer 1980), although in many complex activities they tend to blend into each other as subroutines switch on and off.

Discrete movements

Discrete movements have a clearly defined beginning and ending and possess an all or none character. Examples include single strikes onto the keys of a typewriter or word processor, or in laboratory experiments where the subject has to perform a single on or off response in a reaction time test. Because they are rapid, short range, movements of decisional choice they are committed movements that cannot be modified by sensory feedback before they are completed.

A more complex form of discrete movement that depends upon central programming for its effective deliverance is the quick, acceleratory movement of punching, or throwing an object, which occurs in about 150 to 200 milliseconds. These movements are not voluntarily correctable once initiated and accelerating into the terminal half of their trajectory. Such movements are also classified as *open movements*, *open ended movements* or *ballistic* movements. Jumping is another example. At the end of the acceleratory lower limb extension phase the body is ballistically propelled into momentary free orbit.

In terms of cybernetic theory, the ballistic movement is executed as a *feed forward* sequence that is initiated by central programming. Analysis of any end result error can only be undertaken after movement completion (magnitude of output error compared to reference target), the movement sequence is then reprogrammed in consequence and the ballistic sequence is repeated and reassessed.

Continuous movement

Many everyday activities, such as dressing, washing up, walking, steady running, driving a car, writing, and so on, are examples of continuously controlled activities which, although preprogrammed in intention, can be modified by sensory return because they proceed slowly enough to incorporate error correction during execution. Such movements are classified as *closed* movements, or *closed circuit* movements because they have no open, or non-correctable, end outcome. Examples of continuous movements that specifically use sensory return for error correction are laboratory experiments requiring target tracking, and similar skills employed in amusement arcade and home computer games. Any activity that follows the changing movement of a target is known as *tracking*.

Serial movement

This term can be applied to both continuous movement, where sequential repetition of the same, or similar, movement is an integral part of any continuously changing activity, and to discrete movement, where staccato repetition may have a rhythmic beat, as in touch typing or drumming, or be phased in irregular sequences without strict timing.

An interesting point about tracking is that although the movement does not stop, and is therefore continuous, the conscious adjustments of movement required to track target direction changes are, in fact, *discontinuous*. This is because the decision to change the direction of movement takes about 300 milliseconds, which means no more than three changes per second. Repeated experience of tracking manoeuvres reduces error because central programming leads to an increasing range of anticipatory strategies, reduces response times, and increases subliminal awareness of just perceived cues to possible changes of direction (perceptual anticipation). Nevertheless, there is still a discontinuity of decisional change.

Box 5.5 Task 5.5

Motor skill classification
In pairs, or threes, devise and practise your own examples of *discrete*, *serial* and *continuous* movement, analysing them in terms of their differences in sequencing and end goals of performance.

Cratty (1973) lays particular emphasis upon the importance of sensory learning in motor skill learning. In the early stages of skill practice, conscious level processing of incoming sensory information is relatively generalised with poor discrimination of skill relevant cues. Repetitive practice results in a learning curve rise in the ability to ignore incoming information that is not directly skill relevant and increasing ability to interpret the significance of skill relevant information. As performance moves from peripheral to central programming, the developing motor engram includes a centrally programmed sensory component that has a selective sensitivity for skill relevant clues and a raised threshold against non-relevant input. In effect, a discriminatory sensory cue profile is formed that is specific to the skill. This runs in parallel with anticipatory feedforward strategies so that the motor implications of any change in sensory cueing is perceived at a fast subliminal level, is instantly computed, and the compensatory movement strategy put into effect. This focusing upon those sensory cues that are relevant to skill performance forms the basis of skill concentration.

Body image projection — extending the boundaries

In everyday life most people use a wide range of different tools. Common examples are combs, hairbrushes, hair dryers, cups, jugs and eating utensils. Common occupational tools are screwdrivers, hammers, saws, spades, lawn mowers, strimmers and power tools. People ride bicycles and motor bikes, and drive cars, vans and lorries. Sporting equipment includes tennis rackets, squash rackets, baseball bats, golf clubs, fishing rods, skis, skates and roller skates.

During physical rehabilitation, patients manually hold and use walking frames, crutches and walking sticks. They move and use artificial legs or artificial arms that are firmly attached to the limb stumps by sockets and straps. During prosthetic ambulation, patients usually use walking aids, and those with upper limb prostheses use them to grasp and use other equipment.

The normal three-dimensional body image of conscious awareness is determined by its physical shape, posture and physical boundary of sensorily innervated skin. This awareness is continually updated and reinforced by sensory input. The body is the effector mechanism of the mind which acts upon the external world by using the motive power of the musculoskeletal system. Tools which are manually held or otherwise attached to the body become physical extensions of the body during the period of contact. The reason for holding or attaching tools to the body is to make them available for use as mechanical extensions of the body during purposeful activity. In effect, they become 'add on' effector mechanisms of the mind during use.

Successful tool use is dependent upon a truly remarkable spatial ability of the brain and its conscious mental processes. When using a tool, the mind converts the incoming bodily sensation, which normally reinforces the boundary image of the body itself, into a new body image *plus* tool image extension of personally controlled space. The sensations arising from tool use are projected into this new functional body image as if they were coming from the surface boundary of the tool itself. During the 1950s and 1960s Bekesy pioneered research into this ability of the mind to project sensation into space beyond the body and made the important observation (Bekesy 1967) that 'every well trained machinist projects his sensations of pressure to the *tip of the drill*, and it is this projection that enables him to work rapidly and correctly' (my italics). When using a knife and fork the kinaesthetic sensations arising from the palm and fingers are interpreted *as if* they are coming from the point of contact between the tips of the knife and fork, the food, and the surface of the plate.

This ability to mentally extend the body image into a functionally integrated body/tool unity is a learned skill in its own right, but tends to be overlooked and taken for granted. The experience of a patient who has recently undergone an above knee amputation is a case in point. When the healed surface of the stump is exposed, all sensation from it during handling is related to the body image of the stump itself. When the patient first wears a prosthesis and starts weight-bearing through it, the pressure sensation that is generated in the stump is felt in the body image of the stump. With repeated practice, the stimuli from the stump is integrated with the visual and postural senses of transfer from sitting to standing, and standing to walking, and the stump stimuli lose their tissue boundary location and become interpretatively fused with the position, shape, weight, movement and contact boundaries of the prosthesis during use. Every time that the prosthesis is put on this body *plus* prosthesis body image is put on as well and stays in place until the prosthesis is removed. It then reverts to the skin boundary body image.

Another example is the learned sensorimotor skills of driving a car. Study of the complex skills involved in learning to drive a car tend to concentrate upon integrating the separate subroutines of gear selection, clutch, brake and accelerator pedal control. The external context of safe driving depends upon three-dimensional visual awareness, comparative speed computations, use of mirrors, signalling and so on. Steering the car within the continually changing external environment of roads, pavements, parked cars and moving traffic is crucially dependent upon the creation of a spatially accurate driver *plus* car body boundary image which is the functional extension of the body image during the act of driving. In effect, the car and driver become a unified spatial entity from the moment that the driver sits in the driving seat and holds the steering wheel, and it is within this context that the motor skills of driving are first developed from peripheral programming, where the car is the extended periphery that must be learned and controlled, to central programming, which takes the car and driver as a functional unit. No matter how skilled the driver is in the use of clutch, brakes and gears, if the ability to form this new spatial image is poor, then the risk of accident is high.

Box 5.6 Task 5.6

Body image projection
Practise and analyse in terms of body image projection the difference between a beginner and a professional in sports that require the use of striking equipment such as in tennis and golf.

MOTOR LEARNING — THE NEUROLOGICAL BACKGROUND

New techniques of non-invasive monitoring of brain activity, such as positron emission tomography (PET), can now correlate particular areas of brain activity with particular acts of mental activity as in, for example, the voluntary performance of a simple movement upon request while the brain is being monitored. It has been found that when a subject undertakes to learn a new movement combination, the pattern of brain areas that metabolically 'light up' during performance changes over time as the subject passes through the early stages of trial and error learning, which require sustained conscious control throughout the practice period, to the skill competence of central programming, when the only conscious input is to start, maintain and end performance.

In general terms, the foci of intense brain activity move from the prefrontal lobes and associated anterior cingulate gyrus on the medial aspect of the hemispheres (intention and emotional desire to learn), together with the primary sensorimotor cortex (sensory input concentration and movement output), and regions of the cerebellum and subcortical nuclei involved in primary motor learning, to the secondary sensory association areas, secondary motor areas and subcortical centres that store the movement as a learned whole, or motor engram, that is now centrally programmed. This is a gross simplification, and Roland (1993) should be consulted for detail of known brain mechanisms and their areas of activation during different forms of intentional mental activity, but the principle is clear.

Despite the use of computer programmes to simulate probable neural mechanisms, and the widespread adoption of computer science and cybernetic terminology as models of neural activity, we do not know *how* the brain really works. Nor do we understand the relationship between mind and brain. It is important to realise that we do not, as yet, know how the patterns of neuromuscular activity that we faultlessly perform every day of our lives are represented and stored within the CNS. Brain imaging techniques can correlate brain activity with mental activity but not the actual nature of the interaction.

Some important clues have been revealed by recent research (Welsh et al 1995) using multiple electrode recording of synaptic activity from neurones in the inferior olivary nuclei and related Purkinje cells in the cerebellum. This investigation was done with rats that had been trained to protrude their tongues out to their full extent so that they could just lick an object. During this particular sequence of tongue movement it was found that separate small groups of olivary neurones rhythmically 'drive' the Purkinje cells so that they, in turn, control the tonic excitatory output from cerebellar nuclei by selective inhibition. The outcome is that the pattern of motor neurone to motor unit activation of muscle fibres in the tongue of the rat relates to the tongue protrusion and licking movement.

What the researchers found was that the physical movements of the tongue were represented in the inferior olive by low frequency (circa 40 Hz) electrical oscillations that linked the mosaic of physically separate groups of neurones involved in this movement into a common electrical network. These oscillatory patterns were locked into the same time and spatial sequence as the movements of the tongue. The inferior olivary nucleus in the brain stem was chosen for this experiment as damage to it causes 'cerebellar type' ataxias, and the integrity of the inferior olive is crucial to cerebellar control of movement. The authors referred to the inferior olive as 'an electrically malleable substrate from which unique motor synergies can be sculpted', and McCormick (1995), in his review of this research, referred to the inferior olive as a 'conductor of the symphony of cerebellar neuronal activity'.

The implications of this research are that *if* such oscillatory electrical patterns throughout the CNS were found to be the neuronal equivalent of the physical movements of the musculoskeletal system in space, then it may be possible, in principle, and probably in the far distant future, to record and recognise electrical patterns in, say, the inferior olive, and correlate them with observed physical movement. At the moment it is theorised that when such electrical patterns are not directly coupled to ongoing movement they are maintained during the decoupled periods by continuous refreshment of synaptic activity by low frequency, self-firing, neurones, analogous to the sweeping of the electron beam across the television screen that continually refreshes the image intensity of the individual dots, or pixels, that form the television picture.

Mental practice — skill simulation by mental imagery

As it is normal practice for anyone to think about, and mentally rehearse, a new skill that they intend to learn, or are in the process of learning, the question arises as to whether repeated mental imaging of the desired activity before commencing practice, or during the intervals between practice sessions, would have any outcome effect upon the rate at which the skill is acquired and / or the level of proficiency attained.

The answer is important because if such imaginative rehearsal can be shown to have a measurable effect upon performance, then patients should be encouraged to repeatedly visualise what they are trying to learn, and physiotherapy students should repeatedly visualise the techniques that they need to learn.

Evidence obtained from monitoring the brains of volunteer subjects when they are imagining different physical tasks clearly shows that the same areas of brain are active during the imaging as during performance (Decety et al 1994, Roland 1993), with the important exception of the primary sensory reception areas as they are directly activated by incoming sensory stimuli. For example, when a subject imagines running a marathon, the secondary motor area 'lights up' while the primary motor cortex remains silent, and the secondary and primary sensory cortices act likewise. The important point here is that the cortical and subcortical areas that are activated by mental imaging, including the emotion drive of the limbic system, are the same areas that are directly involved with the processes of memory formation. Kosslyn and Koenig (1992) have shown that the processes of mental imagery closely follow actual experience. For example, subjects who have memorised a map take the same mental time to travel from one imagined map reference to another as they do with the map in front of them. Again, subjects asked to rotate the mental image of a large geometrically shaped object like a capital letter *E*, take a proportionally longer time to rotate it than they do when asked to rotate the image of a small *E*-shaped object. This again parallels real life, and Decety (1993) considers that the known neurophysiological basis of mental imaging forms a strong *prima facie* case for the use of mental simulation of a physical activity before practice, and as a memory reinforcing agent between practice sessions.

Maring (1990) studied the effect of mental practice intervals versus non-mental mental practice intervals upon a simple repetitive activity and found that mental imaging of the activity in the intervals between each set of repetitions facilitated motor learning. In this experiment the subjects sat on a straight-backed hard chair in a standardised sitting position. Each had a splint attached to their forearm that extended to just beyond their hand and ended in an upward-facing cup. An assistant placed a Velcro-covered tennis ball into the cup and the subject first slightly extended the elbow and then quickly flexed to toss the ball forward towards a target mark which was 1.5 m to their front on a Velcro-adherent floor mat. Each subject in the experimental imaging group (N = 13) and the control group (N = 13) threw 10 times in succession, had a 2-minute break, threw another set of 10, and repeated this procedure to a total of 50 individual throws. During each of the four, 2-minute, intervals the experimental group mentally rehearsed the activity with encouragement from the assistant, whereas the control group were given a poem to memorise to block out any mental practice.

Electromyographic (EMG) recordings of motor unit activity from the agonist biceps brachii muscle and the antagonist triceps brachii muscle of each subject showed that the experimental group showed rapid improvement in the timing and amplitude of their respective on/off contractions to control the velocity and effective arc of throw, thus leading to improved accuracy in target area clustering compared to the poem distracted control group. Averaged experimental group accuracy improved from around 30 cm target scatter to 10 cm target scatter over the 50 throws compared to an averaged control group improvement from about 38 cm scatter to 24 cm scatter.

Although much more research needs to be done to quantify the efficacy of mental practice in the acquisition of motor skills, there does seem to be strong grounds for including the planned use of mental imaging before practice and during the intervals between practice, and such periods of mental practice may be of considerable value in rehabilitation.

Box 5.7 Task 5.7

Mental practice of a skill
Using Maring's experiment as an example, divide into small groups and devise a simple, stereotyped movement that could be used to test the hypothesis that mental practice of a skill during the intervals between fixed sets of repetition practice improves performance compared to no mental practice during the same interval durations.

Structured skill instruction — sources of learning

When anyone is acting in the role of skill instructor, he or she must select and use the source, or combination of sources, that will be most effective in ensuring that the learner can understand what the skill involves and how it should be practised.

Again, this seems hardly worthy of comment, but the choice of source(s) presupposes that the intending skill learner can understand and use the source(s) of explanation and instruction to form, in the first instance, a mental skill ghost strong enough to act as a source of self-reference during practice. In rehabilitation, for example, the physiotherapist may need to teach a patient, and/or carers, the use of an inhaler, or regaining postural balance after a stroke, or walking with a frame, or safe transference from bed to chair, and so on. From the viewpoint of the learner, none of these skills will seem very easy to learn, so the physiotherapist must make a personal assessment of the probable ability of the learner to understand what must be done, and use that assessment to choose the most appropriate means of instruction.

Three important factors are involved here:

- the judgement of the skill instructor
- the choice of sources, or means, of skill instruction
- the ability of the instructor to explain, teach and correct.

If, in rehabilitation, the physiotherapist understands what must be done but is unable to convey this understanding to the patient and/or carers, then only frustration and failure can follow, and the same would apply to any other situation where the instructor is unable to explain the skill in terms that the learner can understand and follow.

Possible sources of skill instruction

Demonstration only. Whether in person, or by video, the demonstrator must sequence the skill into subroutine stages so that the learner can follow through. The task for the learner is to successfully internalise the observed demonstration into a skill ghost that matches what is observed, practise the movements by mimicking the demonstrator, and practise in the absence of the demonstrator using the skill ghost as the self-reference standard.

Demonstration with explanation. This is the more usual format, and the purpose of including a verbal explanation to accompany the skill demonstration is to clarify what is involved by using language as an additional source of explanation and comment. This requires considerable skill in presentation. If the two sources of instruction are running in parallel, the attention of the learner is split between the two, possibly in an alternating way, in an attempt to simultaneously understand them both and synthesise them into a unified understanding. If there is any ambiguity between the two sources, the learner will be confused and lose continuity. It must be remembered that conscious attention is a single channel activity, and it may be more effective to sequence the demonstration into successive stages of explanation, demonstration and recapitulation. Whichever method is chosen, it is important that the explanation is simple, concise and complements the demonstration.

Verbal explanation and instruction only. Whether this is delivered in person, or by prerecorded tape, telephone message or direct telephone conversation, the words must accurately describe what is to be done, and the instructional sequence must follow the motor sequence of the skill. It should include advice on how to assess performance and how to correct for error. This is not an easy source to prepare and deliver, nor to understand and put into effect. The task of the learner is to convert word imagery into movement imagery and this is where the main difficulty lies.

Written explanation and instruction. As in the previous source, the words must be chosen with great care. Concise description, correct grammar and correct punctuation are very important as it is suprisingly difficult to be clear and unambiguous. Essential points must be emphasised. The task of the learner is to convert the written words into internalised speech, to understand their meaning as intended, convert the internalised speech into a motor imagery that will act as self-reference, and then practise. Anyone who has ever tried to assemble a do-it-yourself (DIY) kit by following instructions, or has attempted to set up and use a word processor by following the instruction manual, will know how difficult it is to convert the written source into constructive action.

Written source plus illustration. From the point of view of the learner, a combination of writing and key point illustrations make the task of converting source imagery to motor imagery very much easier than just the written word, *assuming* that word and illustration agree with each other. Most instruction manuals include line drawing illustrations to help convert their information into motor imagery.

Any of these sources can be combined to improve clarity of presentation and make the task of the skill learner less open to error and uncertainty. Video presentation presents the opportunity for repeated playback as necessary, and interactive video has considerable potential as a highly effective learning aid.

Box 5.8 Task 5.8

Skill instruction
Choose a task, such as using crutches for non-weight-bearing walking on one foot, and prepare instructions for first time practice of the task, using each of the sources listed, and then implement each one to compare their relative effectiveness.

CONCLUSION

The one certain factor in achieving success in the learning and practising of a skill to the level of competence required is adequate repetition of error-free movement until a long-term motor engram of the whole skill is formed. The formation of this motor memory is coincident with transfer of the learning process from the stage of conscious control of movement, or peripheral programming, which is dependent upon sensory input, to non-conscious control of movement, which unrolls from central programming as anticipatory feedforward and is not immediately dependent upon sensory feedback of any error.

The principles and methods of cybernetics and computer science provide many of the models of probable neurological organisation and function regarding movement control and skill acquisition. They provide a useful explanatory framework and are a valuable source of hypotheses and experiments but they are not the whole answer. This will only exist when neuroscience can explain the brain and its relationship to mind without recourse to models provided by other disciplines.

The emphasis in this chapter has been on structured learning as this is the most common form of learning after early childhood. The most important elements of structured learning are sustained motivation, a clear understanding of the skill as a whole, the experience of attempting to practise the whole skill, practise of clearly defined subroutines of the skill if it is complex, distinguishing the difference between correct performance and error, and a final merging of the subroutines into a seamless whole. The role of the skill instructor is to sustain motivation, to ensure that the learner knows what to do by choosing the method(s) of instruction with care, to recognise and correct error in performance, and reinforce correct movement by repetition and encouragement.

REFERENCES

Bekesy G 1967 Sensory inhibition. Princeton University Press, New Jersey

Cratty B J 1973 Movement behaviour and motor learning, 3rd edn. Lea & Febiger, Philadelphia

Decety J 1993 Should motor imagery be used in physiotherapy? Recent advances in the cognitive sciences. Physiotherapy Theory and Practice 9: 192–203

Decety J, Perani D, Jeannerod M, Bettinardi V, Tadary B, Woods R, Mazziotta J C, Fazio F 1994 Mapping motor representations with positron emission tomography. Nature 371: 600–602

Kosslyn S M, Koenig O 1992 Wet mind: the new cognitive neuroscience. The Free Press, Macmillan, New York

Kottke F J 1980 From reflex to skill: the training of coordination. Archives of Physical Medicine and Rehabilitation 61: 551–561

Kottke F J, Halpern D, Easton J K M, Ozel A T, Burrill C A 1978 The training of coordination. Archives of Physical Medicine and Rehabilitation 59: 567–572

McCormick D A 1995 The cerebellar symphony. Nature 374: 412–413

McGuigan F J 1994 Biological psychology: a cybernetic science. Prentice Hall, New Jersey

Maring J R 1990 Effects of mental practice on rate of skill acquisition. Physical Therapy 70: 28–35

O'Brien C, Hayes A 1995 Normal and impaired motor development. Chapman and Hall, London

Roland P E 1993 Brain activation. Wiley-Liss, New York

Rosenbaum D A 1991 Human motor control. Academic Press, San Diego

Rothwell J 1994 Control of human voluntary movement, 2nd edn. Chapman and Hall, London

Schmidt R A 1982 Motor control and learning: a behavioural emphasis. Human Kinetics Publishers, Champaign, Illinois

Singer R N 1980 Motor learning and human performance, 3rd edn. Macmillan, New York

Welsh J P, Lang E J, Sugihara I, Llinas R 1995 Dynamic organisation of motor control within the olivocerebellar system. Nature 374: 453–457

Whiting H T A 1975 Concepts in skill learning. Lepus Books, London

Wiener N 1961 Cybernetics, 2nd edn. M I T Press, New York

CHAPTER CONTENTS

6

Posture and balance

T. Howe J. Oldham

OBJECTIVES

At the end of this chapter you will be able to:

1. **Define posture and balance**
2. **Describe the ideal posture in lying, sitting and standing**
3. **Discuss the importance of balance and posture and the mechanisms of their interaction**
4. **Discuss the control of balance and posture**
5. **Describe abnormalities of posture and balance**
6. **Discuss the re-education of posture and balance.**

INTRODUCTION

The evolutionary process has resulted in modern man (homo sapiens). One of the early phases of human evolution was the adoption of an erect posture allowing the development of bipedal gait and freeing the arms for other uses, e.g. tool making. Modern man is designed for mobility: muscles generate forces which are exerted about their attachments to bone resulting in movement. However, compared to their four-legged counterparts, humans are precariously balanced animals. Four-legged animals are quite stable and difficult to unbalance when pushed. On the other hand, two-legged animals (man) easily fall over. Two major factors account for this instability: the small contact surface area of the feet relative to the rest of the body, and a high position of centre of gravity relative to the ground.

Gravity is constantly pulling the body towards the ground. This force has to be counteracted by the muscles exerting a continuous pull on the skeleton in the opposite direction to these forces (Thibodeau & Patton 1993). The role the muscles play in this respect cannot be overestimated. In addition, the bones of the skeleton are too irregularly shaped to stand upon each other and maintain an upright position alone (Thibodeau & Patton 1993). The presence of ligaments surrounding joints is still insufficient to afford stability. As such, a passive stability is insufficient and active mechanisms must be employed to maintain balance and posture. These mechanisms do not operate when a person is asleep, hence the inability to sleep unsupported or standing up.

The ability of the muscles to counteract gravity by graded levels of contraction is carefully controlled by the nervous system. The mechanisms by which this is achieved will be described in detail below. Many other systems contribute to the ability of muscles to maintain posture. These include the respiratory, digestive, circulatory, excretory and endocrine systems (Thibodeau & Patton 1993). These will not be considered further in this chapter but the important contribution of other body systems must not be forgotten.

DEFINITIONS

Before proceeding any further, balance and posture need to be defined. The two concepts cannot be considered in isolation as they are interdependent. The term *posture* means simply position or alignment of body parts (Thibodeau & Patton 1993). It is usually thought of in terms of the spine, but it should be remembered that all body parts have a role to play in postural alignment. Furthermore, posture cannot be separated from movement, but should be regarded as 'temporarily arrested movement' as it is in a constant state of change (Bobath 1978), as anyone trying to stay still for any length of time will know. Defining 'good' posture is less easy, however, and can refer to a position that requires the least effort to maintain, puts the least strain on ligaments, bones and joints or maintains the centre of mass over the base of support (Thibodeau & Patton 1993).

Balance, on the other hand, can be defined as 'a state in which the body is in equilibrium' (Galley & Forster 1987). This concept refers to the series of physiological mechanisms that exist to inform the body that a compromise in posture has taken place and undesired movement will occur, and the resulting mechanisms that then restore that posture and prevent the undesired movement. Posture is a result of balancing each body part with respect to the other body parts.

'NORMAL' POSTURE

Normal posture is difficult to define as every person has a unique anthropometric and physiological profile. Morphological body types (somatotypes) may be classified according to three extremes: ectomorphs (long and thin), endomorphs (short and fat) and mesomorphs (athletic and muscular) (Carter et al 1983). Most people are a combination of all three extremes. These anatomical differences, particularly the inborn length of the ligaments, account for many of the greatest differences in posture. People with loose ligaments tend to stand with hyperextended knees and hips (resting on their iliofemoral ligaments) and with flexible, exaggerated curves of the spine that are much greater than those seen in people with tighter ligaments (Larson & Gould 1974). Furthermore, the majority of the population rarely adopts good posture, with slouching being particularly prevalent. In addition, posture may alter throughout the day and may be related to fatigue and emotional state. The number of postural variations makes it difficult to define what constitutes normal posture, particularly in a population that rarely adopts a good one. An ideal posture tends only to be observed in trained individuals and the therapist must aim for a posture that suits the circumstance, i.e. body shape and interaction with the environment, rather than a typical posture.

The ideal static standing posture

When assessing posture, observations should be

undertaken with the subject barefooted as pressure distribution over the feet will differ depending on whether shoes are worn or not (Galley & Forster 1987). The ideal erect posture should be assessed in three dimensions, i.e. laterally, anteriorly and posteriorly, by comparing parts of the body with a plumb or imaginary vertical line. When viewing an ideal posture from the lateral aspect, this standard reference line should pass just anteriorly to the lateral malleolus of the ankle, immediately anterior to the midline of the knee and then directly through the greater trochanter, bodies of the lumbar vertebrae, shoulder joints, bodies of the cervical vertebrae and lobe of the ear (Kendall at al 1993) (Fig. 6.1).

The head should be erect and well balanced on the neck and not held too far anteriorly (poking chin) or posteriorly (Galley & Forster 1987). Incorrect positioning of the head may lead to neck and eye pain and associated headaches. The chest should be erect without tension or overexpansion, and the abdomen should be flat and relaxed without sagging or retraction (Galley & Forster 1987).

Finally, the lumbar area of the back should be slightly hollow (slight lumbar lordosis) (Norris 1994). When viewed from the front, the bare feet should be placed approximately 8 cm apart and the line should bisect the body into two equal halves. Body weight should be evenly distributed between the two halves (Galley & Forster 1987). In this position the pubic symphysis and anterior superior iliac spines and the point of the shoulders should be on the same level in the horizontal plane (Galley & Forster 1987, Kendal et al 1993). When observing the posterior view, other anatomical landmarks including the knee and buttock creases, iliac crests, dimples over the sacroiliac joints, inferior angle of the scapulae, acromion processes, ears and external occipital protuberance can also be used for horizontal alignment (Norris 1995). See Table 6.1 for a posture check-list.

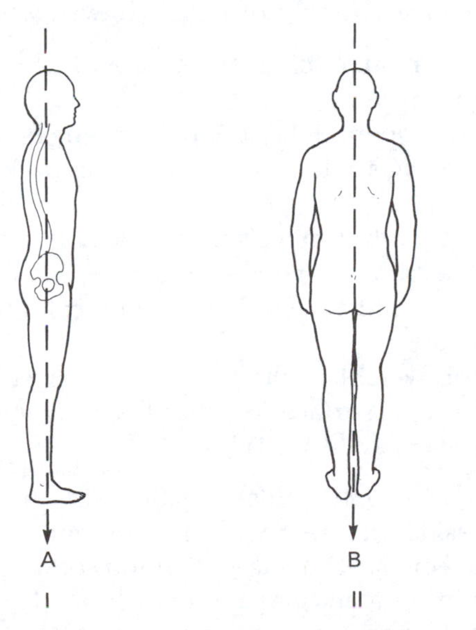

Figure 6.1 A well aligned standing posture: a) lateral view, b) posterior view.

Table 6.1 Posture check-list

Anterior view	
Head alignment	central/L or R lateral tilt/L or R rotation
Shoulders	level/L or R elevation/L or R protracted/ L or R retracted
Waist skin creases	equal/L or R side flexion
Pelvis ASIS	level/L or R elevation
Patellae	level/L or R/medially positioned/ L or R laterally positioned/L or R elevated
Feet	neutral/L or R laterally positioned/L or R medially positioned/L or R flat
Posterior view	
Head alignment	central/L or R lateral tilt/L or R rotation
Acromion processes	level/L or R elevation/L or R protracted/ L or R retracted
Inferior angle of scapulae	level/L or R elevation/L or R abduction/ L or R adduction/L or R winging
Thoracic spine	neutral/L or R scoliosis/kyphosis
Lumbar spine	neutral/L or R scoliosis
Waist skin creases	equal/L or R side flexion
Iliac crests	level/L or R elevation
S-I joint dimples	level/L or R elevation
Buttock creases	level/L or R elevation
Knee creases	level/L or R elevation
Knee position	L or R extension/L or R hyperextension/ L or R flexion
Feet (weight bearing)	equal/L or R diminished
Lateral view	
Head alignment	neutral/anterior poking/extended
Shoulders	level/L or R elevation/ L or R protracted/L or R retracted
Thoracic spine	normal/kyphosis/flattened
Lumbar spine	normal/increased lordosis/flattened
Pelvis	neutral/anterior tilt/posterior tilt
Hip position	neutral/L or R extended/L or R flexed
Knee position	L or R extension/L or R hyperextension/ L or R flexion

> **Box 6.1** Task 1
>
> **Postural assessment**
> Assess the posture of some of your fellow students. Select a sample of males and females and try to include different extremes of somatotypes. Use the posture check-list to help you assess their standing posture. They will not be exactly 'normal' for the reasons discussed previously. Note the differences in posture between males and females.

The ideal lying posture

So far the discussion in this chapter has related to the standing position. However, the strict definition of posture also relates to other body positions. A lying position is known as a recumbent posture. If this position is face down, it is referred to as a prone posture and if face up, a supine posture. In these cases, the pull of gravity is not through the longitudinal axis of the body but through each segment of the body that is in contact with the supporting surface (Larson & Gould 1974). Lying is the least energy consuming of all positions. Whatever the lying position adopted, localised pressure points develop and the position has to be adapted constantly. This is all too obvious when lying on a very hard surface, e.g. the floor. The development of pressure points is a particular problem in people with altered skin sensation, e.g. following spinal injury or peripheral nerve lesions, as they are unable to detect their development and skin breakdown may result. Such individuals must be made aware of the dangers of assuming the same posture for long periods of time and special mattresses to reduce the risk may be provided.

The ideal lying position is one in which rigid body segments sink down just enough to allow remaining segments to accept support from the mattress and essentially maintain the same body alignment as in standing (Larson & Gould 1974). The condition of the mattress is very important. A soft mattress offers little support: when lying supine, the lumbar spine is in a position of flexion and when lying prone, the lumbar spine is forced into an exaggerated lordosis or hyperextension. In side lying in particular, the head will also need to be supported by a pillow to retain its line with the body (Galley & Forster 1987). The number of pillows used and their type also influence posture. Too many pillows or very firm pillows force the head into flexion when supine, and side flexion when side lying. Sleeping posture is important in people who have neck and back injuries as poor posture may aggravate their symptoms. Recently, special pillows and mattresses have become available that offer a more ideal position for the head and neck and afford greater support for the lumbar spine during lying.

The ideal static sitting posture

The sitting posture is more relaxing than that of standing. This position provides a greater supporting surface and allows relaxation of the muscles of the lower limb (Galley & Forster 1987). In the correct sitting position, the centre of mass should extend through the ischial tuberosities and just in front of the eleventh thoracic vertebra (Larson & Gould 1974). Without the additional support to the thighs and back common to most chairs, this is a highly unstable position.

The ideal sitting posture is achieved when:

- the ischial tuberosities provide the major base of support
- the upper thighs add to the sitting base without placing undue pressure on the back of the knee joint
- the lumbar spine is in mid-flexion
- the entire spine is supported via a backrest with a slight backward inclination from the perpendicular
- the weight of the legs is transferred to the supporting surface of the floor by the feet (Larson & Gould 1974).

Unfortunately, this position is practically impossible to attain with many modern seating arrangements. Seats are often too soft and deep, short or long and have too much of a backward slope. Furthermore, it is very common for individuals to slide their pelvis forwards (slouch) resulting in a centre of mass behind the ischial tuberosities. This results in a convex curvature of the lumbar spine (loss of the lumbar lordosis) (Larson & Gould 1974) and concave curvature of the thoracic spine, the former placing excessive strain on the posterior spinal ligaments and causing posterior bulging of the intervertebral discs. Different sitting postures are also achieved depending on whether the subject is sitting on a chair or the floor (Fig. 6.2).

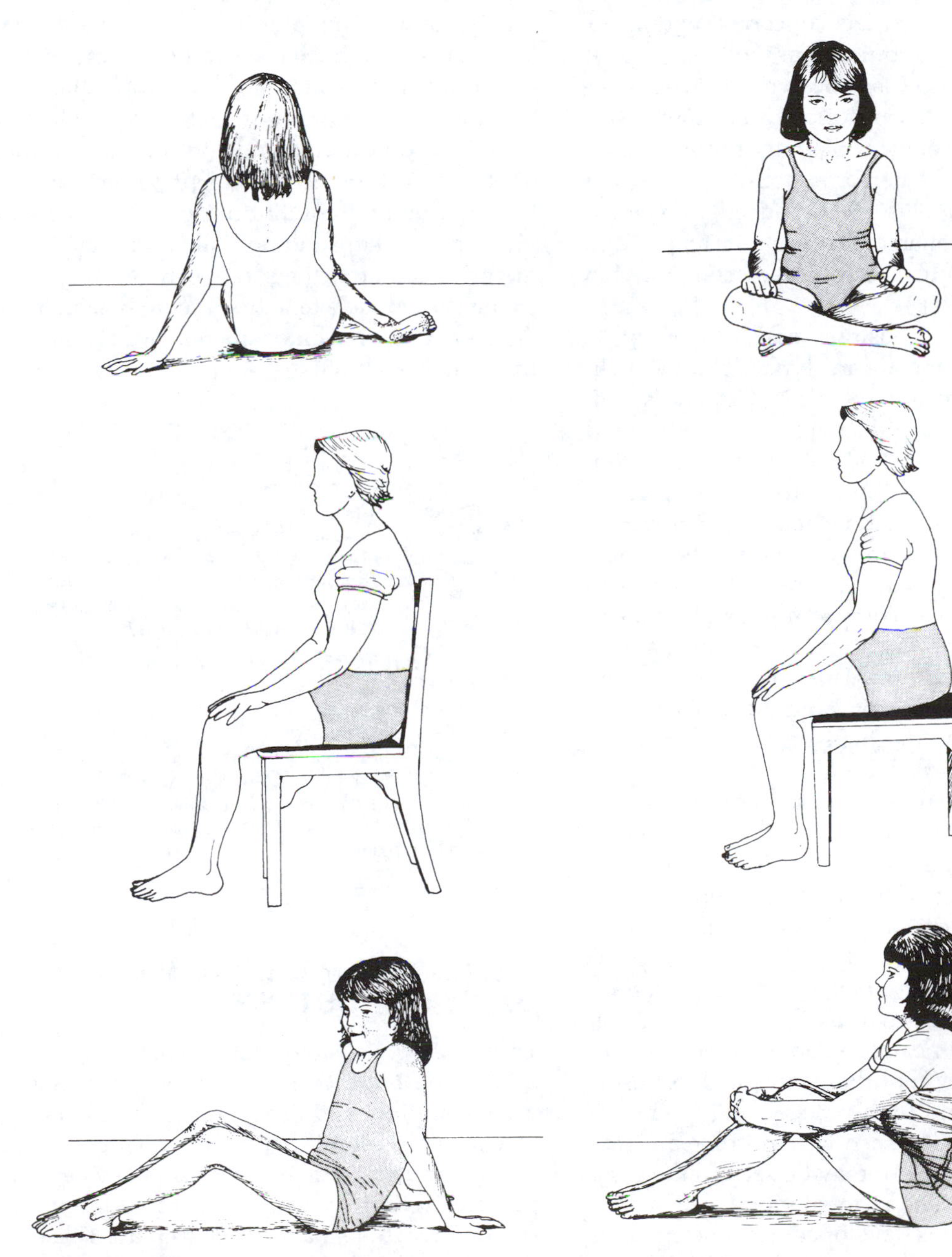

Figure 6.2 Different sitting positions.

THE IMPORTANCE OF BALANCE AND POSTURE

The importance of balance and posture cannot be overemphasised as demonstrated in the following quote by Carpenter (1984): 'Movement begins and ends in posture and for the most time the motor system is not in fact concerned with moving the body at all but rather with keeping it still'. This definition is, however, a gross oversimplification with postural adjustment (i.e. maintenance of balance) playing a fundamental role throughout the performance of a movement (Peterkin 1969, Bobath 1969). Indeed, movement occurs against a pliable postural tone adjusted to maintain the equilibrium of the person (Peterkin 1969, Willard 1992).

In the standing position and throughout a movement, humans are in dynamic rather than static equilibrium (Green 1978). Gravity is continuously trying to pull us towards the ground (Thibodeau & Patton 1993). Any slight deviation from the upright posture must be counterbalanced to maintain that position. A number of complex mechanisms are involved in this process. These range from those systems that tell us when our posture has been compromised and we are in danger of falling to those mechanisms that restore our equilibrium. This chapter focuses on these two mechanisms and describes how through balancing mechanisms we can maintain posture.

Balance is essential for every activity carried out during waking hours and skilled movements are dependent upon the ability to maintain equilibrium in a variety of positions and under many conditions (Davies 1985). The lower limbs must be able to support the body on its base of support and withstand any unexpected perturbations (Stein 1982). Muscles make continual adjustments to maintain balance and equilibrium. This requires contraction of the flexors and extensors synergistically and with precise timing. Thus, it can be seen that balance and posture are intimately related. Posture may be seen as the starting point and the end point of movement.

Bobath (1978) has developed the concept of the normal postural reflex mechanism (NPRM) providing the background for all skilled movements (it should be noted that this concept presupposes an intact adult brain). The NPRM is dependent upon three factors: normal muscle tone, reciprocal innervation and inhibition, and automatic movement patterns. Normal muscle tone must be sufficiently high to support the body against gravity and to initiate and control movement, but not so high that it impedes movement. Reciprocal innervation and inhibition allow stabilisation of certain body parts while allowing selective, coordinated and controlled movement of others. Automatic movement patterns include the righting and equilibrium reactions that provide the background for all voluntary movements. These reactions range from very small changes in muscle tone that are not visible to the naked eye to gross movements of the limbs and trunk (Davies 1985).

Box 6.2 Task 2

Effects on balance
Stand on a wobble board with your feet approximately 10 cm apart then with your feet together. Then try standing barefoot on a large medicine ball. How do these alterations in base of support alter your ability to balance?

Stand with the aid of elbow or axillary crutches so that your feet and the crutches are in a straight line. Attempt to maintain your balance while a fellow student pushes you in anterior and posterior directions. Now place the crutches slightly in front of you so that you form an approximate triangle with the crutches and your feet. What difference does this make to your ability to balance?

BALANCING MECHANISMS TO MAINTAIN POSTURE

Three major physiological mechanisms exist to inform the body that a compromise in posture has taken place and to elicit a series of balancing reactions to restore that posture. These include: the vestibular system, the visual system and pressure receptors in the feet. In addition, neck righting reactions have a role to play. These systems do not operate in isolation but offer an integrated approach to maintenance of posture involving the entire nervous, muscular and skeletal systems.

Pressure receptors in the feet

Pressure receptors in the feet provide the body with information about the distribution of body support. Differences in pressure at different points on the sole of the foot tell us the position of the vertical projection of the centre of mass relative to body supports (Carpenter 1984). This is very important, for in order for the body to remain upright, its centre of mass must always pass through its base of support, i.e. the feet in the upright position. When the line of gravity falls outside of this position the body will fall over (Norris 1993). To prevent such a catastrophe, the position of the body must change to restore the status quo (Carpenter 1984).

If position is shifted from an equal pressure distribution under both feet to an increased pressure under one foot, the body responds by increasing the tone of the extensor muscles in that limb with a corresponding increase in the flexor muscles in the opposite limb (Green 1978). As a result, the body is prevented from falling over and an upright posture is maintained (Carpenter 1984). These reactions are seen in a more dramatic form if the body is pushed from side to side. The body has to constantly maintain an upright position by reversing its reactions in response to the direction of force. These mechanisms form the *postural sway reaction*. These reactions are not solely related to external force but also apply in normal standing when the body is never completely still but swaying constantly (Fig. 6.3).

The postural sway reaction assumes that some support is already present. Other movements may be required, however, if the position of the centre of mass of the body extends outside its base of support. These include *stepping reactions* where two feet are involved and *hopping reactions* where there is only one foot involved. In cases where postural support is required, but not present, *placing reactions* are employed.

Proprioceptive input from the skin, and pres-

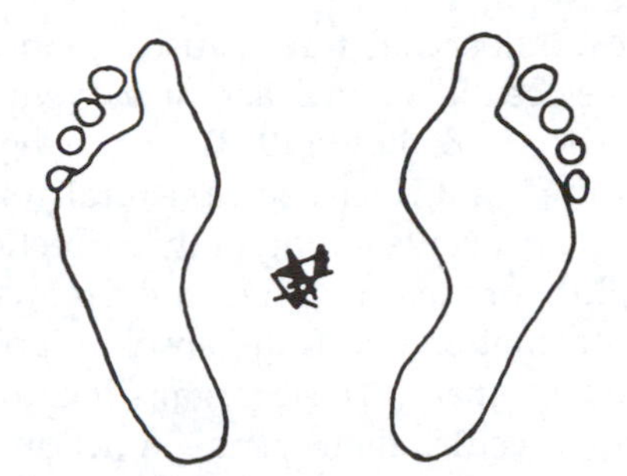

Figure 6.3 The oscillating path of the body's centre of mass during standing.

sure and joint receptors of the foot, have been observed to play a significant role in minor perturbations of the supporting surface, but are of minor importance during rapid displacements of the supporting surface (Deiner et al 1984). It should be noted that a complete loss of proprioceptive input from the lower limbs results in a severe loss of postural stability and a pathognomonic body tremor (Deiner et al 1984). Pressure receptors are to be found not only in the feet but also throughout the body. These receptors contribute to the total knowledge of body position but for simplicity are not considered in detail here.

Vestibular system

The vestibular system provides the body with two sets of information. Rotation of the head activates sensory receptors of the:

- semicircular canals to provide information regarding the angular acceleration of the head (Berne & Levy 1993).
- otolith organs to provide information about the effective direction of gravity.

Stimulation of the semicircular canal system provides information about the rate rather than direction of movement (Seeley et al 1992). Such information results in postural adjustments mediated by commands transmitted through the spinal cord via the lateral, medial and vestibulospinal tracts and reticulospinal tracts (Berne & Levy 1993, Rutishauser 1994). The lateral vestibulospinal tract activates the extensor muscles throughout the body that control posture. The medial vestibulospinal tract causes contraction of neck muscles that counteract the movement of the head (Berne & Levy 1993). If the head is moved to the right, increased postural tone on the right side prevents falling in that direction.

The otolith organs are the only organs that provide information about the absolute position of the head in space. These organs include the utricle and saccule; their prime function is to keep the head upright despite changes in the position of the body. This is achieved through changes in tone of the neck muscles via mechanisms known as the *head righting reflexes* (Carpenter 1984). These mechanisms restore the position of the head to its neutral balanced position (Seeley et al 1992).

The vestibular system is poorly developed in man and balancing in daytime is mainly carried out via impulses from the eyes (Green 1978). In people with visual impairments, however, balance has to be maintained by the pressure of the feet and the vestibular mechanisms described above.

Tonic neck reflexes

Changes in neck muscle tone are not only mediated in response to information received via the vestibular system but from the muscle spindles of the neck itself. One of the largest concentrations of muscle spindles exists in the neck (Berne & Levy 1993). Stimulation of these spindles evokes tonic neck reflexes. Bending the head to the left will result in the contraction of the extensor limbs on the left side and relaxation of the flexor limbs on the right (Berne & Levy 1993).

The visual system

Other receptors in the head that assist in the maintenance of posture by providing information about the position of the head are those of the retina (Carpenter 1984). The visual system tells us that the image of an object has moved relative to the retina. The ability of this system to distinguish between an object moving, the eye moving relative to the head, or the head and eye moving together is of paramount importance to the maintenance of posture. How this is achieved, however, is still subject to debate, though the end point of the reflex results in contraction of the neck muscles to right the head (Keel & Neil 1966).

The important role of the eyes in the maintenance of balance, however, can be illustrated by asking a person to stand on one leg initially with the eyes open and then closed. Once the eyes are closed, the person tends to fall over as the visual reference point for maintaining that position is lost. In addition, the visual system enables the precise timing and control of a movement in relation to the environment (Galley & Forster 1987), for example, the foot striking a step when alighting a bus.

Box 6.3 Task 3

Factors affecting balance
Find a partner and ask him/her to time how long you are able to stand on your non-dominant leg while barefoot and with your eyes open. Repeat this procedure with your eyes closed. Note the differences in your ability to balance.

With your eyes closed spin yourself around in a circle several times then try and stand still. What factors are contributing to your altered ability to balance?

Interaction of postural information to maintain balance

The various postural reactions described above in the normal person are coordinated in definite cortically controlled patterns common to everyone (Bobath 1969). These reactions are automatic and involve predictable changes in muscle tone according to the position of a person's head in relation to the trunk. This results in a corresponding increase in flexor or extensor activity to restore balance. These *automatic postural reaction patterns* form the background against which automatic and voluntary movement patterns are based (Bobath 1969). Figure 6.4 illustrates the interrelations between the various mechanisms involved in the maintenance of posture.

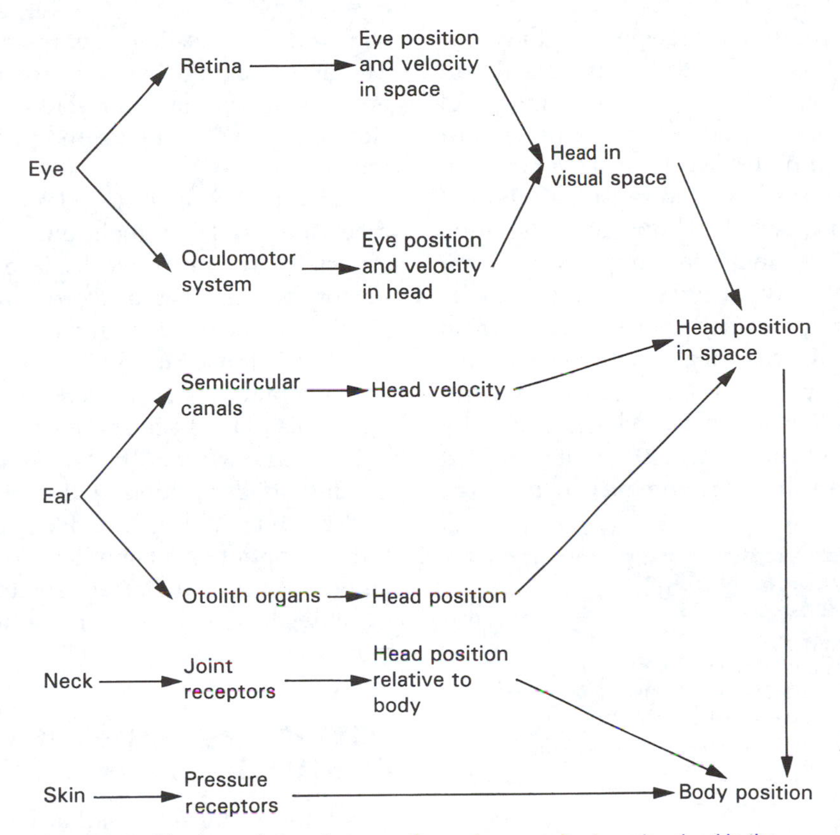

Figure 6.4 The interrelations between the various mechanisms involved in the maintenance of posture (Carpenter 1984).

FACTORS THAT CONTRIBUTE TO POOR OR ALTERED POSTURE

There are many factors that contribute to poor posture. These include: pain, decreased range of movement and flexibility, muscle imbalances, altered joint biomechanics, pathological conditions, joint hypermobility and ligament laxity, altered sensation and proprioception, psychological state, adaptations to the environment, and finally and more commonly, persistent adoption of poor posture (habituation). Poor posture puts abnormal strains on muscles, bones, joints and ligaments. In some cases this may lead to deformity which may interfere with various bodily functions, such as respiration, digestion, circulation (Thibodeau & Patton 1993) and mobility.

Pain causes reflex contraction (spasm) of the muscles surrounding the source of the pain. This is the body's mechanism of restricting movement (splinting) of the injured body part to protect it from further damage. Obviously this causes an alteration in posture. For example, following an injury to the hand, a natural response is to hold the arm close to the body and, usually, the other arm is wrapped around it for added protection. Patients who have a prolapsed intervertebral disc in the lumbar region often present with muscle spasm in their paravertebral muscles. This is often unilateral and causes a lateral tilt of the pelvis. This altered posture disappears when the cause of their pain is resolved.

Decreased range of movement occurs for the following reasons: decreased flexibility of the muscles or ligaments surrounding a joint or altered joint biomechanics following trauma, pathology or congenital malformation. Muscle tightness occurs following muscle imbalances where one muscle group is stronger than another. Excessive strengthening of a muscle group leads to hypertrophy (enlargement) of the muscle fibres, but this could lead to increased soft tissue apposition and a decreased range of movement results. Weakness in the antagonist muscle group (muscles working in opposition) means that they are unable to resist the action of the stronger muscles that are causing an alteration in posture. It is, therefore, important to incorporate flexibility exercises into muscle strengthening regimes and that care is taken to avoid creating muscle imbalances between muscle groups (agonists and antagonists). Injury to bones or joints that cause a mechanical or bony restriction to movement may also result in altered posture due to an altered alignment of body parts.

Box 6.4 Task 4

Altered posture
Get your partner to simulate a leg length discrepancy by getting them to place one foot on a book approximately 3–4 cm high. Note the differences this makes to their standing posture.

Laxity of ligaments surrounding joints allows the movement of the joints beyond normal limits: this is known as *hypermobility*. This causes an alteration in posture as the end of joint range is increased. This can be seen in female gymnasts who are able to hyperextend their lumbar spine. The length of ligaments is predetermined at birth but they may also become elongated by persistent overstretching and may become lax following injury.

Alterations in sensation and proprioception mean that awareness of the position of a body part in space and its relationship with a supporting surface are diminished. This may affect posture as the individual is unable to determine the position of their body parts with respect to others.

Psychological state has a large influence on posture (Morris 1982). Compare, for example, the postures adopted by: a footballer who has just scored a goal and the players on the opposite team; a parent scolding a child and the child being scolded; a person who has just had an important honour bestowed on them and a person who is feeling depressed or anxious. Therapists should be aware of how feelings of anger, fear, elation, depression, anxiety, submission and power can alter the posture of an individual.

Many people have to adapt their posture to meet the needs of their environment. This is especially true in the work place. For example, a person working at a word processor spends much of their time looking at a VDU. If the position and height of the VDU are incorrect, unnecessary strain may be placed on the muscles and ligaments of the neck and back due to this habitual altered posture. A factory worker may spend all day in a position that facilitates the task undertaken while leaning over a production line. If this adoption of an abnormal posture is a regular occurrence, anatomical structures may become overstretched and pain and long-term damage may result.

DEVIATIONS FROM THE IDEAL POSTURE

The major posture types that will be considered in this section are: scoliotic, kyphotic, lordotic, kypholordotic, sway back and flat back. These posture types will be considered with respect to the body segment alignment, and those muscles that are elongated and weak or short and strong respectively.

Scoliosis

Scoliosis is a lateral curvature of the lumbar and/or thoracic spine and may be either mobile or fixed (Fig. 6.5). Mobile scoliotic postures may result from:

- persistent adoption of such a posture
- from a painful pathology, i.e. a prolapsed lumbar vertebral disc
- as a compensation for lower limb problems, i.e. leg length discrepancy or abnormal pelvic tilt
- reduced muscle strength or altered muscle tone of the paraspinal muscles unilaterally, i.e. following head injury or CVA (stroke).

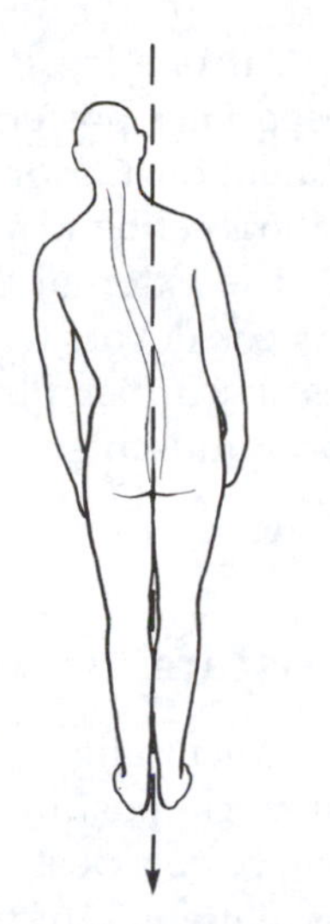

Figure 6.5 Posterior view of a subject with a thoracolumbar scoliosis.

A fixed (structural) scoliosis does not disappear with alterations in posture. The vertebral bodies rotate towards the convexity and the spinous processes towards the concavity of the curve. Secondary curves develop to counteract the effects of the initial scoliosis but these often become fixed later (Apley & Solomon 1982).

Kyphosis

Kyphosis is an increase in the convexity of the thoracic spine when viewed laterally. A kyphosis may be either mobile or fixed (Apley & Solomon 1982). Mobile kyphosis may be the result of:

- persistently adopting such a posture, i.e. in obese individuals or during pregnancy and immediately following childbirth
- association with other postural defects, e.g. flat feet
- weak erector spinae muscles
- a compensation to hip deformity, i.e. fixed flexion or congenital dislocation of the hip.

Fixed kyphosis occurs in patients with ankylosing spondylitis and Scheuermann's disease and senile kyphosis exists in the elderly as a consequence of intervertebral disc degeneration.

Lordotic (hollow back) posture

In normal posture the lumbar spine should be slightly hollow. This hollowing (the lumbar lordosis) is influenced by the tilt of the pelvis. The pelvis balances on the hip joints like a see-saw. Control of this see-saw is maintained by the abdominal, spinal and hip muscles and the surrounding ligaments (Norris 1994). The abdominal, gluteal and hamstring muscles work together to tilt the pelvis backwards and flatten the lumbar spine. At the same time the hip flexors and spinal extensors increase the lumbar curve by tilting the pelvis forwards (Norris 1994). If an imbalance in these muscles results in excessive lengthening and weakness of the abdominal and gluteal muscles and tightening of the iliopsoas and spinal extensor muscles, the person assumes a 'pot belly' type posture (Norris 1994). The pelvis is tipped forwards and increases the curvature of the lumbar spine (Fig. 6.6a). This muscle imbalance is known as the *pelvic crossed syndrome* (Norris 1994). In addition to the problems described above, the hamstrings attempt to compensate for the weakened gluteal muscles during walking. Since the hamstrings are not as strong as the gluteals, hip extension is

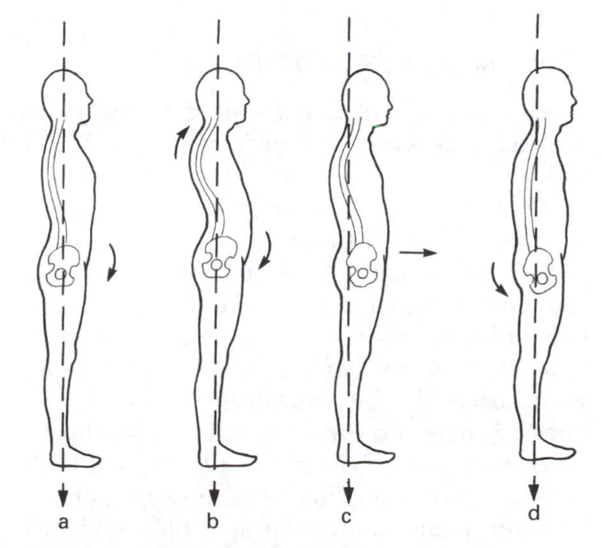

Figure 6.6 Lateral views of different posture types: a) lordotic, b) kypholordotic, c) sway back, d) flat back (adapted with permission from Kendall F P, McCreary E K 1983 Muscles testing and function, 3rd edn. Williams & Wilkins, Baltimore).

weaker. The body tries to compensate for this by extending the lumbar spine resulting in further undue stress in this area (Norris 1994).

Kypholordotic posture

The body segment alignment of an individual with a kypholordotic posture will be altered in the following way (Fig. 6.6b). Their head will be the most anteriorly placed body segment. It will be held forwards but their cervical spine will be extended, i.e. they will have a 'poking chin'. Their scapulae may be abducted. They will have an increased thoracic kyphosis with an increased lumbar lordosis and their pelvis will be tilted anteriorly. They will stand with their hips flexed but their knees will be hyperextended. They will have elongated and weak neck flexors, with weakness also in the upper portion of erector spinae and the external oblique. Conversely, the neck extensors and hip flexors will be short and strong. If the scapulae are abducted, there will be weakness of the middle and lower fibres of the trapezius but the serratus anterior, pectoralis major and/or minor and upper fibres of trapezius will be short and strong.

The sway back posture

The body segment alignment of an individual with a typical sway back posture is altered in the following way (Fig. 6.6c). The pelvis is positioned in neutral or often is tilted posteriorly and is positioned to be the most anterior of the subject's body segments. Subjects have a kyphosis of the thoracic region with a flattening of the lumbar lordosis. This results in the hip joints being positioned in front of the standard posture line with observable hyperextension of the hip and knees. Where the subject stands predominantly on one leg, the pelvis will be tilted down to the non-favoured side. This gives the appearance of a longer favoured leg but this is only evident in the standing position.

Individuals with a sway back posture will have weak and elongated hip flexors with short and strong hamstrings due to standing with hyperextended hips. If one leg is favoured, the gluteus medius (especially posterior fibres) will be weak and elongated, the tensor fascia lata will be strong and the iliotibial band will be tight on the favoured side. The external oblique, upper back extensors and neck flexors will be elongated and weak whereas the upper fibres of internal oblique and low back musculature will be short but not strong.

The flat back posture

The altered body segment alignment of an individual with a flat back posture is as follows (Fig. 6.6d). They will have a posterior tilt of their pelvis and a loss or flattening of the lumbar lordosis. They will stand with hyperextended hip and knee joints and their head will be positioned anteriorly resulting in increased flexion of the upper thoracic spine. They will have elongated and weak hip flexors with short and strong hamstrings. Their erector spinae (back muscles) may be slightly elongated and weak but their abdominal muscles may be strong.

RETRAINING POSTURE AND BALANCE

As we have seen, there are many factors that contribute to poor posture. The therapist should identify whether there are any pathological or biomechanical abnormalities and whether any anatomical structures are tight or weak. They should then initiate a programme of stretching for tight structures and strengthening for those muscle groups that are weak. This should be monitored to avoid developing a further muscle imbalance or instability. Pathological or biomechanical abnormalities will require further investigation and may necessitate involvement from orthotists or surgeons.

Many patients are unaware that their posture is incorrect and may in fact be contributing to their physical symptoms. For the correction of habitual posture it is essential that the patient is aware of what is correct and what is bad and has become 'habit'. It is only then that correction may begin to take place: patient education is, therefore, essential. This may include advice on the height of furniture and work surfaces, types of mattresses, pillows, etc. and lifting and handling techniques.

Long length mirrors are frequently used in physiotherapy departments to make patients more aware of their static posture either in sitting or standing and dynamic posture during movement. Mirrors are also employed in gait retraining, for example, following the provision of a prosthesis after lower limb amputation.

A further method of posture retraining is biofeedback. This is a technique involving the use of biological signals. Electromyographic activity (EMG) is produced by muscles when they are active. These signals may be detected by an electronic device. Any muscle contraction results in an increase in electrical activity, i.e. an increase in the amplitude of the EMG signal. EMG may be used to detect when unwanted movements or positions are occurring and may be identified to the patient by the sounding of an alarm or by the illumination of a light switch. The miniaturisation of such devices allows them to be worn by subjects in their normal environments while performing normal activity. This is especially useful in occupational health, where habitual adoption of a bad posture may cause injury. The identification of an incorrect posture or unwanted muscle activity allows the subject to reassess their posture and, if appropriate, alter it.

Patterns of muscle activity (dynamic posture and balance) may also be retrained using visual feedback by interfacing the EMG with a computer and appropriate computer software. Patients may be able to play games by using the intensity of muscle contraction or using activity in different muscle groups. This technique is especially useful in posture and balance education in children.

Balance may be affected by a person's ability to move quickly from one position to another. Several factors may have an influence on this: pain, musculoskeletal pathology, muscle weakness, biomechanical abnormalities, proprioceptive loss, reduced neuromuscular control and age. *Static balance* refers to a person's ability to maintain their balance when in a specific posture and *dynamic balance* refers to keeping the body's segments under control during movement to prevent falling. Obviously, both types of balance are important and may require retraining. Static balance is perhaps the easiest to retrain as it involves maintaining the body's equilibrium in a particular posture. This may be achieved by starting with a large base of support or by increasing the amount of support above normal levels. As balance improves this additional support may be gradually reduced. For example, in standing extra support may be achieved by holding on to a rigid support, e.g. wall bars, with the arms. A larger base of support is achieved by standing with the legs wide apart, a smaller base of support is obtained by standing with the feet together.

Dynamic balance may be facilitated by using moveable supports instead of rigid ones, e.g. gymnastic balls instead of wall bars, or by using more unstable bases of support, e.g. wobble boards. Advanced training may involve a person standing on a wobble board and attempting to maintain their balance while throwing and catching a medicine ball. More functional tasks may include walking over an uneven surface and turning round corners and negotiating obstacles. As mentioned earlier, methods of biofeedback may also be useful.

SUMMARY

Balance and posture are interrelated and a series of physiological mechanisms exist to inform the body if posture has been compromised and the body is in danger of falling. The body responds by altering its posture (moving) in an attempt to maintain or regain its balance. These mechanisms form a feedback loop and are constantly informing the body of changes in position and pressure distribution. Normal static posture is difficult to define as every person has a unique anthropometric and physiological profile. However, there are some common types of deviations from the normal. Some examples of these different types of posture have been described and methods for retraining balance and posture have been introduced. It should be remembered that balance and posture are vital for everyday activities and it is often not until a failure of one or more components of these complex systems that we realise how sensitive the system is. The physiotherapist plays a major role in posture education and the retraining of balance and posture following pathology or injury.

Box 6.5 Study questions

1. What are the main physiological systems involved in balance?
2. What factors contribute to poor posture?
3. What are the common types of poor habitual posture?
4. What methods may be used to retrain posture?
5. What factors should be taken into account in patients who have sensory impairments?
6. What implications do visual impairments have during recovery from injury?
7. What advice may you offer patients when teaching them to use crutches or walking sticks?
8. What methods may be used to retrain balance?

Acknowledgements

The authors would like to thank Mr Chris Norris for his contribution to this chapter.

REFERENCES

Apley A G, Solomon L 1982 Apley's system of orthopaedics and fractures, 6th edn. Butterworths, London
Berne R M, Levy M N 1993 Physiology, 3rd edn. C V Mosby, St Louis
Bobath B 1969 The treatment of neuromuscular disorders by improving patterns of co-ordination. Physiotherapy 55: 18–22
Bobath B 1978 Adult hemiplegia: Evaluation and treatment, 2nd edn. William Heinemann Medical Books London
Carpenter R H S 1984 Neurophysiology, 1st edn. Edward Arnold, London
Carter J E L, Ross W D, Duqet W, Aubry S P 1983 Advances in somatotype methodology and analysis. Year Book of Physical Anthropology 26: 193–213
Davies P M 1985 Steps to follow. Springer-Verlag, Berlin
Deiner H C, Dichgans B, Guschlabauer, Mau H 1984 The significance of proprioception on postural stabilisation as assessed by ischemia. Brain Research 296: 103–109
Galley P M, Forster A L 1987 Human movement: an introductory text for physiotherapy students, 2nd edn. Churchill Livingstone, Edinburgh
Green J H 1978 An introduction to human physiology, 4th edn. Oxford University Press, Oxford
Keel C A, Neil E 1966 Samson Wright's applied physiology, 11th edn. Oxford University Press, London
Kendall F P, McCreary E K, Provance P G 1993 Muscles testing and function, 4th edn. Williams and Wilkins, Baltimore
Larson C B, Gould M 1974 Orthopaedic Nursing, 8th edn. C V Mosby, St Louis
Morris D 1982 The pocket guide to manwatching. Triad / Granada, London
Norris C M 1993 Weight training principles and practice, 1st edn. A & C Black, London
Norris C M 1994 Flexibility principles and practice, 1st edn. A & C Black, London
Norris C M 1995 Spinal stabilisation 4: Muscle imbalance and the low back. Physiotherapy 81: 20–31
Peterkin H W 1969 The neuromuscular system and re-education of movement. Physiotherapy 55: 145–153
Ruitshauser S 1994 Physiology and anatomy — a basis for nursing and health care, 1st edn. Churchill Livingstone, Edinburgh
Seeley R R, Stephens T D, Tate P 1992 Anatomy and physiology, 2nd edn. Mosby Year Book, St Louis
Stein J F 1982 An introduction to neurophysiology. Blackwell Scientific Publications, Oxford
Thibodeau G A, Patton K T 1993 Anatomy and physiology, 2nd edn. C V Mosby, St Louis
Willard F H 1992 Medical neuroanatomy. J B Lippincott, Philadelphia

CHAPTER CONTENTS

7

The effects of age on human movement

M. Trew

OBJECTIVES

When you have completed this chapter you should be able to:

1. **Understand why physical ability may deteriorate with increasing years**
2. **Understand the basic physiological changes that can occur with age**
3. **Recognise age induced changes in posture and gait**
4. **Understand the importance of recognising that some age changes are reversible**
5. **Understand the concept of the threshold of ability.**

INTRODUCTION

The way in which people move and their ability to undertake physical activity inevitably changes with the passage of years. Whilst a reduction in movement capability and the ability to perform functional activities can have a catastrophic effect on lifestyle and quality of life, it does not always have to be so. It is clear that a number of age induced changes can be prevented, slowed down or even reversed.

This chapter looks at those aspects of the physiological and pathological responses to ageing which have a direct effect on movement and the ability to undertake everyday activities. The age induced changes in posture and gait are used as examples of how age may affect movement or function, and the final part of this chapter considers how exercise and activity may be beneficial.

Box 7.1 Task 1

Before reading this chapter, write down your ideas of what an elderly person in their eighth or ninth decade would be like; use the headings of 'appearance', 'functional abilities', 'physical abilities' and 'mental capabilities' to help you. It may also be helpful to compare your expectations of the abilities of an elderly person with your expectations of someone in their third decade.

Discuss with your fellow students the conclusions you have reached and keep a note of your discussions for further reference.

The loss of the ability to move easily and freely is a common problem associated with ageing. The onset is often insidious and it is sometimes difficult to decide whether the reduction in movement ability is caused by age induced changes and disease or whether it is a consequence of an increasingly sedentary lifestyle. Elderly people tend to avoid or have no need to undertake activities which require strength, speed or extreme ranges of joint movement and many spend increasing periods of time in sedentary pursuits (Fiatarone 1990). In a number of cultures there is an expectation that life should slow down with the passage of years and that old people deserve to take life easy. Where these social expectations exist it is almost inevitable that many older people's lifestyles will become increasingly sedentary and will eventually contribute to their physical deterioration. A sedentary lifestyle, over many years, will result in a gradual decline in physiological and, sometimes, psychological functions. This, if carried to the extreme, will eventually result in a reduced ability to perform even the most basic physical and cognitive tasks associated with activities of daily living. In these cases there will inevitably be a reduced quality of life and a substantial dependence on carers.

Although the general population usually accepts a slower pace of life and the loss of ability to undertake some functional activities as an inevitable consequence of growing old, it is clear that at least some of the changes are caused by inactivity and are therefore both preventable and reversible (Fiatarone 1990, Raab et al 1988). Good health education and health promotion schemes which explain the importance of maintaining a reasonably active lifestyle in the later years are clearly important. Elderly individuals who exercise regularly are less likely to suffer from movement problems and those who take up exercise late in life benefit from an improvement in function in a number of body systems (Aniansson et al 1984, Grimby 1988, McArdle et al 1991, Meusel 1984).

Human movement in elderly people can be compromised in three ways:

- through structural changes which are a direct result of the ageing process
- because of the effects of disease which may or may not be associated with growing old
- as a consequence of the effects of inactivity.

This chapter is concerned with the effects of the ageing process and with the effects of inactivity on human movement. The degeneration of a number of body systems is an inevitable part of ageing, but when considering the functional ability of many old people it is important to be aware that some of the changes may not necessarily be the result of the inevitable and irreversible ageing process. The individual may be reduced to a lower than necessary level of function by societal and psychological constraints and these may be reversible. It is clear that a number of conditions and symptoms commonly associated with elderly people are not a consequence of ageing so much as a consequence of inactivity. This illustrates why it is important to take a positive approach when working with older people and why it is quite acceptable to have expectations of improvement.

PHYSIOLOGICAL CHANGES ASSOCIATED WITH AGEING

Cardiovascular changes

The function of the cardiovascular system is significantly reduced with increasing age and this has a negative effect on both the intensity and duration of any physical activity that is performed. Resting heart rate changes little with age, but the maximum heart rate attainable on activity drops significantly causing a concurrent reduction in exercise tolerance. An approximate guide to the expected maximum heart rate for older people can be obtained by using the formula: 220 minus the age in years = maximum heart rate.

By the age of 85 the resting stroke volume is reduced by about 30% and the myocardium is noticeably hypertrophied. Also, at this age, the resting cardiac output may be less than half the value it was in the third decade (Fitzgerald 1985).

These changes mean that the ability of the heart to deliver blood to the tissues is impaired; oxygen uptake in the muscles is also reduced, not because of a loss of muscle oxydative enzymes but as a consequence of the actual loss of muscle fibres. It seems likely that the oxydative enzymes are little affected by age, though there may be a reduction in the resynthesis of ATP due to a reduction in phosphogens (Grimby 1986). With both a reduced delivery of blood to the muscles and also a reduction in the number of muscle fibres available to use that oxygen, there is a general reduction in movement capacity. A further adverse effect of ageing on the cardiovascular system is the loss of elasticity in the arterial walls. This leads to an undesirable increase in blood pressure on strenuous activity due to the arteries' inability to accommodate the increase in blood flow.

These cardiovascular changes make activity more difficult with advancing years but in themselves should not have too serious an effect on the ability to perform normal, everyday functions. However, if the natural age changes are combined with a long-term loss of fitness due to an increasingly sedentary lifestyle, then the consequences may start to become serious and essential functional activities may require so much energy that they take an unacceptable length of time to complete.

Respiratory system changes

Increasing age may result in degenerative changes in the costovertebral and costosternal joints and may be combined with calcification of the costal cartilages. Eventually these changes will lead to a reduction in thoracic mobility in both the anteroposterior and lateral directions. Elastic recoil of the chest wall, which is necessary for expiration, becomes reduced as a consequence of a generalised loss of elastic tissue and this puts major emphasis on movements of the diaphragm as a means of changing thoracic diameter and lung capacity. All these changes mean that a disproportionate amount of effort is needed for respiration during strenuous activity and these age changes can eventually lead to a 20% increase in respiratory energy requirements (Fitzgerald 1985). By the age of 70, vital capacity can also have decreased by 50% yet, despite all these changes, it appears that a decrease in respiratory function is not a major factor in the limitation of exercise tolerance in elderly people, particularly if they are fairly active. The changes in the heart, the peripheral circulation and the muscles appear to be far more significant in reducing movement capabilities.

Muscle changes

With increasing age there is a reduction in both aerobic and anaerobic capacity affecting the endurance, strength and speed attributes of muscle. As with the other body systems the most marked changes occur in extreme old age, particularly after the eighth decade (Fiatarone 1990, Grimby 1986, Young 1986).

In elderly people beyond their eighth or ninth decade the reduction in muscle bulk has occurred to such an extent that it can easily be noted on visual examination. This reduction in muscle mass is mainly a consequence of loss of muscle fibres, though there may also be a reduction in size of the Type II fibres (fast twitch), the latter probably caused by disuse atrophy. Although the visible signs of loss of muscle bulk are not always apparent until old age, computer tomography shows that muscle fibres are being lost from at least the age of 30 and are being replaced with intramuscular fat. The fat masks visual signs of muscle fibre loss until a substantial number of fibres have disappeared (Fiatarone 1990).

Despite the significant loss of muscle fibres,

there is normally sufficient spare capacity for this not to be a major constraint on function until extreme old age is reached. Once again it is the combination of age changes and disuse atrophy which is likely to be the cause of the 'weakness' exhibited by many old people.

The lifestyle of many middle-aged and elderly individuals puts little demand on the Type II fibres which are responsible for power and speed activities and this inevitably leads to disuse atrophy.

Box 7.2 Task 2

Start observing children, young adults, middle-aged and elderly people. Watch the speed with which they tackle everyday tasks and the ranges of joint movement that they employ. Are you able to note any differences between these age groups?

Interestingly, the pattern of atrophy of Type II fibres is not constant throughout all regions of the body. The quadriceps are more subject to atrophy than muscle groups in the upper limb, which may be a reflection of a reduction in lower limb activities rather than an age induced change in muscle structure (Grimby 1988). The muscles of the upper limb remain in constant use even when the individual has lost the ability to walk and are often used in fairly rapid movements. This continued use probably contributes to the retention of a greater proportion of Type II fibres in upper limb muscle groups when compared with lower limb muscles.

Most authorities feel that the decrease in muscle function and muscle mass is mainly due to the loss of muscle fibres and a reduction in oxygen delivery. Whilst there may be a reduction in the enzymic capacity of muscle, this is not a necessary consequence of old age and therefore can be slowed down by the use of exercise (Fitzgerald 1985, Grimby 1986, McArdle et al 1991). An individual muscle fibre in an elderly person is structurally and functionally similar to one in a young person and will therefore respond to physical demands and training in the same way. It is the reduction in number of fibres and the reduced delivery of oxygen which reduces the overall power and endurance capacity of the muscle, not age changes affecting the structure of the muscle fibre itself (Young 1986).

Neurological changes

The neurological system is not protected against age induced changes although in the absence of disease or trauma it can, in some individuals, continue to function remarkably well into extreme old age. Commonly, ageing can cause a reduction in the effectiveness of the neurotransmitters, a change in nerve cells, especially in the number and effectiveness of synaptic connections, and a reduction in the number of nerve cells (Pickles et al 1995). Reaction times are significantly increased with age, probably due to the reduction in nerve conduction velocity which can be altered by up to 15% (Fitzgerald 1985). The number of motor units decreases so that activation of muscle becomes more difficult. There will eventually be a reduction in sensory capacity as the number of sensory nerve endings declines and thresholds for transmission of sensory information increase. All these factors may lead to the control and performance of human movement being reduced (Fitzgerald 1985, Grimby 1986).

Skeletal changes

With increasing age there is often a significant loss of bone mass, particularly in elderly women, which may be due to the ageing process or caused by disease. The situation can be exacerbated by lack of exercise, a reduction in dietary calcium, poor diet and also genetic factors (Pickles et al 1995). This osteoporosis and the increased incidence of falls in older people make them particularly vulnerable to fractures.

Reduced range or quality of movement in joints is a common problem associated with age. In the later decades connective tissue and muscle become less elastic, altering the quality of joint movement. Individuals who adopt a sedentary lifestyle are vulnerable to soft tissue contractures. Any loss of range of movement in the major joints will inevitably reduce the ability to move normally and undertake functional activities with the same efficiency as a younger person. Arthritis starts to develop from an early age and after the sixth decade it is a common cause of movement problems. Research indicates that there is a gender effect on the incidence of musculoskeletal disorders, with women being more likely to have joint problems than men (Arber & Ginn 1994). This, coupled with the fact that in most countries women live longer than men, inevitably leads to musculoskeletal problems being a significant factor in reducing movement capabilities in older people. Joint disease, combined with a reduction in function of other body systems, greatly reduces the functional ability of older people and has an adverse effect on their quality of life. Providing there are no serious degenerative changes, exercise can increase the range of movement and a small improvement in joint range may cause a disproportionately large functional improvement (Fitzgerald 1985).

Box 7.3 Task 3

Look through your grandparents' photograph album. Can you identify the physical changes that are occurring through their life? Can you suggest the reasons for these changes?

Ask your grandparents about the lifestyle they were leading when the photographs were taken and evaluate whether their answers support the theories of ageing discussed in this chapter.

Postural changes with age

Usually an individual's posture is assessed in relation to an imagined 'ideal' static posture similar to that exhibited by healthy, normal young adults. It is almost impossible to measure dynamic posture because it is changing all the time. So presumptions have to be made that deviations from normal static posture will be reflected in dynamic posture.

With increasing years there are some common deviations from normal posture which can be seen in most elderly people; these deviations occur at different times for different people but are most usual in the eighth and ninth decades. Static standing posture shows typical changes which originate in the spine; other body segments then have to readjust to compensate for the altered position of the centre of gravity (Fig. 7.1). In many elderly people there is an increase in their thoracic kyphosis and either a flattening of the lumbar curve with posterior pelvic tilt or a compensatory increase in lumbar lordosis. The hip and knee joints become increasingly flexed and the ankle joint is dorsiflexed to counter these changes in body alignment which have tended to displace the centre of gravity anteriorly in relation to the individual's base. Because of the thoracic kyphosis, the upper limbs no longer hang by the sides but now hang in front of the body when the individual is relaxed.

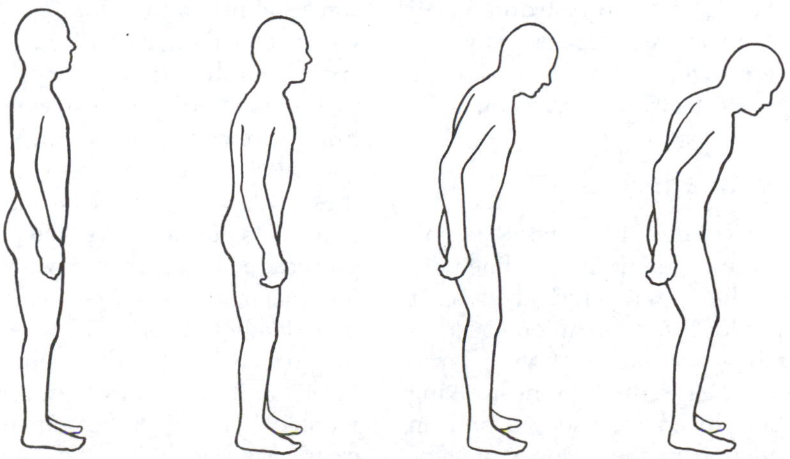

Figure 7.1 Typical postural changes associated with ageing.

This displaces the centre of gravity even further anteriorly and can stretch the individual's balance ability to the limit. To compensate, extension of the shoulder joints usually occurs and in order to reduce the energy requirements of holding the upper limbs behind the body, the fingers may be interlocked behind the back. There are a number of reasons why these changes occur but most commonly they result from degenerative changes of the intervertebral discs, osteoporosis of the vertebrae and trunk muscle weakness. Following on from these changes will be compensatory positioning of other joints to counter the alteration in position of the centre of gravity (Kauffman 1987, Pickles et al 1995).

Such changes in posture have a significantly adverse effect on movement, particularly as the alteration in the positions of the centre of gravity will lead to feelings of insecurity in standing and walking. The need to clasp the hands behind the back in order to counter the anterior displacement of the upper trunk limits arm swing in walking and in turn reduces the natural rotation of the whole spine. Step length is directly affected by spinal rotation and if rotation is prevented then step length will be reduced, inevitably resulting in a slower walking velocity. In addition, the increase in thoracic kyphosis and the flattening of the lumbar spine will tend to compress the abdominal contents against the under surface of the diaphragm, reducing its capacity to descend and somewhat impairing respiratory function. The flexed posture of the hip, knee and the dorsiflexed position of the ankle joint reduces step length in walking and all these factors contribute to slowing down or reducing the fluency of movement.

Gait changes with age

It is easy to observe the differences in gait between old and young people but, in line with other age changes, the age at which they occur varies between individuals. As mentioned earlier there is usually a loss in spinal rotation and arm swing. This leads to a reduction in walking velocity due to a reduced step length and, in some cases, a reduction in the number of steps taken per minute can be expected. There is also a noticeable increase in the length of the double support phase of walking and a reduction in the propulsive force generated at the push off phase (Murray et al 1969, Whittle 1991, Winter et al 1990). Gait changes associated with age are, on the whole, due to altered posture, a reduction in spinal movement, muscle weakness and a loss in balance ability which is rooted in neurological changes (Pickles et al 1995, Winter 1990).

IMPROVING MOVEMENT AND FUNCTION IN OLDER PEOPLE

It is far more worthwhile that people approach old age with a good level of physical fitness and an expectation that they will continue to be physically active than that they slide into such an inactive lifestyle that they eventually need intervention from a therapist. To a certain extent an individual's expectations of old age will be based on cultural norms and financial constraints. All cultures ascribe certain roles to old people and in many cases these are of reducing involvement in society and a reduction in activity levels generally. If changes in society's expectations are to be effected, then they will have to occur from childhood upwards. Youngsters need to be brought up recognising that it is normal for their grandparents to be active and taking a full and interested part in life, rather than it being so unusual that the active elderly person is labelled 'super gran' or 'super grandad'. Good health education can establish and reinforce these expectations, which can then be carried throughout the life span. This should give a greater likelihood of the maintenance of physical well-being and, hopefully, a concurrent good quality of life.

Financial constraints also influence lifestyle and those with low incomes are more likely to have poor diets and less access to a range of enjoyable physical activities than those who are affluent. In the major industrialised countries women are more likely to have an income near to the level of the poverty line than men and also a greater level of physical impairment (Arber & Ginn 1994). As women live longer than men, these factors are clearly significant but the solution is not easy to find and changes in society may be necessary.

For those individuals who reach old age in poor physical condition and who live a sedentary lifestyle, intervention from the health care team is likely to be necessary. The combination of inactivity and ageing results, amongst other things, in a loss of muscle mass, poor exercise tolerance and some limitation in joint movement. If the worst circumstances prevail, then by the eighth decade many sedentary people may be so frail that the limit of their physical ability is reached simply by the performance of normal activities of daily living. This leaves them with little or no physical capacity in reserve and it takes only the slightest deterioration for their level of ability to drop below the requirements of daily living. Such a drop, which can be caused by a minor illness requiring a few days bedrest, often proves catastrophic and results in loss of independence. Young (1986) describes a threshold of ability which is necessary for the performance of activities of daily living; if an individual falls below that threshold then self-care becomes impossible. Many very old people, because of age changes, disease and inactivity, live on the border of that threshold. For them, generating sufficient torque in their quadriceps and hip extensor muscles in order to rise from a low chair or toilet may require them to work to their maximum ability and getting dressed may well cause the heart rate to rise to its age-predicted maximum. Under these circumstances it is no wonder that frail elderly people find the simplest activities difficult and take an inordinately long time to complete even the most basic tasks.

When old people reach the stage when the simple and essential activities of everyday life require maximum ability, their frailty usually results in increasing dependency on the help of others and also a greater likelihood of accidents. The quality of life for these frail, old people is often poor, and their dependency places a consequent strain on their relatives and carers (Fiatarone 1990).

For their own self-respect and for the benefit of their carers, the elderly person needs to be able to undertake, at the very least, the basic activities of daily living. They need the physical attributes to be able to move around their living area unassisted and to attend to their eating and toileting needs.

Even at an extreme age there is evidence that training can induce positive changes and a small strength gain may represent a very large improvement in the quality of life. The beneficial effects of exercise for young people are well known and it is generally accepted that exercise for people in the older age range will, at the very least, maintain their existing condition, if not improve the function of all major body systems. These include the heart, lungs, musculoskeletal, metabolic and central and peripheral nervous systems and some authors have noted that the progress of vascular disease, diabetes, hypertension, chronic obstructive airways disease, osteoporosis and arthritic conditions is retarded in subjects who undertake exercise programmes (Fisher & Pendergast 1995, Fitzgerald 1985, Thompson et al 1988).

Unfortunately, conclusive proof that exercise for old people is beneficial is lacking. Some authors have shown benefits: Aniansson et al (1984) found that training elderly people two times a week for 10 weeks using a strength training programme produced up to 13% increase in quadriceps strength and a general feeling of well-being. Conversely, other authors have not been able to show improvement. Thompson et al (1988) exercised a small group of elderly subjects for 16 weeks at a level which should have induced a training effect. At the end of the exercise programme, however, no significant improvement had been shown. Whether statistically significant improvement equates well with functional improvement is not known. If the subjects were near that all-important threshold of ability, then a small improvement might not be statistically significant but might make a large difference in the ability to undertake activities of daily living and to the quality of life. Consensus of opinion suggests that exercise is beneficial to older people and that the very least effect that can be expected following a programme of exercise is an increased feeling of well-being and an enthusiasm for maintaining an adequate level of physical activity (Fitzgerald 1985, McArdle et al 1991, Meusel 1984, Thompson et al 1988).

Endurance exercises have been shown to increase the maximum oxygen uptake and reduce the heart rate, blood pressure and blood lactate in elderly people (Fitzgerald 1985). With appropriate, long duration training, muscle strength and endurance can increase with the same order of magnitude as that seen in younger people on an identical training programme. Obviously no amount of exercise can replace lost muscle fibres, but for those that remain it is possible to induce hypertrophy, better recruitment of motor units and improved delivery and uptake of oxygen (Grimby 1986).

A major concern is that if frail or very old people are to see a substantial improvement in their physical capacity, they may need exercises of an intensity which they are unable to tolerate. The physiotherapist has to achieve a balance between exercise and activity which is so gentle as to be ineffective and exercise which might be effective but which will place a dangerous stress on the heart (Shephard 1984). Some studies have recently shown that, providing there are no disease processes which would contraindicate exercise, resistance training is both safe, desirable and effective (Pickles et al 1995). Careful monitoring and a gradual build-up of the intensity of exercise should be sufficient to ensure the safety of the patient and there are many examples of elderly people achieving activity levels well above that of younger members of the population (McArdle et al 1991).

Box 7.4 Task 4

Go to your library and obtain a journal or magazine which publishes the results of athletics meetings. Look at the results for the veterans and note their ages. These athletes don't fall into the stereotyped category of old people and are able to perform to high levels and obtain great satisfaction.

As not all people enjoy competitive sport, it is necessary to be aware of other sorts of activities which could maintain or increase levels of physical ability. Talk to a range of middle-aged and older people about what sort of activities they might find enjoyable and list the results of your discussions.Consider whether the activities you have listed are likely to have a beneficial effect.

Ideally, as people age they should be encouraged to develop the habit of exercise so that the age changes which might reduce their ability to perform everyday activities may be avoided or delayed. Unfortunately, this does not happen in many cases and the physiotherapist may have no contact with the patient until they become dependent following a long period of inactivity. Under these circumstances, motivation to exercise is often difficult but if the exercise programme is specifically planned for the individual's needs and if achievable goals are set, significant improvements may be seen. Patients should be made aware of the treatment goals and should, where possible, take an active part in their setting. They should be taught how to monitor their own progress and records should be kept so that improvements can be seen and can be used as motivators.

Exercises for very frail people who have a history of falls or who have significant lower limb problems may initially have to be non-weight bearing, chair based. But the principles of exercise are the same whether for young people, fit elderly people or people who are old and frail. The exercise programme should include exercises which will increase:

- joint range
- muscle strength
- muscle endurance
- heart and respiratory rate
- the ability to move groups of muscles and joints quickly
- coordination
- balance
- the ability to perform functional activities.

For those elderly people who are more fit, encouragement to take part in any physical activity is important: brisk walking, jogging and dancing are excellent methods of undertaking weight bearing exercise. In all three examples the participant will be using major muscle groups and joints and there will be some beneficial stress placed on their balance mechanisms and their cardiovascular and respiratory systems. Swimming utilises major muscle groups in both the upper and lower limbs and moves the joints through a wide range of movement and stresses the heart and lungs satisfactorily, but it is non-weight bearing. Non-weight bearing activity can be beneficial if the individual has joint pain but swimming does not require or stimulate balance reactions and the effects of weight bearing through the spine and long bones are lost. As such it should be seen as an adjunct and not an alternative to other weight bearing activities.

Very little research has been undertaken into the most effective duration of exercise programmes for elderly people. Presumptions must be made that the rules that apply to the design of exercise programmes for the young population will be appropriate across the age range. Exercises should be undertaken not less than two times a week. Initially, the duration of the exercises will depend on the ability of the individual and may have to start at a very low level of perhaps a few minutes per day. The exercises, whether for the power or endurance capacity of muscle, should show progression and, if possible, the level of endurance exercises should take the heart rate into the cardiac training zone (60–80% of the predicted maximum heart rate for the individual).

The fear of falling

In addition to those factors already considered in this chapter it is necessary to address the effect of the fear of falling on human movement and the ability to perform functional activities. With ageing can come an increase in falling frequency and this phenomenon presents a major problem both to the patient, their carers and the rehabilitation team. With every fall is the possibility of musculoskeletal damage which may result in the faller being placed on bedrest. For those people with an already reduced threshold of physical ability, any period of inactivity can prove very serious and may lead to dependency. Falling also leads to a loss of confidence when undertaking weight bearing activities and this in turn leads to a reduction in those activities. Currently, there is no consensus on why elderly people fall, but it is likely that it is a consequence of a number of factors, some of which are intrinsic to the individual and some of which are related to their environment (Pickles et al 1995). A thorough assessment of the individual and their environment may reveal potential contributing factors on which a rehabilitation programme may be built. If not, then a programme which improves lower limb joint range and muscle function, especially around the ankle joint, combined with exercises to improve postural stability, may be successful.

The future for old people, in terms of their ability to move and function normally, can be excellent. The maintenance of a reasonable level of physical activity into old age is important and, where fitness levels have dropped significantly, hope should not be given up because there is always capacity for improvement.

Box 7.5 Task 5

Look again at your description and attitudes towards older people. Having read the chapter, have your opinions and expectations changed? Are your thoughts towards ageing more positive or more negative?

REFERENCES

Aniansson A, Ljungberg P, Rundgren A, Wetterqvist H 1984 Effect of a training programme for pensioners on condition and muscular strength. Archives of Gerontology and Geriatrics 3: 229–241

Arber S, Ginn J 1994 Women and aging. Reviews in Clinical Gerontology 4: 349–358

Brown M, Rose S J 1985 The effects of aging and exercise on skeletal muscle: clinical considerations. Topics in Geriatric Rehabilitation 1: 20–30

Fiatarone M A 1990 Exercise in the oldest old. Topics in Geriatric Rehabilitation 5(2): 63–77

Fisher N M, Pendergast D R 1995 Application of quantitative and progressive exercise rehabilitation to patients with osteoarthritis of the knee. Journal of Back and Musculoskeletal Rehabilitation 5: 33–53

Fitzgerald P L 1985 Exercise for the elderly. Medical Clinics of North America 69: 189–196

Grimby G 1986 Physical activity and muscle training in the

elderly. Acta Medica Scandinavica, Supplement 711: 233–237
Grimby G 1988 Physical activity and effects of muscle training in the elderly. Annals of Clinical Research 20: 62–66
Kauffman T 1987 Posture and age. Topics in Geriatric Rehabilitation 2: 13–28
Mahler D A, Cunningham L N, Curfman G D 1986 Aging and exercise performance. Clinics in Geriatric Medicine 2: 433–452
McArdle W D, Katch F I, Katch V L 1991 Exercise physiology, energy, nutrition and human performance, 3rd edn. Lea & Febiger, Philadelphia
Meusel H 1984 Developing physical fitness for the elderly through sport and exercise. British Journal of Sports Medicine 18: 4–12
Murray M P, Kory R C, Clarkson B H 1969 Walking patterns in healthy old men. Journal of Gerontology 24: 169–178
Pickles B, Compton A, Cott C, Simpson J, Vandervoort A 1995 Physiotherapy with older people. W B Saunders, London
Raab D M, Agree J C, McAdam M, Smith E L 1988 Light resistance and stretching exercise in elderly women: effect upon flexibility. Archives of Physical Medicine and Rehabilitation 69: 268–272
Shephard R J 1984 Management of exercise in the elderly. Canadian Journal of Applied Sport Science 9: 109–120
Thompson R F, Crist D M, Marsh M, Rosenthal M 1988 Effects of physical exercise for elderly patients with physical impairments. Journal of the American Geriatrics Society 36: 130–135
Winter D A, Patla A E, Frank J S, Walt S E 1990 Biomechanical walking pattern changes in the fit and healthy elderly. Physical Therapy 70: 340–347
Wittle M 1991 Gait analysis, an introduction. Butterworth Heinemann, Oxford
Young A 1986 Exercise physiology in geriatric practice. Acta Medica Scandinavica, Supplement 711: 227–232

CHAPTER CONTENTS

8

Joint mobility

T. Everett

OBJECTIVES

At the end of this chapter the student should be able to:

1. **Describe the structure and function of joints**
2. **Discuss range of movement**
3. **Describe how movement is produced at joints**
4. **Discuss normal facilitation and restriction of joint range**
5. **Discuss the abnormal restriction of joint range**
6. **Give the classification of joint movement**
7. **Discuss the rationale of the use of movement to increase mobility.**

INTRODUCTION

The aims of this chapter are to introduce the student to the structure and function of joints and discuss the concepts and classification of joint mobility.

Biomechanically the body can be considered as composed of eight segments divided between the axial and appendicular skeleton. The head and neck, and trunk, make up the two segments of the axial skeleton. The other six segments are equally divided between the upper and lower limbs; the upper limb consisting of arm, forearm and wrist and hand, whilst the lower limb consists of the thigh, leg and foot and ankle.

All movements, including locomotion, involve the motion of these bony segments, be they in the appendicular or axial skeleton. Junctions between these segments are provided by the joints (juncturae, articulations or arthroses) which are classed into three groups: fibrous or fixed (synathroses), cartilaginous (amphiathroses) or synovial (diarthroses), the last being the only freely moveable joint (Williams et al 1989). It is at these joints that the motion actually takes place. Movements of these segments are produced mainly by the internal forces provided by muscles. This combination of the bones that form the core of the segments and the muscles that produce the force that provides movement is described as the musculoskeletal system. It is vital that this musculoskeletal system is intact for functional movement to occur. The role of the muscles producing the movement is covered in Chapter 9 of this text, but the vital component of this system, the synovial joints, will be looked at below.

The major characteristics of a synovial joint include the surface of opposing bones being in contact, but not in continuity, and covered in hyaline cartilage. These bony ends are joined together via ligaments and the whole joint complex, which may or may not include the ligaments, is surrounded by an extensive synovial lined fibrous joint capsule. The viscous synovial fluid secreted by this synovial membrane not only provides the articular cartilage with nutrition but acts with it to decrease the coefficient of friction within the joint to a level that is low enough to reduce the possibility of joint surface destruction. Intracapsular structures are usually covered by synovium. Intra-articular discs or menisci may be found within the synovial joint helping congruity and acting as a shock absorber. Labra and fat pads may also be found within the joints, increasing joint surface area and absorbing shock, respectively.

There are a large number of different types of synovial joints, classified according to their shape. They include plane, saddle, hinge, pivot, ball and socket, condylar or ellipsoid (Palastanga et al 1994), but movement at all these joints can be considered as either physiological or accessory. A physiological movement is the movement that the joint performs under voluntary control of the muscles or is performed passively by an external force but still within the available range of the joint. An accessory movement is an integral part of the physiological movement that cannot be isolated and performed actively by muscular effort.

Although there may seem to be a large number of directions in which joints may move, a system of description has been devised to make the analysis of movement simpler. From the anatomical position (standing upright with the arms at the side and palms and head facing forward), movements can be described as occurring in three planes and around three axes. The frontal plane splits the body into front and back halves, the sagittal plane splits the body into right and left halves, and the transverse plane splits the body into top and bottom halves.

Movements within these planes take place around three axes. The axes can be described as being perpendicular to the plane of movement. Therefore, there are two horizontal axes and one vertical axis. The sagittal axis is at 90° to the frontal plane and therefore allows movements within that frontal plane. These movements consist of abduction, adduction, deviation and lateral flexion. The frontal axis allows rotations of the segments within the sagittal plane and consists of the movements of flexion and extension. These are both horizontal axes. The vertical axis is at 90° to the horizontal plane and movements around this axis give rotatory motion.

The actual movements performed can be described in the degrees of freedom the joint allows. A uniaxial joint will possess only one degree of freedom, i.e. rotation about only one axis. An example of this is flexion and extension at the elbow. A two axis joint, biaxial, has two degrees of freedom, such as the radiocarpal joint which has flexion and extension at the wrist about one axis and ulnar and radial deviation about the other axis. The movement available at a multiaxial joint can be described as having three degrees of freedom. This type of movement can be described at the shoulder where flexion, extension, abduction, adduction and internal and external rotation take place.

Box 8.1 Task 1

Movements of the body segments are described in terms of their planes and axes. For each of the major joints of the body describe the planes in which the segments move and the axes that they move around.

RANGE OF MOVEMENT

With the combination of uniaxial, biaxial and multiaxial joints, the body may adopt a multitude of functional positions. When analysing these positions or movements, the components are broken down for each individual joint and the range of movement (ROM) of that joint described. This movement may not be the maximum movement that the joint may achieve, i.e. its full range of movement (FROM), but only a functional component of it.

Movements of the joints are dependent on many factors and the description of these factors is largely dependent on the discipline by which the movement is being studied. One such discipline is arthrokinametics. This is the intimate mechanics of the joints and is dependent, for its description, largely on the shape of the joint surfaces. Most of the synovial joints are complex in their formation, having more than one axis within the joint. This is brought about by the joints being ovoid in different axes of the same joint. This means that, although joints have roughly reciprocally shaped surfaces, the maximum congruity of the articular surfaces occurs at specific positions within the range of movement and does not necessarily equate with the end of range of the physiological movement. This position of maximum congruity is called the close pack position and is the position of greatest joint stability. At this close packed position not only is there greatest joint surface contact, but the ligaments are often taut. Loose pack position is where the apposition of the joint surface is the least; muscle, ligaments and capsule are usually lax and the joint is in its least stable position (Hall 1995).

Physiological movements are rarely pure movements but usually contain different elements. These may be a combination of physiological movements, such as side flexion of the cervical spine which, if examined closely, will be seen to involve both side flexion and rotation in combination. By studying the arthrokinematics it has also been shown that there is a combination of accessory type movements as previously described. These accessory movements are considered to be of three types: spin, roll and glide. A roll refers to one surface rolling over another as a ball rolling over a surface. An example of roll is seen when the femoral condyle rolls over a fixed tibial plateau during knee extension. Gliding, on the other hand, is a pure translatory movement, one fixed point sliding over the other joint surface. A slide usually takes place in an anterior-posterior or medial-lateral direction and this type of movement is again seen when the femur slides forward on a fixed tibia at the knee joint. Spin is like a top spinning, a pure rotatory motion. These movements enable the range of movement to be increased at the joint and also improve the congruency of the joint surfaces to improve stability (Norkin & Levangie 1992). Descriptions of these movements and their combinations for specific joints can be found in many anatomical texts.

So it can be seen that movement is not the straightforward unidimensional process that it first appears to be. Movement at the joint is caused by a force acting on the bony segment which in turn produces the movement at the joint. This force may be an internal force, the concentric or eccentric work of the muscle, or an external force, for example the force of gravity.

When the joint is moved by the force of muscle contraction, either concentrically or eccentrically, the range of joint movement may be described in terms of the excursion of the muscle. This excursion consists of the full range of the muscle, that is, the inner, middle and outer range, each being roughly a third of the full range of movement. The inner range is where the muscle is at its shortest, middle range is the middle third of the muscle excursion and the outer range is where the muscle is at its longest. This is graphically illustrated in Figure 8.1.

The actual range through which the joint

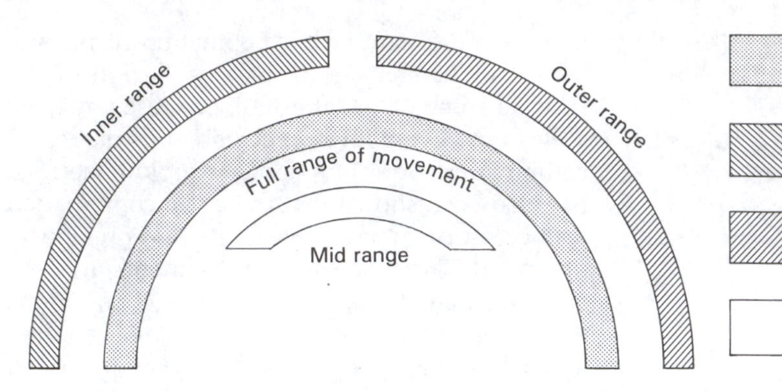

The maximum amount of movement of the muscle

From maximum contraction to approximately half range

From approximately half range to maximum muscle elongation

From approximately midway through inner range to midway through outer range

Figure 8.1 Diagrammatic representation of range of movement.

moves, either actively or passively, is measured in degrees of a circle. Range of movement is usually measured by goniometry. This gives an accepted objective measurement that may be used when analysing joint motion as a part of movement analysis or used as an objective marker when assessing patients. Chapter 14 will give a full description of the process of goniometry.

> **Box 8.2** Task 2
>
> Work in small groups and choose some simple activities. One person performs the activity whilst the others observe the movement very closely. Try to estimate the range of movement of each joint and identify the muscles that are working to produce the movement. Describe the range of movement of the muscle and the type of muscle contraction throughout the activity.

FACILITATION AND RESTRICTION/LIMITATION OF MOVEMENT

Anatomical, physiological and arthrokinaesiological factors combine to give normal facilitation to, and restriction of, movement within a joint.

Normal facilitation

ROM is facilitated by the following:

- bony shape
- hyaline cartilage
- capsule supporting synovial membrane
- additional structures
- elastic ligaments.

The reciprocal convex and concave shape of the joint surfaces combined with the accessory movements of roll, spin and glide provide a greater surface area over which the two bone ends can move. Hyaline cartilage, present on the articulating surfaces of the bone ends, has the dual effect of providing a smooth surface over which the bone ends can glide and affording the joint protection from wear and tear. Both the effects of the hyaline cartilage are enhanced by the presence of synovial fluid within the joint. The fluid layer over the two joint surfaces will reduce the amount of friction between the bone ends when movements occur and act as a form of shock absorber to reduce the trauma of constant impact, particularly in weight bearing joints. Synovial fluid produced by the synovial membrane also provides the joint with some of its nutrition. Most of the capsule surrounding joints is lax, thus permitting a large range of movement by the joint. Many joints have additional structures within the joint; examples of these are the menisci in the knee or the glenoidal labrum within the shoulder. Both these structures, and others like them, can act as a mechanism for increasing joint surfaces, increasing congruity or affecting stability within the joint. Occasionally, the ligaments surrounding the joints contain yellow elastic fibres other than the usual collagen, enabling the ligament to allow the joint an increase in range by providing more flexibility (Williams et al 1989).

Normal limitation/restriction

Normal limitation of range of movement is brought about by:

- articular surface contact
- limit of ligament extensibility
- limit of tendon and muscle extensibility
- apposition of soft tissue.

The two main factors that limit joint mobility are the shape of the joints and the type of structures that run over them. Articular surface contact could mean that the joint is in the close pack position, as in the elbow, where the joint is actually prevented from extending beyond approximately 180° by the olecranon of the ulnar impinging onto the humerus. Flexion of the elbow, on the other hand, is limited by the bulk of the biceps pressing against the forearm, this being an example of soft tissue apposition.

Most ligaments and all tendons are primarily composed of white fibrous collagen. One of the properties of collagen is that it is fairly inelastic and stretching achieved by deformation requires strong forces. Therefore, within normal activities, if the ligaments or tendons are at their maximum length, no more movement is possible at that joint. The structure of muscle offers the opportunity of more stretch as it crosses the joint and thus affords greater mobility, but it still has a limit of extensibility. The limitation that muscle offers to joint mobility is seen particularly when the muscle stretches over two joints, such as the hamstrings stretching over the hip and knee. If the hip is flexed, then the amount of knee extension is limited as the muscle is already partly stretched. If the hip is extended, then the hamstrings are relaxed allowing for a greater range of knee extension.

Although both normal facilitation and limitation are important factors to consider when discussing joint range, for physiotherapists, joint range becomes an issue when there is an abnormal limitation in that range. Abnormal limitation of joint range is usually brought about by either injury or disease to its structure, surface or surrounding soft tissue, i.e. the muscles producing the movement or their functioning.

Abnormal limitation

The above factors can be summarised as:

- destruction of bone and cartilage
- bone fracture
- foreign body in joint
- tearing or displacement of intracapsular structures
- adhesions / scar tissue
- muscle atrophy or hypertrophy
- muscle tear, rupture or denervation
- pain
- physiological factors.

Any disease that destroys the articular cartilage, such as osteoarthritis or rheumatoid arthritis, will impair the functioning of the joint and thus limit movement. This may be for two reasons. Either the destroyed surface will physically prevent the movement, or the pain produced when the two exposed surfaces grind together may produce a reduction in range or a deterioration in the quality of the movement. Either singly or together these may actually prevent movement altogether. A fracture near to, or within, the joint will also prevent movement via mechanical obstruction or pain. The same applies to a foreign body within the joint complex. If there has been an injury to the soft tissue surrounding or within the joint, then repair to that tissue usually takes place by the formation of fibrous or scar tissue which does not have the same extensibility as the tissue it is replacing. Range of motion will, thus, be limited.

The joint itself may be intact, but if the muscles that produce the movement have a dysfunction, the net result is a decrease in ROM. If the muscle is atrophied to a large degree, it would not create sufficient force to move the joint through its full range. Conversely, if there was a large amount of muscle hypertrophy, ROM would also be decreased due to the increased amount of soft tissue apposition. The muscle itself may be intact, but its neural control may be impaired. This could range from total denervation, causing flaccidity of the muscle, to lack of higher centre control, which may cause spasticity. Local spinal reflexes may also have the effect of limiting movement by causing the muscle to be in spasm.

The body's response to pain is usually to keep the part still and avoid movement. This may be only short term, but if the pain becomes chronic, adaptive shortening may occur. The pain may disappear, but the pattern that the brain has adopted due to the memory of the pain that occurred on movement may continue. Other psychological problems such as depression, lack of motivation or self-confidence may also be responsible for the subject not moving. This may also be transient and cause no physical limitation of movement, but if the condition persists then adaptive shortening may also occur.

It is important to remember that what has so far been discussed describes a decrease in movement (hypomobility), but the opposite may also occur. This is termed hypermobility where the range of movement exceeds that of the expected physiological range. This could be due to pathological change either at the joint or elsewhere within the musculoskeletal or neuromuscular systems. It may, however, be a natural phenomenon caused mainly by laxity of ligaments or a congenital joint deformity, but it can also result from a deliberate attempt to increase the joint range well beyond that which is functionally acceptable, in a gymnast or ballet dancer, for example. As Lewit (1993) states, this may be an advantage to these sportspeople but with increased mobility there may be a decrease in stability.

Box 8.3 Task 3

In groups, look at the major joints of the body and visually assess the differences in the range of movement between each person. Can you discover what is limiting the movement for each joint? Can you find anyone with hypermobility? Are there any differences between males and females in ROM or hypermobility?

Before physiotherapists can treat any decrease in the range of movement it is obvious from the above that the cause of the decrease will have to be known. It has been shown that the cause may be in the joint structure (surface or intracapsular), the structures surrounding or running over the joint (ligaments or tendons), or the neuromuscular system that produces the movements. So, to establish the pathological changes that have occurred, it is vital that the therapist performs a full and detailed assessment. Once the pathology is known, the therapist can choose a method of treatment whose physiological effects alter the pathological changes that have occurred to limit joint movement.

Once the assessment has been made it is important to know the physiological effects of the possible treatment options and match them with the effects they will have on the pathological changes. This is the rationale of the treatment.

Limitation of movement, from whatever cause, impairs function of the joint and the muscles producing the movement. Measures that increase the range of movement must also include methods that strengthen the muscles in their new, lengthened position. The degree of ROM gained must be able to be controlled, and the stability of the joint maintained, or further injury may result.

It is important that the details of the anatomy and arthrokinaesiology are understood as well as the pathology of the joint, as these have an effect on the rationale of the treatment choice when there is a pathological reduction on joint range.

Many studies have looked at actual joint ranges of physiological ROM and presented tables of values (Norkin & White 1995). It is accepted, however, that each ROM is specific to each individual and discrepancies may even exist when comparing both sides of an individual. Two of the factors that have an effect on the ROM obtained are age and gender. Younger children appear to have a greater amount of hip flexion, abduction and lateral rotation, and a greater amount of ankle dorsiflexion, than an adult. Elbow movements are also shown to be greater than those of an adult, whilst there is found to be less hip extension, knee extension and ankle plantarflexion. Older age groups seem to have a generalised appendicular and axial joint decrease. Gender appears to have different effects on different joints depending on what movement that joint is performing (Norkin & White 1995).

There are many physiotherapy modalities to increase the ROM. The most obvious is the use of movement itself. The main classifications of these movements are described below.

TYPES OF JOINT MOVEMENT

Movement is one of the main methods that therapists use to increase joint range. The therapist will use the different ways in which the joint moves as a basis for these different methods. There are two main types of movement:

- **passive movement**
- **active movement.**

Passive movement is defined as those movements produced entirely by an external force, i.e. no voluntary muscle work. These can be subdivided into:

- relaxed passive movements
- stretching
- accessory movements
- manipulations.

Active movements are those movements within the unrestricted range of a joint produced by an active contraction of the muscles crossing the joint. These can be subdivided into:

- active assisted exercise
- free active exercise.

PASSIVE MOVEMENTS

Relaxed passive movements

These are movements that are performed within the unrestricted range by an external force and involve no muscle work of the particular joint, or joints, at which the movement takes place.

These movements can be performed in three ways:

Manual relaxed passive movements are movements that are performed by another person, usually the physiotherapist, within the unrestricted range.

Auto-relaxed passive movements are performed within the unrestricted range by the person themselves, i.e. with their unaffected limbs.

Mechanical relaxed passive movements are performed by a machine but still occur within the unrestricted range.

Relaxed passive movement has been a core skill of the physiotherapist for many years and is still widely used today. As with many of the traditional skills, little evidence of its clinical effectiveness has been published (Basmajian & Wolf 1990) but the following is the accepted rationale for its use.

Indications

It is indicated when the patient is unable to perform an active full range movement. The reason for this inability may include: unconsciousness, weak or denervated muscle, spinal injury, pain, neurological disease or enforced rest.

Effects

- maintain ROM
- prevent contractures
- maintain integrity of soft tissue and muscle elasticity
- increase venous circulation
- increase synovial fluid production and therefore joint cartilage nutrition
- increase kinaesthetic awareness
- maintain functional movement patterns
- reduce pain.

If muscle is not moved through its full range, it will adapt to the demands being placed upon it. The actin and myosin protein filaments (the contractile element) will be reabsorbed and thus the area for cross bridge formation will be decreased. This will cause muscle weakness and the inability to perform the movement. The muscles will adopt the new position and will be shortened, i.e. they will have adaptive shortening. The noncontractile elements within the muscle, the connective tissue, will add to this effect by increasing the collagen turnover rate, which is the balance of collagen production and destruction (Basmajian & Wolf 1990). If more collagen is produced, it increases the stiffness of the muscle and decreases its propensity to stretch. If no movement takes place, the muscle will adapt to the new position and thus contractures will occur.

Other soft tissues, such as ligaments and tendons, will also be similarly affected. As they have a greater proportion of collagen, the increase and change in consistency will also lead to stiffness and eventually to contractures.

By placing stresses on these tissues the collagen turnover rate is normalised and the elasticity of the tissues is maintained. There is, however, no increase in the strength of the muscle as this requires a greater physiological demand.

If a limb, particularly the lower limb, is not moved, venous congestion may occur. This is because the muscle pump does not work to aid venous return. Pooling occurs and is increased through dilation of the vessels. Perfusion is also decreased and this can lead to an increased risk of deep vein thrombosis. Passive movements will act as a prophylaxis to prevent stagnation. This is achieved by physically compressing the veins and one-way flow is achieved via the valves within the veins themselves. Lymph is also encouraged to move. Compression of the tissues increases the hydrostatic pressure and thus encourages tissue perfusion and fluid reabsorption. This may be useful in reducing oedema.

If the synovial fluid is swept over the articular cartilage, it will provide nutrition and help prevent the deterioration of the surface. Production and absorption of the synovial fluid by the synovial membrane is stimulated by movement of the joint. This becomes more important if the immobility is due to injury as one of the consequences of injury is inflammation and repair by fibrosis, which in itself will increase the risk of adhesions forming within the joint. Thus, with passive movement, this risk will be decreased.

The brain is said to recognise gross movement patterns, so if these patterns are not able to occur, there is the possibility of the memory of the pattern being lost. Passive movement in these patterns will decrease this risk. Kinaesthetic awareness is maintained when performing passive movements by stimulating the nerve endings within the joint complex.

Rhythmical movements are said to reduce pain by causing a relaxation effect within the muscles (Gardiner 1981). This may partly be achieved by the removal of waste products from the area (through the increased circulation) and any chemical irritants. Stimulation of the mechanoreceptors may also subserve the sensations from the pain nerve ending and thus decrease their effect (see accessory movements for further explanation).

Contraindications

- Immediately post-injury as this may increase the inflammatory process.
- Early fractures where movement may cause disruption of the fracture site.
- Where pain may be beyond the patient's tolerance.
- Muscle or ligament incomplete tears where further damage may occur.
- Where the circulation may be compromised.

Principles of application

Passive movements may either be performed in the anatomical planes or in functional patterns. The choice and type of movement will depend on the findings of the assessment and the aims of the treatment. The same basic principles of application need to be considered whichever movement is chosen. These include:

- The part should be comfortable, supported and localised.
- The patient should be comfortable, warm and supported.
- Hand holds should be as near the joint as possible.
- The motion should be smooth and rhythmical.
- Speed and duration should be appropriate for the desired effects.
- Range should be the maximum available without stretching or causing pain.

- Muscles that stretch over two or more joints must be relaxed (after Hollis 1989).

Auto-relaxed passive movements

Although the rationale is the same, the method of application must be modified. Patients who have to perform their own passive movements are usually those with a long-term problem. People with spinal injuries, for example, must retain their joint range and muscle length if they are to perform the functions necessary for daily living (Bromley 1991).

Mechanical relaxed passive movements

Unlike manual or auto-passive movements which, by their nature, have to be carried out intermittently, mechanical passive movements may be carried out continuously. Mechanical devices for producing continuous passive movement (CPM) were first used by Salter in 1970 (McCarthy et al 1993). Although their designs and protocols of use may alter, they all have essentially the same function.

The rationale is the same as for any relaxed passive movement but the benefits of continuous movement are particularly evident following surgery (Kisner & Colby 1990). Basso and Knapp (1987) found that CPM decreased joint effusions and wound oedema whilst increasing range of movement and decreasing pain in postoperative knee patients.

STRETCHING

Stretching differs from relaxed passive movement in that it takes the movement beyond the available range. This available range may be limited due to disease or injury. Stretching may also take the joint beyond the normal physiological range. Whereas relaxed passive movements are designed to maintain length of soft tissue and hence joint range, stretching should result in a change in length of the soft tissue structures crossing over the joint, with the consequent increase in joint range. Passive stretching is not the only method of increasing joint range via the soft tissues. This can also be attained via active stretching which will be discussed later.

Stretching of biological material

The response to stretching of biological material will depend on the composition of that material. Some tissues show great elasticity, for example skin and others, such as ligaments, have little elasticity and will deform under stress. Bone shows very little elasticity and tends to fracture under stress. A composite stress/strain curve of biological material is shown in Figure 8.2.

The *toe region* occurs as the wavy collagen is straightened out. This leads into the *elastic range* where the material will return to its original shape before reaching the *elastic limit*. Beyond this point the material will not return to its original shape or size. Beyond this point the material will enter the *plastic range* where permanent deformation occurs and the material no longer obeys Hooke's law. This means that the stress is no longer proportional to the strain. In the *necking phase* considerable strain occurs with only minimal stress leading to *failure* of the material, when there is complete rupture.

Stretching of muscle initially occurs in the series elastic components (actin and myosin) and the tension within the tissue increases sharply. A mechanical disruption of the cross bridge formation occurs as the filaments slide apart. The sarcomeres then 'give' with an abrupt lengthening. After a while the sarcomeres return to their resting length. If this stress is maintained, the muscle will adapt to its new length. This adaptation is only transient (Kisner & Colby 1990).

Stretching of collagen is mainly due to plastic deformation as the material is fairly inelastic. Connective tissue will reorganise itself in response to a sustained stretch provided the stress is not too great or applied for too long (Basmajian & Wolf 1990). Therefore, any stretch will be fairly permanent. It can be seen that care must be taken not to reach the necking phase when stretching as this would compromise the integrity of the tissue.

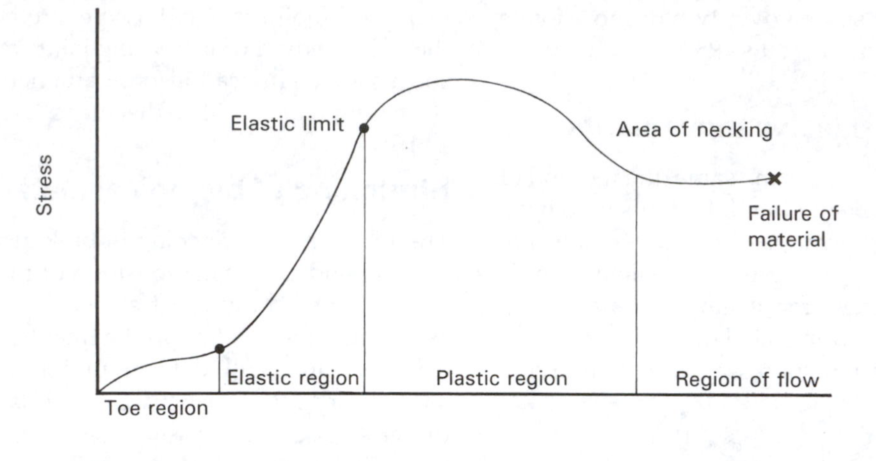

Figure 8.2 Stress/strain curve for biological material. From the end of the toe region to the elastic limit, the material obeys Hooke's law and the stress is proportional to strain.

Smith (1994) makes the suggestion that muscles and tendons are also viscoelastic in that they are time or rate change dependent and it is this viscoelasticity that is responsible for the stretch.

To summarise, the effects of stretching:

- increase joint ROM
- increase soft tissue length
- relieve muscle spasm
- increase tissue compliance in preparation for an athletic event (Vujnovich & Dawson 1994).

Passive stretching can, like passive movements, be performed as manual, auto- or mechanical stretching.

In manual stretching the therapist produces the sustained stretch at the end of range with the patient relaxed. The stretch should be held for at least 15 seconds and repeated several times. Patients themselves produce the stress required to elicit stretch with auto-stretching. Mechanical stretching involves a low stress being applied by a machine over a prolonged period.

The above text refers to static stretching, which is defined as a slow sustained stretch at the end of range. This is in contrast to ballistic stretching which is defined as small end of range bounces. This method of stretching has been shown to increase the risk of injury and should not be performed (Vujnovich & Dawson 1994).

ACCESSORY MOVEMENTS

Maitland (1986) defines accessory movements as those movements of the joints which a person cannot perform actively, but which can be performed on that person by an external force. Although these movements form an integral part of the normal physiological movement, and occur throughout the ROM, they cannot be physically isolated by the person themselves. However, if the accessory movement was being performed by the physiotherapist, the person would be able to stop the movement from taking place. These movements take the form of glides (medially, laterally, longitudinally), compressions, distractions or rotations.

Anatomists group accessory movements into two categories: type one are those that cannot be performed unless resistance is provided to the active movement; and type two are the movements that can be produced only when the subject's muscles are relaxed (Williams et al 1989). The latter are the type that are used for therapeutic purposes.

Accessory movements are used to increase the

range of movements at joints and also to decrease any pain that is present. The mechanism for decreasing pain depends, of course, on the cause of that pain. If the pain is caused by the decrease in ROM then the rationale of treatment will be the same as that for increasing the range. If not, then the usual effect of small amplitude movements, applied rhythmically to the joints, is to have an inhibitory effect by the afferent impulse traffic from articular receptors, blocking the pain (Grieve 1988).

Twomey (1992) suggests that articular cartilage facilitates the ROM of joints. If there is joint immobility, the articular cartilage will degenerate more quickly as it requires movement and loading to ensure adequate nutrition. This nutrition is facilitated by the synovial fluid which is swept over the joint surface. As the stimulation of the synovial membrane decreases, the amount of synovial fluid produced is decreased and thickens with decreased osmolarity. Accessory movements assist in the maintenance of synovial production and, by the small oscillatory movements they make, produce the washing effect of the synovial fluid over the joint surfaces.

Twomey (1992) also suggests that movement is important for the nutrition of all collagenous tissue as well as the prevention of adaptive shortening. Small amplitude movements at the end of range will elongate connective tissue, ligaments, joint capsule and other periarticular fascia via the processes described in the section dealing with stretching. End range movements will also break down any intra-articular adhesions that have been formed (Grieve 1988).

If joint stiffness is caused by fibrocartilaginous blocks and subsequent connective tissue shortening, Threlkeld (1992) claims that the relative positions of the joint surfaces can be altered to restore normal accessory movement and restore displaced material.

MANIPULATION

Joint manipulations differ from all other types of therapeutic movement because the patient has no control over the procedure. These movements are potentially dangerous and should therefore only be performed by skilled professionals with much experience in the mobilisation of joints. Manipulations are small amplitude forceful movements that take the joint past the available physiological range. As Maitland (1986) states, manipulations are performed very quickly before the subject has time to prevent the movement from taking place. Lewit (1993) describes manipulation as a technique for treating end range blocking of joints. Displaced intra-articular material may be one reason for the restriction of movement. Joint misalignment may be another. This blocking, or restriction of movement, is claimed to have two effects: one is the restriction of the subject's functional movement; the other is the effect on the accessory movement or joint play. Restrictions caused by meniscal or other material within the joint are termed loose bodies.

Lack of movement of a joint may lead to adaptive shortening of the soft tissue structures surrounding it; capsule, ligaments, tendons or muscles, for example. This secondary consequence will in itself cause a restriction of movement and possibly pain when movement is attempted. Thus, a vicious circle is set up. The initial immobility may be as a consequence of pain. If the pain is caused by trauma to the joint, then the problem may be compounded by the presence of adhesions within and around the joint that are the natural consequence of the inflammatory process.

Shortened soft tissue structures may also be manipulated to physically break the structures.

Effects

The primary effect of manipulating the joint is to restore its mobility. A secondary effect may be to decrease pain, if the pain was a direct result of abnormal tension on structures due to incorrect functioning of the joint.

These effects are brought about in a variety of ways. As the joint is manipulated the surfaces are caused to gap, creating space. The movement of the surfaces, together with the greater space created, possibly causes any physical obstruction between the joint surfaces to be moved clear. Joint surfaces may then be realigned, causing the correct afferent information to be sent to the spinal cord. Fibrous adhesions caused as the result of the inflammatory exudate, the organisation of the synovial fluid or the adaptive shortening of any soft tissue structure will be physically torn. This takes the structure through the plastic phase and very rapidly past the breaking point. The hopeful consequence of this procedure is to free the joint to perform its full functional excursion.

Contraindications of the above technique are the same as those for passive movements.

ACTIVE MOVEMENT

Active movement can be thought of as movements of the joint within the unrestricted range that is produced by the muscles that pass over that joint. The movement may be active assisted or free active. Therapeutically, these movements are performed as exercise.

Active assisted exercise is exercise carried out when the prime movers of that joint are not strong enough to perform the full ROM of that joint. Assistance may be given by any external force but it is important that the external force provides assistance only, the movement is augmented and must not become a passive movement. The external force has to be applied in the direction of the muscular action but not necessarily at the same point. External assistance may be:

- manual assistance
- mechanical assistance
- auto-assistance.

Manual assisted exercise is the assistance to the subject's muscular effort by the physiotherapist. The therapist is able to change the assistance as the muscles progress through their ROM compensating for such factors as angle of pull and length–tension relationships. The amount of assistance may also be changed as the muscle strength increases.

Mechanical assistance may be provided by a variety of means. Isokinetic equipment has the facility for active assisted movement provided the trigger forces are set low enough. The most useful mechanical assistance is that of sling suspension, where the assistance is given in two ways. Firstly, the resisting force of friction is reduced by physically lifting the segment clear of its resting surface, which also helps to counteract the force of gravity by supporting the segment. Secondly, depending on the point of fixation of the sling suspension, gravity may be used to assist the movement, provided the desired movement occurs on the downward arc of the curve produced by the segment within the sling suspension.

Auto-assisted exercise may be performed by the subject themselves using the same principles as those of manual assistance. More common, however, is for the assistance to be a combined form of auto- and mechanical assistance. This is seen in the use of bicycle pedals for lower limb mobility or pulleys for upper limb mobility.

Free active exercise differs from assisted exercise in that the movement is carried out by the subject themselves, with no assistance or resistance to the movement, except that of the force of gravity. There are many ways that free active movements can be performed. These include:

- Rhythmical: this uses momentum to help perform the movement taking place in one plane but in opposite directions.
- Pendular: these are movements performed in an arc and are useful for improving mobility as, on the down curve of the arc, the movement is assisted by gravity.
- Single or patterned: depending on the aims of the intended movement, the choice of single or patterned movements is made. As a general rule, single movements are used to demonstrate or restore actions whereas patterned movements are used for functional activities. The use of biceps to flex the elbow as a pure movement is an example of a single movement. Reaching out to pick up an object (food) and taking it to the mouth also involves flexion of the elbow but this movement also contains other joint movements in a functional pattern (the feeding pattern).

These movements may also be classed as to their effect:

- Localised: designed to produce a local or

specific effect; mobilising a particular joint or strengthening a particular muscle.

- General: gives a widespread effect over many joints or muscles, running, for example (Gardiner 1981).

Exercise affects all the systems of the body and is covered elsewhere in this and many other textbooks. What must be remembered, however, is that the effects produced cannot be isolated to one particular system or even one effect within that system. Exercise, for example, will maintain muscle length and joint range but it may also alter the strength of that muscle and have a more widespread effect on the cardiovascular system.

One of the major local effects of exercise is the increased rate of protein synthesis, thus producing more actin and myosin as a response and facilitating an increase in muscle length. Connective tissue also responds to increased exercise by becoming stronger in order to cope with the increase in function that is required of that muscle (Basmajian & Wolf 1990).

If active exercise is performed regularly and through the available physiological range, it has all the effects of passive exercise previously explained, including maintaining joint range, increasing joint nutrition and decreasing pain. Active movement has the advantage of strengthening muscles to some extent, thus providing stability for the joint with its increased range. Another advantage of performing active movements is that, if they are performed in a rhythmical manner, they may promote relaxation of the muscles surrounding the joint. If this happens, the joint range may well be increased, especially if the restriction was due to muscle spasm.

CONCLUSION

Movement occurring at joints depends upon a variety of anatomical and biomechanical factors which both facilitate and/or limit the range of movement available. How movements are classified and described will depend on which discipline is being studied. For this text, movements are classified as either active or passive, with subdivisions of each. Restriction of movement, caused by pathological changes, can be successfully managed by the physiotherapist using different forms of movement. This is achieved once the rationale of the chosen method is known and correctly applied to the fully assessed patient.

The preceding texts by no means represent an exhaustive survey into the therapeutic modalities to improve joint range and muscle length. They are, however, representative of the basic principles of the techniques that are used based on normal joint movement.

REFERENCES

Basmajian J, Wolf S 1990 Therapeutic exercise, 5th edn. Williams and Wilkins, Baltimore

Basso D, Knapp L 1987 Comparison of two continuous passive motion protocols for patients with total knee implants. Physical Therapy 67: 360–363

Bromley I 1991 Tetraplegia and paraplegia; a guide for physiotherapists, 4th edn. Churchill Livingstone, Edinburgh

Gardiner M D 1981 The principles of exercise therapy, 4th edn. Bell and Hyman, London

Grieve G 1988 Contraindications to spinal manipulations and allied treatment. Physiotherapy 75: 445–453

Hall S 1995 Basic biomechanics, 2nd edn. Mosby, St Louis

Hollis M 1989 Practical exercise therapy, 3rd edn. Blackwell Science, Oxford

Kisner C, Colby L 1990 Therapeutic exercise, foundations and techniques, 2nd edn. F A Davies, Philadelphia

Lewit K 1993 Manipulative therapy in rehabilitation of the locomotor system, 2nd edn. Butterworth-Heinemann, Oxford

Maitland G 1986 Vertebral manipulation, 5th edn. Butterworths, London

McCarthy M, Yates C, Anderson M, Yates-McCarthy J 1993 The effects of immediate continuous passive movement on pain during the inflammatory phase of soft tissue healing following anterior cruciate ligament reconstruction. Journal of Sport and Physical Therapy 17: 96–101

Norkin C, Levangie P 1992 Joint structure and function: a comprehensive analysis, 2nd edn. F A Davis, Philadelphia

Norkin C, White J 1995 Measurement of joint motion: A guide to goniometry, 2nd edn. F A Davies, Philadelphia

Palastanga N, Field D, Soames R 1994 Anatomy and human movement: structure and function, 2nd edn. Butterworth-Heinemann, Oxford

Smith C 1994 Warm-up procedures: to stretch or not to stretch. Journal of Sport and Physical Therapy 19: 12–17

Threlkeld J 1992 The effects of manual therapy on connective tissue. Physical Therapy 72: 61–70

Twomey L 1992 A rationale for the treatment of back pain and joint pain by manual therapy. Physical Therapy 72: 53–60

Vujnovich J, Dawson N 1994 The effect of therapeutic muscle stretch on neural processing. Journal of Sport and Physical Therapy 20: 145–153

Williams P, Warwick R, Dyson M, Bannister L 1989 Gray's anatomy, 7th edn. Churchill Livingstone, Edinburgh

CHAPTER CONTENTS

9

Tension and relaxation

M. Trew

OBJECTIVES

When you have completed this chapter you should be able to:

1. **Understand the difference between positive and negative stress**
2. **Understand the basic physiological changes in tension and relaxation**
3. **Recognise the common changes in movement patterns or posture that occur in situations of tension**
4. **Describe several different methods of inducing relaxation.**

INTRODUCTION

Even minor degrees of tension will affect normal patterns of movement or posture and, where levels of tension are very high, human movement will be noticeably disrupted. This chapter considers stress or tension in general terms and how it may affect movement: methods of inducing relaxation are described in the final part of the chapter.

Though the general population is familiar with the words 'tension' and 'relaxation', a surprisingly large number of people are unable to recognise these two states in themselves. Well developed physical self-awareness is not common, probably because there is a tendency to regard the body as something to be accepted until it starts to go noticeably wrong. Despite this, everyone has the potential to know what tension and relaxation are, though some people may need guidance to recognise these states in themselves. When an individual is able to monitor their own levels of tension and relaxation, they have taken the first step in being able to control and modify their reactions to stress.

NORMAL LEVELS OF AROUSAL

It is quite normal to vary between periods of stress when arousal is high and periods of low arousal and relaxation. In fact, these differences in the level of arousal are essential for efficient activity. When action is needed, the arousal level, and therefore the level of muscle tone, is raised. When there is no call for high levels of mental and physical activity, then it is appropriate for the general level of arousal to be reduced so that energy may be conserved and depleted energy stores may be replenished.

In addition, all individuals have natural periods of greater or lesser mental and physical arousal throughout each day. The most obvious example is the wake/sleep cycle. At a more detailed level this basic rhythm is also evident in the various functions of the body: the rhythmical contraction and relaxation of the heart, the inspiratory and expiratory phases of the breathing cycle, and in movement, where one muscle group contracts whilst the antagonists relax to allow the movement to take place. These alternating rhythms can be speeded up or slowed down in response to both internal and external stimuli, depending on the needs and demands of the particular situation.

Relaxation requires a de-emphasis of arousal states and, in particular, a reduction in muscular activity but, because the mind and the body are inextricably linked, there must also be a reduction in mental activity. Mental activity such as anxiety, emotional tension and fear can all have a deleterious effect on tension and it is not possible when considering relaxation to divorce emotional states from physical states (Elton 1993).

When the natural balance between tension and relaxation is disrupted it can lead to an excessively high level of mental and physical arousal during which the muscular, respiratory and cardiovascular systems start to exhibit inappropriate activity. This is variously described as stress, tension, anxiety or an inability to relax and it leads to feelings of mental and physical tiredness, headaches, abnormal joint postures and a loss of the fine control of movement (Basmajian & Wolf 1990). In severe cases, other systemic manifestations such as dizziness, dysfunction of the digestive system, shortness of breath and disruption of the immunological response may occur (Baker 1987). Stress is defined by Selye (1984) as the non-specific result of any demand upon the body. This result can be mental, somatic or both. The physiological responses to stress are controlled by the autonomic nervous system and the psychological responses are thought to be controlled from the hypothalamus and the limbic lobe of the cerebral cortex.

PHYSIOLOGY OF STRESS AND RELAXATION

The autonomic nervous system is divided into two separate systems, the sympathetic and the parasympathetic, and they control the viscera. Afferent impulses from the viscera are integrated either at spinal level or within the reticular formation of the pons, medulla and midbrain. The two systems frequently act in a complementary manner affected by the quantity and type of synapse transmitter. The sympathetic system releases noradrenaline and the parasympathetic nerve endings release acetylcholine. Noradrenaline is also released as a hormone by the adrenal gland thus enhancing the effect of the sympathetic system.

The cardiovascular centres of the pons are responsible for the regulation of blood pressure. The dorsal vagal nuclei contain the parasympathetic preganglionic cell bodies of the vagus nerve which is responsible for the control of the viscera of the thorax and upper abdomen. The Edinger–Westphal nuclei of the midbrain controls the reflex changes of the eye. These nuclei receive input from the reticular formation and other sensory inputs, particularly from the hypothalamus (Moffett et al 1993).

It is the sympathetic system that is activated in times of stress initiating the 'fight or flight' response. The body becomes ready for action: there are chemical changes in skeletal muscle which result in the development of tension in the contractile components, in the eyes the pupils dilate and there is vasoconstriction of the salivary glands leading to a dry mouth. There is an inhibition of gastric and digestive secretions and a constriction of the sphincters with an increase in sweating. There is also an increase in the rate and force of heart contractions delivering more blood and, therefore, more oxygen to the tissues that need them. Cardiac output is increased and the distribution of blood is adjusted through constrictions of the small blood vessels in the skin and abdominal viscera so that blood is not directed to non-essential tissues. In addition to all this there is a bronchodilatation in the lungs and a general increase in metabolism.

These are natural responses to stress and they only become inappropriate and possibly pathological when the original reason for stress is removed and the responses remain. Because so many systems are affected by the normal stress response, it is not surprising that in individuals who have abnormal levels of tension there may be symptoms involving a vast range of body systems. For example, experimental work has shown that abnormally high and prolonged levels of adrenaline and noradrenaline in animals lead to hyperactivity and enlargement of the adrenal cortex, and atrophy of the thymus and lymph nodes (Selye 1984). It has also been shown that prolonged and excessive stress weakens the effectiveness of the immune response and therefore increases vulnerability to disease. At a subjective level, anyone who has experienced high levels of stress will have noticed the general feeling of tiredness which results from sustained high levels of muscular activity.

Relaxation produces the opposite situation and, amongst other things, results in a noticeable reduction in blood pressure, heart and respiratory rate. Muscle tension is no longer generated and the elastic components of the muscles cause the fibres to return to their resting length — this is muscle relaxation (Astrand & Rodahl 1987). In complete muscle relaxation there is little or no apparent activity in the contractile components, although some tone is always present as a consequence of the elastic components of muscle.

WHEN STRESS BECOMES ABNORMAL

At times of either personal or environmental stress there will be challenges and difficulties to face. These lead to a rise in the level of arousal in the brain and the wide range of physical responses discussed earlier. This normal stress response, which is often experienced by the population at large, is usually followed by fatigue, a reduction in the level of arousal and eventually sleep. Selye (1984) defined this form of stress as positive because it is clear that a certain level of arousal is necessary to meet the challenges of life and living efficiently. Problems arise only when the response to stress becomes excessive and out of control, in which circumstances stress can be described as negative or distressful.

The capacity to cope with stressful events varies from person to person. Some people have the ability to recognise excessive stress and deal with these situations or prevent them occurring (Baker 1987). Others fail to recognise excessive stress or are unable to take avoiding action, which inevitably leads to the development of stress symptoms. At a gross level these symptoms include fatigue, loss of sleep, and work inefficiency which leads to further muscle tension, causing even greater fatigue. The individual will often attempt to make up for the loss of efficiency by trying to increase their level of arousal, but this is self-defeating as it leads to more physical and emotional tension. If this situation is allowed to continue, a self-perpetuating cycle becomes established which can lead to depression and ill health.

There is no clear dividing point between the experience of positive stress and the change into a clinically recognised anxiety disorder because the duration of stress symptoms varies greatly as does an individual's ability to cope. However, there is no doubt that stress is a major problem which can affect any member of the population and this is evidenced by a review by Durham & Allan (1993) which indicated that at least 30% of the population may be driven to seek professional help for stress related problems.

Pattern of stress

The physical symptoms of stress which are easily recognised are often reflected in over-activity of the general body musculature. This excessive muscular tension forms well recognised patterns. An individual may demonstrate inappropriate muscular tension in any of the major muscle groups, but usually one or two muscle groups may be particularly affected. Because the pattern of tension may manifest itself in a variety of ways, it is important to observe the whole body when looking for signs.

The facial muscles often show abnormal tone; corrugator and frontalis, when not relaxed, will produce vertical and horizontal lines on the forehead. Tension headaches may develop if there is excessive tone in the occipitofrontalis and the suboccipital muscle groups. Inappropriate muscle activity in orbicularis occuli and muscles of the eyelids can either result in the eyes being narrowed or held widely open or in ticks, particularly on the inferior surface. Very frequently the masseter muscle responds to stress by contracting to hold the jaws tightly together; on careful observation of the lateral aspect of the cheeks it is often possible to see the muscle twitching. For some people, tension in the muscles of the jaw may also manifest itself in grinding of teeth during sleep. The tongue may be held rigidly on the roof of the mouth and the muscles of the throat may contract abnormally making speech rather harsh and lacking in fluency of pronunciation. There may also be a click as the tongue unsticks itself from the roof of the mouth at the start of each sentence.

The posture of the head at times of stress varies between individuals, but common to all is elevation of the shoulder girdle as a result of abnormal tension in the upper fibres of the trapezius. This is easily recognised because the neck appears shortened and the normally sloping contour of the neck/shoulder region becomes more angular. The upper limbs are often held quite rigidly close to the chest with the elbows flexed. The trunk is generally held stiff and, because of tension in the abdominal musculature, diaphragmatic respiration can be inhibited so that the individual has to breathe using the intercostal and accessory muscles of respiration. In these cases, upper chest movement is observable and the accessory muscles in the neck can be seen contracting.

In general, repeated movements such as finger tapping or swinging the knees from flexion to extension when sitting can be indicators of abnormal muscle tone. Classically, it is also expected that adduction, flexion patterns are symptomatic of stress. This may be seen in a seated subject who is tense and will sit rigidly on the edge of the chair with their lower limbs adducted and flexed, while in a standing subject the arms may be folded tightly and the legs crossed at the ankle.

Though there are many ways in which stress or tension can be manifested, it is rare for one person to demonstrate all these symptoms simultaneously and it may be necessary to use careful observation in order to identify the signs. In addition, some people have learnt to hide the signs of tension and in consequence will make an effort to slump back in their chair or to stand in a relaxed manner. Despite these efforts it is hard to control tension indicators and an experienced observer will usually be able to notice some abnormal movements or postures.

Apart from postural symptoms and general feelings of fatigue, increased stress can show in other ways. There may be an increase in adrenaline production which will result in an increase in heart and respiratory rate. There may also be changes in blood and digestive chemistry which become abnormal if maintained for any length of time (Astrand & Rodahl 1987). Skilled performance of motor tasks is usually decreased with a higher incidence of anxiety-induced errors in people who are tense than in people who are more appropriately relaxed (Basmajian & Wolf 1990). This is probably caused by a combination of reduced muscle control in those muscles with abnormal tone and a disruption of concentration. Tension can also lead to emotional changes which may vary from anger to despair; intellectual performance may also suffer in these circumstances.

Many students suffer from increased tension, which can interfere with their performance. The stress of study, fear of failure, desire to succeed and pace of work, as well as poor environmental conditions and financial worries, may all contribute to this condition.

Box 9.1 Task 1

Make a list of the symptoms that you manifest when tense. Watch other people carefully and see if you can notice examples of tension in the way in which they move or in their general posture.

RELAXATION THERAPY

There are a number of specific techniques that can be used to encourage relaxation but it is clear that treating the symptoms alone is not enough and the underlying cause must also be addressed. It is important to seek a solution for the causes of tension as well as learning methods to remove the symptoms. Relaxation techniques can be either physical or behavioural in approach, but there is controversy over which might be the most successful. Elton (1993) suggests that a combination of physical and behavioural techniques are most successful and that it may therefore be necessary for the patient with severe, chronic symptoms to seek help from those versed in both physical and psychological therapy. There is further evidence of the specificity of effectiveness of relaxation techniques. This means that a cognitive treatment approach will have mainly cognitive effects whereas if a physiological approach is used, most benefits will be seen in the reduction of the physical manifestations of tension (Lehrer et al 1994). If an individual has a generalised response to stress, then a mixed approach to relaxation may be the most appropriate treatment.

In physiotherapy the goal of relaxation therapy is primarily to encourage the patient to gain physical control so that their response to stress is appropriate rather than destructive. This should then lead to the adoption of normal body posture and normal patterns of movement which will be more energy efficient and will, in turn, lead to a reduction in the feelings of fatigue. Some relaxation techniques require the patient to be rather passive, for example a sedative massage or a session in a flotation tank may be enjoyable and temporarily effective but overall the improvements are less likely to be long lasting than when a technique is used which encourages self-control (Elton 1993). To be effective the patient has to recognise tension in themselves and understand its causes and effects. Once this is achieved and they are aware of their abnormal responses to stress, they can be taught to adopt strategies which will lead to more normal reactions and an immediate improvement in physiological function (Elton 1993).

General and local relaxation

Methods of relaxation may involve either learning to relax generally or developing the ability to recognise tension in specific parts of the body and then targeting individual muscles.

General relaxation can be used either as a treatment in its own right or as part of a more extensive treatment plan. Sometimes patients are very tense due to psychological or physical stress, and treatment of a totally unrelated condition may be impossible until general relaxation is gained; in this case relaxation will be preparatory to the main programme.

At other times general relaxation is not appropriate and the treatment objective may be to teach control of abnormal tension in specific muscles. This is often the case where fear of pain leads to protective muscle spasm round a particular joint and in this case local relaxation is often used in conjunction with pain-relieving treatments.

Box 9.2 Case study 1

An executive secretary in her mid-thirties was referred for treatment for unspecified low back pain. On arrival it was noticed that she sat very upright in the chair with her legs tightly wound round each other. Her arms were folded and her shoulder girdle was noticeably elevated. Initially it was thought that she was probably nervous about the impending treatment and that once she realised that the session would be neither painful nor unpleasant, she would relax. This did not happen and it became evident that even undertaking an assessment of her back problem was going to be difficult because of the degree of tension she exhibited. The remainder of the session was spent making her aware of her tension levels and giving her strategies for relaxation. For the next few weeks treatment focused primarily on relaxation and as she became more adept at reducing tension she was able to report a marked reduction in her back problem, a reduction in the incidence of headaches and an increase in general well-being. She now continues the relaxation on her own at home and has not required further treatment.

Developing self-awareness

Whatever technique is used, whether for local or general relaxation, the first step is to gain the patient's cooperation and develop their self-awareness. Until patients can recognise the feeling of abnormal muscular activity they will not be effective in trying to reach a more normal level. They must also learn to recognise the 'feel' of being relaxed so that they can monitor their level of success (Madders 1988). Developing self-awareness can be done by using a variety of biofeedback mechanisms. A mirror is an effective means of demonstrating abnormal posture and the success of attempts to reduce increased tone is immediately apparent. Videotape can also be used to record the patient's posture and movement patterns and it can then be viewed and discussed. Repeat filming will demonstrate the success of the intervention process. Patients can also be made aware of the level of activity in their muscles by using electromyography (EMG). Simple EMG biofeedback machines will indicate the level of electrical activity in those muscles underlying the electrodes. They are encouraged to try and relax while using the EMG output to measure the degree of success. Electronic blood pressure and pulse rate monitors can also be used as effective biofeedback mechanisms and at a simpler level patients can take their own pulse or count their own respiratory rate. With the information gained they can impose conscious control over cardiac or respiratory rate.

The overriding aim is that the patient comes to recognise the feeling of relaxation and that they know what activities will cause a reduction in the symptoms of stress. Emphasis is placed on teaching them to discriminate between abnormal and normal tension so that they can recognise when they have been successful in achieving a relaxed state and are able to produce relaxation at will. The skills of monitoring muscle activity, joint position, respiratory and heart rates have to be learnt and this demands regular practice and concentration.

Preparing for relaxation

When teaching the patient total body relaxation it is advisable in the early stages to work in a fairly quiet, pleasantly warm environment where the patient can concentrate on the task in hand (Mitchell 1990). The choice of starting position should be left to the patient but it must be one of comfort and full support. Side, supine or prone lying with pillows and blankets for support and comfort are the most likely to be successful but should never be imposed upon the patient. Often during the process of relaxation therapy the initial starting position may become inappropriate and the patient should be reassured that they may change position at any time.

As the ability to relax develops, different positions should be adopted and the environment should be made less protective until the patient is able to relax under any circumstances. It is essential that the treatment sessions proceed beyond the application of a technique and the therapist's goal should be to help patients develop strategies which will allow them to relax in situations which have previously generated stress (Crist & Rickard 1993).

The words used in the treatment session are

important. Instructions should be brief, to the point and spoken in a calm, quiet manner so that they give a mental concept of what is required. The same phrases and commands should be used repeatedly so that the words themselves may become a trigger for relaxation. Harsh sounding words should be avoided and where a command to relax is given the tone of the voice should reflect the meaning. The therapist should conceal their own personality so that the focus of the session is on the task and not on the individuals involved.

Timing is also important. Patients must be given ample time to assimilate the instructions and monitor their responses, in particular, whether they have achieved a release of tension. In the introductory sessions this may take considerably longer than might be expected.

SPECIFIC RELAXATION TECHNIQUES

Apart from teaching patients to develop an increased self-awareness and, through that, a measure of personal control, specific relaxation techniques may also be used. There are two main approaches: *progressive muscle relaxation*, which is based on work reported by Jacobsen in 1929, and *imaginal techniques*.

Progressive muscle relaxation depends on the contraction of a muscle or muscle groups followed by a period of relaxation. Imaginal techniques require the focusing of thoughts on an activity or place or on a repetitive sound or movement.

Whilst there is no clear evidence as to which approach is most effective or whether personality type might affect the choice of procedure (Crist & Rickard 1993), the best technique for a specific patient may be that which targets their main manifestations of stress (Lehrer et al 1994). In any event, the therapist needs to be flexible when choosing the relaxation technique and must be prepared to change the approach if success is not being achieved.

Contract-hold-relax or contrast method (Hollis 1988, Gardiner 1981, Ricketts & Cross 1985)

Fox (1996) cites work undertaken by Sherrington in the early part of this century which suggested that a maximal relaxation follows a maximal muscle contraction. This mechanism is mediated through the golgi tendon organ. When a muscle contracts hard, considerable tension is placed on the golgi tendon organs in that muscle and this stimulates the inverse stretch reflex. Impulses from the golgi tendon organs in the contracting muscle pass to the spinal cord and synapse through to alpha motor neurones which transmit the impulses to the same muscle. These are inhibitory impulses and, if the original contraction was strong enough, will result in inhibition of contraction and therefore relaxation (Fox 1996). It is possible that this reflex is utilised in some relaxation techniques.

The contract-hold-relax relaxation technique requires the patient to undertake, in sequence, strong isometric contractions of all major muscle groups and in particular those which are manifesting abnormal tone. The contraction is held and it is then followed by a period of relaxation which should be at least as long as the contraction. The patient should be asked to concentrate on the feeling of relaxation in order to learn to reproduce it. It is usual to work distal to proximal and to begin with one limb and teach contraction of each individual muscle group. When each major muscle group in a limb has been taken through this process, a total limb isometric contraction is undertaken before the process is repeated on the next limb. This process is repeated for all four limbs and the trunk and face. Interspersed at regular intervals through the session should be relaxed diaphragmatic breathing with the emphasis on the expiratory phase.

Commands

Tighten (a muscle) — hold — let go
Breathe deeply in — hold — let the air sigh out of your mouth.

The speed of progression through this technique depends on the patient. It may be necessary to repeat each of the stages many times before relaxation is obtained and it may take many weeks before the whole body is involved.

The physiological or Mitchell's technique (Hollis 1988, Gardiner 1981, Mitchell 1990)

The patient is required to contract the muscles antagonistic to those in tension. This moves the nearby joints out of the position of tension and simultaneously induces reciprocal relaxation in the previously tense muscle groups. After each contraction the patient should be made to concentrate on the feeling of the new, relaxed position.

It is thought that this technique utilises the mechanism of reciprocal innervation. As a muscle contracts, impulses from the muscle spindle will pass via the spinal cord to the antagonistic muscle group causing inhibition (Kandel et al 1991). For example, if the trapezius muscle is exhibiting abnormal tone, the patient will be encouraged to contract the shoulder girdle depressor muscles to cause reciprocal inhibition of trapezius.

It is usual to work proximal to distal and to start with the upper limbs followed by the lower limbs, trunk and finally the head. Relaxed diaphragmatic breathing exercises are interspersed throughout the programme. As with the previous technique, groups of muscles are worked individually after which these movements are summated before moving on to the next limb. The starting position is not important as it will change as the patient relaxes.

Commands

Move in a given direction — stop — let go
Feel the new position
Breathe deeply in — hold — let the air sigh out of your mouth.

Visualisation or the suggestion method (Hollis 1988)

The aim of this method is to encourage the patient not to think of things that worry them but to concentrate on something that is non-threatening and pleasurable. It is a form of auto hypnosis and rather similar to techniques used in certain types of meditation. Whereas the previous two techniques concentrated on reducing muscle tone with the presumption that mental relaxation would follow, this technique tries to reduce the mental activity and replace worrying thoughts with something pleasant. If the technique works, then a reduction in muscle tone should follow automatically. Ricketts and Cross (1985) suggest that this method is most successful with individuals who may be able to relax physically but find it difficult to stop thinking.

The room must be quiet and warm and the patient comfortable and well supported. It may help to cover them with a blanket. In hypnotic tones the physiotherapist guides the patient to think about pleasurable, non-threatening experiences or objects. Patient are encouraged to absorb themselves in what is being said, fully concentrating on the word picture being painted by the physiotherapist. The physiotherapist should persuade the patient to think about repetitive movements or sounds such as waves gently lapping on the shore or the gentle rustle of a breeze in the trees. The physiotherapist should speak slowly and in low tones. No harsh sounding words should be used and long vowel sounds should be drawn out.

Interspersed in this technique can be the suggestion that the patient's limbs feel heavy or that they feel to be gently floating. As before, this technique is improved by the inclusion of deep relaxed breathing exercises.

Before starting this technique it is important to explore the types of experiences that have given the patient pleasure or the nature of the object on which they wish to focus. In this way the technique will be appropriate and meaningful. Individuals who find this technique particularly helpful can learn to trigger relaxation when alone by thinking about their chosen subject. It is also helpful to concentrate on repetitive sounds or actions such as breathing.

Sedative massage (Madders 1988)

Slow, rhythmical stroking, effleurage and kneading can be very relaxing and are a valuable way of introducing the patient to the concept of relaxation. It is essential that the physiotherapist's hands are relaxed and remain so throughout the treatment. The strokes should initially be light but the depth can be increased as the patient adapts to the feel of the massage. For some people being asked to undress or being touched can be stressful and where this is the case it is possible to allow the patient to remain clothed. Once they are suitably positioned for the massage a large blanket is tucked firmly round them and stroking is applied through the material. As the aim of all relaxation therapies is to enable the patient to take control over their own body, massage should only be seen as an interim step; as soon as possible the patient should be introduced to more active methods.

EMG biofeedback (Basmajian & Wolf 1990, Elton 1993)

This method has been shown to be most successful when used in conjunction with other approaches (Elton 1993). Biofeedback gives specific information about the state of tension in individual muscle groups enabling the patient to take control of the reduction of tone in their muscles with confidence. There is instant feedback of success and the impartiality of the machine's response is often seen as reassuring.

The electrodes are attached over muscle groups which are known to be in tension and the patient is placed in a variety of starting positions and encouraged to reduce the electrical activity in the muscles. Once sufficient self-awareness has been developed, the patient is able to reduce inappropriate muscle tone without the help of the machine. Nigl and Fischer-Williams (1980) reported several cases of low back pain associated with abnormal muscle activity which were greatly improved by using EMG to teach specific muscle relaxation. Where there is a generalised response to stress, the electrodes are often placed over the frontalis muscle and patients can be taught to control their stress using input from this muscle.

Low frequency breathing exercises (Leuner 1991, Madders 1988)

Low frequency diaphragmatic breathing exercises induce relaxation and can be used alone or in conjunction with other techniques. In order to breath diaphragmatically, the abdominal muscles must relax to let the diaphragm descend. In the expiratory phase a sighing action is encouraged so that the air is expired through elastic recoil of soft tissue whilst the respiratory muscles and muscles of the throat and mouth relax. The frequency of respiration should be low in order to avoid hyperventilation. As the patient is resting whilst being asked to undertake deep breathing, it is necessary to reduce the respiratory rate to less than the normal resting rate in order not to disrupt the normal blood chemistry. Eight or ten breaths a minute is often successful and pauses may be incorporated both at the end of the inspiratory and the expiratory phases. This technique results in a reduction in muscle tone as well as a reduction in blood pressure, heart rate and electrical activity in the brain (Leuner 1991). The mental relaxation which most patients feel from this method results from both the physiological changes and the concentration needed to maintain a low frequency respiratory pattern. This is a repetitive, non-stressful activity which leads to mental relaxation in much the same way as some meditative techniques.

Box 9.3 Case study 2

Mrs V. was an asthmatic patient with serious social problems. Her husband was in prison and she was trying to bring up three teenaged children in a small flat on the thirteenth floor of an ageing tower block. She was unable to work and had to cope on a very limited income. She was deeply concerned that her children should not follow the example of her husband and turn to crime, but she found it very hard to exert control over them. A major problem was that the lift in the block of flats was frequently out of action and her asthma made it impossible for her to tackle the stairs. Consequently, she felt a prisoner in her home and her children were aware that she had no control over them once they were out of the flat. She was experiencing frequent, serious asthma attacks which were resulting in repeated periods of hospitalisation.

It was decided to try low frequency breathing exercises and general relaxation with this patient and the results were quite dramatic. The incidence of asthma attacks serious enough to warrant hospitalisation dropped to approximately one per year and with the improvement in her health she felt slightly more able to cope with her life in general.

Pendular exercises (Hollis 1988)

Relaxed swinging of a limb is a repetitive rhythmical movement which probably causes relaxation through two mechanisms. Firstly, it raises the threshold of transmission of impulses from sensory receptors in the moving joint and surrounding soft tissues, leading to a reciprocal inhibition of the afferent impulses from the same spinal segment and a reduction in local muscle tone. Secondly, the repetitive nature of the movement tends to reduce mental tension. Any starting position which allows a limb to swing with a minimal amount of muscular effort can be used. Pendular exercises in suspension are particularly effective, but whatever position is chosen, the exercise should start with small range movements and never move beyond what is comfortable.

Rhythmical passive movements

These are similar to pendular exercises in that the limb is subjected to gentle repetitive movements until the threshold of sensory receptors is raised. In this technique the patient should attempt to relax whilst the physiotherapist moves the limb. It is important to undertake this technique with constant velocity and range of movement of the joints as any variation in these factors will disrupt the rhythmicity of the procedure.

Physical activity

Several authors have shown that physical activity reduces stress levels. It has also been shown that when an individual develops a high level of physical fitness their response to stress is improved (Wilfley & Kunce 1986, Roth & Holmes 1987, Brandon & Loftin 1991). It does not appear that the form of the physical activity is crucial to the technique and swimming, running, ball sports and fast walking have all been shown to be successful.

Although it is possible to teach relaxation, the causes leading to excessive tension should not be ignored and the patient should be encouraged to seek solutions for the original cause of stress. This is likely to involve other members of the health care team or the social or welfare services. While the original cause of stress continues, the ability to move with an economy of effort and therefore less fatigue will be reduced.

REFERENCES

Astrand P O, Rodahl K 1987 Textbook of work physiology, 3rd edn. McGraw-Hill, New York

Baker G H 1987 Psychological factors and immunity. Journal of Psychosomatic Research 31: 1–10

Basmajian J V, Wolf S L 1990 Therapeutic exercise, 5th edn. Williams and Wilkins, Baltimore

Brandon J E, Loftin M J 1991 The role of fitness in mediating stress: a correlational exploration of stress reactivity. Perceptual and Motor Skills 73: 1171–1180

Crist D A, Rickard H C 1993 A fair comparison of progressive and imaginal relaxation. Perceptual and Motor Skills 76: 691–700

Durham R C, Allan T 1993 Psychological treatment of generalised anxiety disorder: a review of the clinical significance of results in outcome studies since 1980. British Journal of Psychiatry 163: 19–26

Elton D 1993 Combined use of hypnosis and EMG biofeedback in the treatment of stress-induced conditions. Stress Medicine 9: 25–35

Fox S I 1996 Human Physiology, 5th edn. W C Brown, Boston

Gardiner M D 1981 The principles of exercise therapy, 4th edn. Bell & Hyman, London

Hollis M 1988 Practical exercise therapy, 3rd edn. Blackwell Scientific Publications, Oxford

Jacobsen E 1929 Progressive relaxation. University of Chicago Press, Chicago

Kandel E R, Schwartz J H, Jessell T M 1991 Principles of neural science, 3rd edn. Elsevier Science Publishing, New York

Lehrer P, Carr R, Sargunaraj D, Woolfolk R L 1994 Stress management techniques: are they all equivalent, or do they have specific effects? Biofeedback and Self Regulation 19: 353–401

Leuner H 1991 Ein neuer weg sur tiefenentspannung: das respiratorisch feedback. Kranken Gymnastik 43: 246–253

Madders J 1988 Stress and relaxation. Macdonald & Co, London

Mitchell L 1990 Simple relaxation: the physiological method for relieving tension. John Murray, London

Moffett D F, Moffett S B, Schauf C L 1993 Human physiology and foundation frontiers. Mosby Year Book, St Louis

Nigl A J, Fischer-Williams W 1980 Treatment of low back strain with electromyographic biofeedback and relaxation training. Psychosomatics 21: 495–499

Ricketts E, Cross E 1985 The Whitchurch method of stress management by relaxation exercises. Physiotherapy 71: 262–264
Roth D L, Holmes D S 1987 Influence of aerobic exercise training and relaxation training on physical and psychological health following stressful life events. Psychosomatic Medicine 49: 357–365
Selye H 1984 The stress of life. McGraw-Hill, New York
Wilfley D, Kunce J 1986 Differential physical and psychological effects of exercise. Journal of Counselling Psychology 33: 337–342

CHAPTER CONTENTS

10

Function of the lower limb

M. Trew

OBJECTIVES

When you have completed this chapter you should be able to:

1. **Discuss the importance of walking**
2. **Identify the characteristics of normal gait**
3. **Describe the gait cycle**
4. **Recognise the joint movements and muscle activity that occur in normal gait**
5. **Recognise the normal patterns of movement for ascending and descending stairs**
6. **Recognise the normal patterns of movement for rising from a chair.**

INTRODUCTION

The main functions of the lower limbs are to support the body when standing and to enable locomotion. To be able to stand and move is essential to leading a normal, active life, and it is important for physiotherapists to have a detailed understanding of the function of the lower limbs in order to plan purposeful rehabilitation programmes. In this chapter three major functions of the lower limbs are explored: walking, stair climbing and getting out of a chair. The similarities in the patterns of movement of these activities are obvious and will lead to an understanding of other, less common, functions of the lower limbs.

WALKING

Walking is the most common means of moving about and is an essential part of daily life. Other methods of locomotion include running and, less commonly, crawling, hopping and jumping. All these methods of locomotion have common patterns of movement and by studying walking it becomes easier to understand the rest.

Walking can be defined as a rhythmical, reciprocal movement of the lower limbs where one foot is always in contact with the floor. People normally walk for a purpose, perhaps because they want to reach a certain place at a certain time, although walking for the sheer pleasure of the activity is also common. Increasingly, walking is being advocated as a safe and effective way of maintaining fitness, particularly in the later years of life (Arakawa 1993, Hardman & Hudson 1994), and in many countries walking for pleasure and health is a popular pastime. Providing walking speeds of more than 6 km/hr are maintained and hills are incorporated into the route, it is possible to maintain reasonable lower limb joint range and a functional level of cardiorespiratory fitness.

Walking requires the simultaneous involvement of all lower limb joints in a complex pattern of movement. In addition, there is movement of the joints of the vertebral column, including the cervical spine and, when unrestricted, the upper limbs swing in a reciprocal pattern. This involvement of all the body segments requires considerable neural control which explains why, at birth when the nervous system is not fully developed, walking is not possible. It is not until the infant has gained control over all parts of their body and is able to balance that the first uncertain steps can be taken. Even then the child is unable to walk while carrying objects and it is some years before the activity becomes mature and fully automatic.

In the health care professions it is quite common for the term *gait* to be used in preference to walking. 'Gait' means the manner or way in which walking takes place and implies a detailed consideration of the way in which the joints and muscles are involved. Gait re-education requires careful assessment of how the patient is deviating from the normal patterns of joint movement and muscle activity. Precise objectives then need to be set to achieve as near normal sequence of movement as possible. Normal gait is remarkably efficient in terms of energy and, when considering gait re-education, it is important to strive to imitate a normal pattern of movement. The more gait deviates from normal, the more energy will be required and the less the patient will be able to achieve.

Basically, all people walk in the same way, with the lower limbs moving reciprocally to provide support and propulsion. Throughout the whole of the body those joint movements which occur in the sagittal plane are very similar between individuals (Murray et al 1964). If the upper limbs are unencumbered, they demonstrate a stereotyped pattern of reciprocal movement in phase with the lower limbs. The differences in gait between one person and another occur mainly in movements in the coronal and transverse planes, for example the foot angle can vary dramatically between individuals as can the amount of trunk lateral flexion. The range of head and trunk rotation in the transverse plane also varies greatly between individuals, with some people having almost imperceptible rotation and others rotating through such a wide range that they appear to be swaggering (Murray et al 1964, Smidt 1990).

Box 10.1 Task 1

It is important to develop good observational abilities
Next time you are in a suitable place, watch the way in which people walk and see if you can work out how the gait of one individual differs from another. In particular you should observe the degree of knee extension in the stance phase and the rotation and lateral flexion of the trunk. Are the differences in gait between individuals simply a consequence of changes in the velocity of walking?

Walking is a smooth, highly coordinated, rhythmical movement by which the body moves step by step in the required direction. The forces that cause this movement are a combination of muscle activity to accelerate or decelerate the body segments and the effects of gravity and momentum.

Walking has often been described as a fall followed by a reflex recovery of balance and, to a certain extent, this is true. To initiate the first step the anterior muscles at the ankle contract to move the centre of gravity forwards in relation to the feet and this causes a loss of balance anteriorly. Once the centre of gravity has been displaced, gravity and momentum continue the movement initiated by the muscles. In order to avoid a fall there is a reflex stepping reaction which causes one of the lower limbs to be moved forwards. This alters the base so that the centre of gravity is once again above the feet. If walking is to continue, the centre of gravity must again be displaced anteriorly, though on this second step the propulsion comes from a small contraction of the calf muscles.

Very little muscle activity is required to cause this second forward displacement because the momentum acquired on the first step carries the body forwards. Balance is lost again and a second reflex step occurs. This continues until the purpose of walking has been achieved and, because the momentum of walking removes the need for much muscle activity, the whole process is energy efficient. Once a steady walking cadence has been reached, the actual process of walking requires low energy expenditure and it is possible to walk for long periods of time with surprisingly little fatigue (Smidt 1990). There are two instances in walking on level ground when energy expenditure is high. Energy expenditure is relatively high on the initiation of movement when inertia has to be overcome so that the body weight can be displaced forwards. Conversely, at the end of walking an increase in energy expenditure is required to stop forward movement of the limbs and trunk.

The energy efficient nature of walking is lost if the utilisation of momentum is restricted. Re-education programmes should aim to develop the rhythm of gait and the smooth swinging motion of the upper and lower limbs.

Box 10.2 Task 2

Analyse the effects of using a walking frame
Consider the gait of a patient using a quadruped walking frame. Is the smooth, rhythmical, reciprocal movement of normal gait preserved? Will using a walking frame conserve or increase energy expenditure? How do you think the energy expenditure needed for walking with a frame would compare with the energy needed for walking, fully weight bearing, with crutches?

Terminology of gait

Walking is a complex activity and, if it is to be fully understood, it needs to be broken down into named parts. Some of the terminology of gait relates to the period of time during which events take place and some refers to the position or distances covered by the limbs. When analysing gait it is essential to consider both the temporal and spatial components because disease or trauma can affect either. These components are illustrated in Figures 10.1 and 10.2.

Gait cycle. This is the period of time during which a complete sequence of events takes place. This could be from the moment the right foot strikes the floor until the right foot strikes the floor again. It is usual to consider the gait cycle as beginning when the heel of one foot strikes the floor (initial foot-floor contact), but it may be measured from any moment in the gait cycle. The gait cycle is subdivided into a stance phase and a swing phase, which describe the periods of time when the foot is either in contact with the floor or swinging forward in preparation for the next step.

Swing phase. This is the period of time when the limb under consideration is not in contact with the floor.

Stance phase. This is the period of time when the limb under consideration is in contact with the floor. In walking there is always a period of time when both feet are in contact with the floor, this is called 'double stance'.

At an average speed of walking, the stance phase takes about 60% of the time of the gait cycle and the swing phase about 40% (Murray 1967). In slow walking the stance phase can constitute more than 70% of the gait cycle with the swing phase being less than 30%. As the velocity of walking increases, the length of time in the stance phase decreases until on very fast walking the stance phase may be reduced below 57% of the cycle (Smidt 1990). The period of double stance also decreases with increasing velocity. When walking very slowly the double stance may last as long as 46% of the total gait cycle, yet on very fast walking the double stance period may be reduced to 14%.

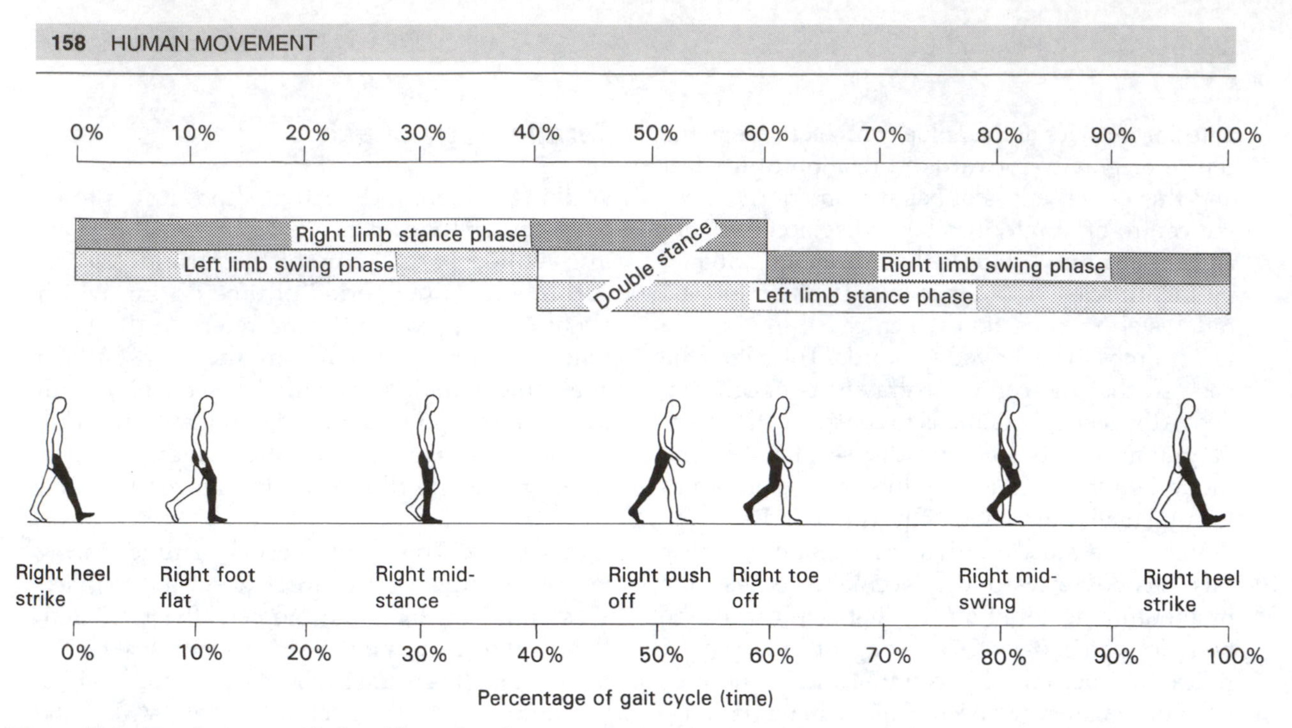

Figure 10.1 Terminology and timing of the gait cycle.

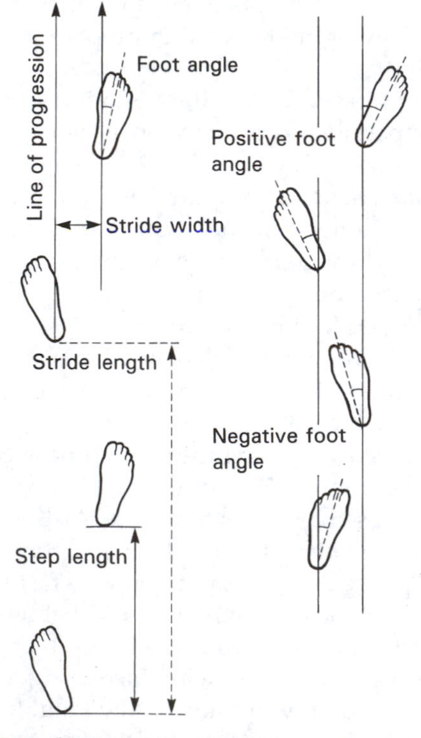

Figure 10.2 Characteristics of gait that can be measured from footprints.

Stance phase

The stance phase provides the stability of the gait and is necessary if an accurate swing phase is to take place. It can be subdivided into the following stages.

Heel strike. In normal walking this is the first moment of foot-floor contact for the leading limb. At the moment of heel strike the following limb is also in contact with the floor, giving a position of double stance. This is the moment when the whole body centre of gravity is at its lowest and the walker is at their most stable.

Foot flat. At foot flat the whole foot comes into contact with the floor and is followed rapidly by the mid-stance phase. During heel strike and foot flat, strategies to absorb the impact of foot-floor contact are normally seen.

Mid-stance. In mid-stance the body is carried forward over the stance limb and the opposite limb is in the swing phase. The whole-body centre of gravity passes from behind to in front of the stance foot during this phase and it is in mid-stance that the centre of gravity rises to its highest position in relation to the supporting surface. This is the position when the walker is at their least stable.

Push off. This stage can be further divided and it is important to understand the difference between these substages. Initially, there is 'heel off', usually followed by a propulsive stage that is called 'push off' which leads to the moment of 'toe off' when propulsion ends and the swing phase starts.

Swing phase

During the swing phase, the swinging limb moves in front of the stance limb so that forward progression may take place. This phase can be divided into three stages.

Acceleration. The force generated by the hip flexors, and to a lesser extent by the plantar flexors, accelerates the non-weight bearing limb forwards.

Mid-swing. This corresponds with mid-stance and at this moment the swing phase limb passes the stance limb.

Deceleration. In this final stage of the swing phase, the lower limb muscles work to decelerate the swing limb in preparation for heel strike. The muscle action in this phase is usually eccentric and requires less energy than those times in the gait cycle when concentric activity is needed to accelerate a limb.

When considering pathological gait, a knowledge of the step rate is important and the term 'cadence' is used to indicate the number of steps taken per minute. The cadence mainly depends on the velocity of walking. In slow walking the cadence may be 40–50 steps per minute, whereas moderate walking will cause an increase in cadence to around 110 steps per minute and this figure will rise with increasing velocity until running occurs. If a patient has pain, joint stiffness, muscle weakness or poor balance, the cadence will be reduced. An increase in natural cadence is often taken to indicate an improvement in a patient's walking ability, but it might be more appropriate to use the patient's ability to vary cadence as an indicator of walking skill.

There are a number of spatial components which it is useful to measure as part of the analysis of gait (Fig. 10.2).

Stride length. This is the distance between successive foot-floor contact with the same foot. This might be the distance between the first point of contact of the right heel on the floor and the second point of contact of the right heel.

Step length. This is the distance between successive foot-floor contact with opposite feet. In this case, it could be the distance between the point of right heel strike to the point of left heel strike. There are two steps to every stride.

Step and stride length are dependent on several factors including the length of the lower limb, the age of the subject and the velocity of walking. Short lower limb length, increasing age and decreasing velocity will all reduce the step and stride length.

Foot angle. This is the degree of in-turning or out-turning of the foot: if the foot turns in, there is said to be a negative foot angle; if the foot turns out, the angle is positive. The majority of the population walk with a positive foot angle of up to 30°. This angle is mainly associated with the degree of rotation at the hip joint and to a lesser extent the rotation between the tibia and the femur. In some cases tibial or femoral torsion will influence foot angle.

Stride or step width. This is the distance between the two feet and it is normally measured from the mid-point of the heels. This distance varies between individuals but on average it is about 7 cm; however, when a person has poor balance they tend to increase their stride width to give themselves a greater base of support. It is interesting to note that on slow walking the stride width tends to be greater than on rapid walking.

Joint and muscle activity in the stance phase

As shown in Figure 10.3, the pattern of joint movement is less complicated at the hip joint than the knee or ankle joints. Whilst the hip joint has only one phase of extension and one phase of flexion, the other two joints have two phases of each movement in each gait cycle.

Muscle activity, as indicated by electromyography (EMG), shows variability between subjects and also when different walking velocities

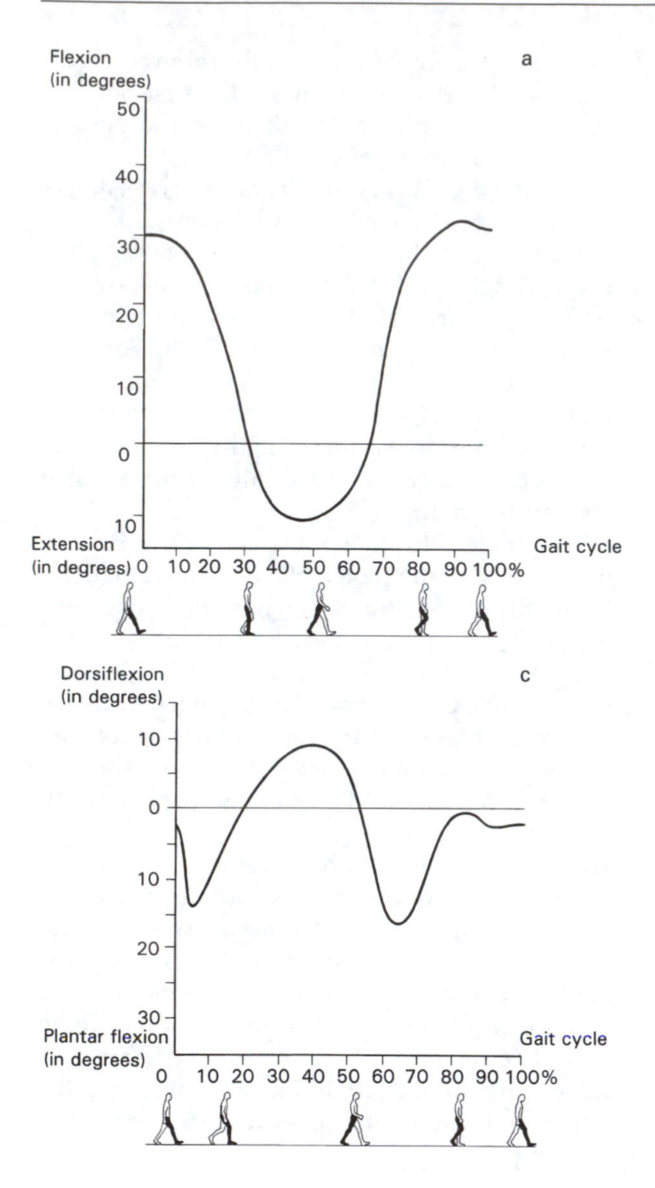

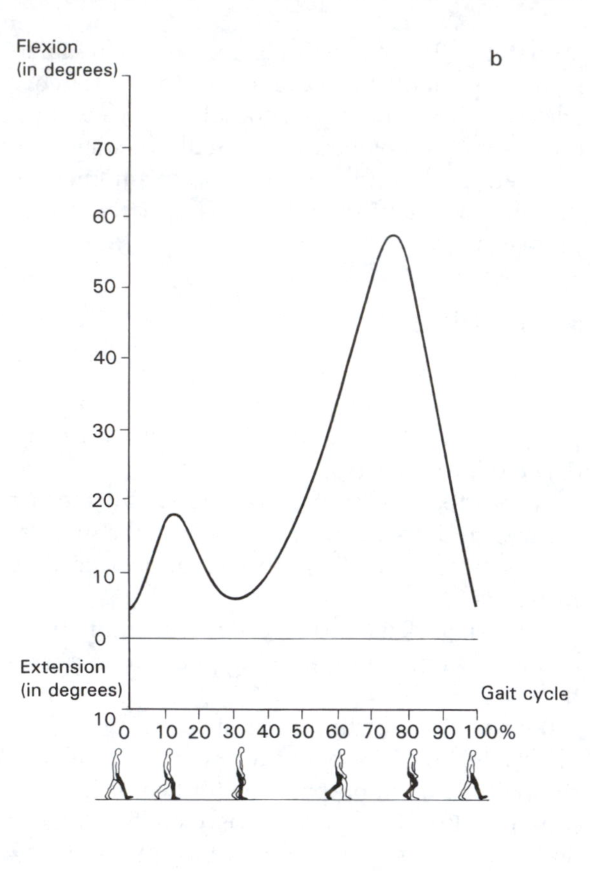

Figure 10.3 Pattern of hip, knee and ankle joint movement during a single gait cycle (walking velocity — 3.5 km/hr). a) hip joint; b) knee joint; c) ankle joint.

are chosen. A guide to the common patterns of major muscle activity is given in Figure 10.4; the data for this was gathered from subjects walking at a moderate pace. It is interesting to note that in substantial portions of the gait cycle there is little or no muscle activity occurring in the majority of the muscle groups. This explains why gait is energy efficient.

Heel strike. At the instant of heel strike the hip joint is partially flexed and gluteus maximus and the hamstrings contract immediately to initiate hip extension. The knee joint will either be in full extension or flexed to about 5° and the quadriceps will be working eccentrically to control the knee flexion which follows immediately after heel strike. At the ankle joint there is almost full range dorsiflexion so that heel strike is possible. This position is produced prior to heel strike by concentric action of the dorsiflexor muscles which will then change to eccentric activity to lower the forefoot to the floor. At the metatarsophalangeal joints there is a similar pattern of

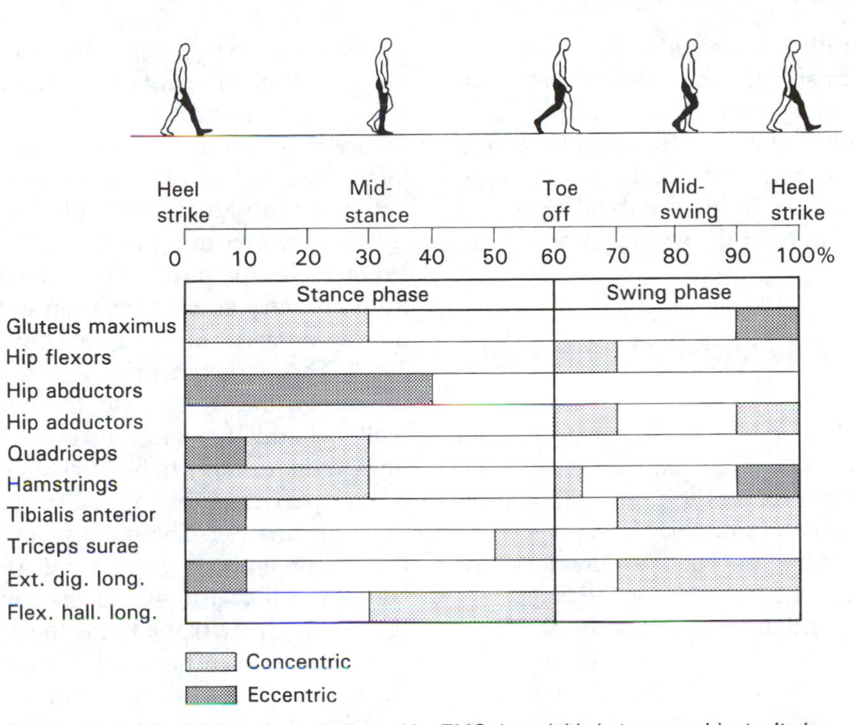

Figure 10.4 Muscle activity, as indicated by EMG, is variable between subjects. It also varies with velocity and the faster the velocity, the more muscle input will be required. This figure shows the type and duration of muscle activity which might be expected in moderate velocity walking.

activity with the muscles positioning the joints in extension ready for heel strike and then lowering the toes into floor contact.

Foot flat. The hip joint is beginning to move into extension by concentric action of the hip extensors, but the knee joint has flexed further in order to cushion the effect of heel strike and also to reduce the vertical displacement of the centre of gravity that would otherwise occur as the body passes over the stance limb. This knee flexion is controlled by eccentric work of the quadriceps group; measurements taken in our laboratory show it can often be as great as 30°. At the ankle joint there is controlled plantar flexion to lower the foot to the floor which is undertaken by eccentric work of the dorsiflexors. Without this controlled plantar flexion the foot would 'slap' down uncomfortably. As the foot achieves good foot-floor contact there is a small amount of eversion to transfer the body weight across from the lateral border of the foot towards the great toe.

Mid-stance. Hip extension continues but by now it is being produced by momentum so the muscles are no longer active. While the contralateral limb is in the swing phase, the pelvis is unsupported on that side and the hip abductors on the stance limb initially contract to control pelvic levels and lower the pelvis towards the swing side by eccentric muscle action. The knee joint remains in slight flexion. There is sometimes a minor burst of activity in the ankle dorsiflexors to pull the tibia forwards over the foot, but once this movement has been initiated, momentum and gravity will take over and, depending on the velocity of the movement, the calf muscles may need to exert a slowing influence through eccentric muscle action.

Heel off, leading to push off. At the beginning of this phase, the centre of gravity is in front of the stance foot so the force of gravity will increase the range of hip extension and ankle dorsiflexion. As full range dorsiflexion is reached, the heel will rise off the floor and the plantar flexors will contract concentrically to provide the propulsive component of push off. This contraction is not usually very large as momentum contributes to moving the body forwards. In normal walking the hip extensors are inactive at this stage and, in fact, the hip and knee joints are usually starting to flex in preparation for the swing phase (Inman et al 1981).

Joint and muscle activity in the swing phase

Acceleration. Minor forces generated on push off by the hip flexors and the plantar flexors accelerate the limb forwards in the swing phase assisted by momentum and gravity. The hip and knee joints are both flexing and there is dorsiflexion to ensure that the toes do not catch on the floor.

Mid-swing. In mid-swing, flexion of the knee and hip joints continues to keep the foot sufficiently raised above the floor for the toes not to drag.

Deceleration. The hip continues to flex but the hamstrings start to work eccentrically to slow down the movement at the hip joint. The knee joint moves from flexion to extension; it is interesting to note that the quadriceps play no part in this movement. The whole of the lower limb is being moved forwards by flexion of the hip and the resulting momentum causes knee joint extension. Towards the end of the deceleration phase, knee joint extension may have to be slowed down and this is achieved by eccentric action of the hamstrings. In preparation for heel strike, the dorsiflexors contract quite strongly to ensure that the foot is in the optimum position for heel strike.

Movement in the trunk, shoulder girdle and upper limbs

It is possible to walk with little movement of the trunk and no movement in the upper limbs, but in these circumstances gait is awkward and tiring (Murray et al 1967). Flexion and extension of the hip joints causes some rotation in the lumbar spine and, in order to keep the head facing forwards, the thoracic and cervical spine rotate in the opposite direction. Reciprocal movements of the upper and lower limbs occur with the right upper limb flexing at the shoulder joint simultaneous with flexion at the left hip joint. Normally the shoulder joint starts to flex or extend slightly before the same movement is seen in the elbow joint. The range of movement varies greatly between individuals and in the same individual it will vary according to the velocity of walking. Whereas the patterns of movement in the lower limb are very similar between individuals, they are more varied in the upper limb (Murray et al 1967). This is not surprising as the movement of the upper limb is not essential for the process of walking and the range of movement of upper limb joints is likely to be affected by the momentum imparted by the lower limbs and the degree of trunk rotation. There are several reasons why the upper limb moves during gait. It has been suggested that arm swing may impart momentum through the trunk to the lower limbs; subjective reports from fatigued walkers indicate that they have reduced the effort of walking by deliberately increasing their arm swing. It is also possible that arm swing acts to correct over rotation at the lumbar spine (Murray et al 1967).

Any vertical or lateral displacement of the centre of gravity will reduce the economy of walking. In the double stance phase, the lower limbs are quite far apart and this will cause a lowering of the centre of gravity. In the mid-stance phase, the centre of gravity will be displaced vertically but, by having a semiflexed knee joint during this phase, the amount of vertical displacement is reduced.

RUNNING

The main way in which the movement pattern of running differs from walking is in the absence of double stance and its replacement by a period when there is no foot-floor contact at all. In general terms, running and walking are very similar, though in running the movement is much quicker and the stride is lengthened. The force of push off is greater and heel strike may be replaced by toe strike. The trunk remains upright, although the line of gravity falls further outside the base than in walking. Movement of the upper limbs becomes essential and the elbows are more flexed than in walking.

Box 10.3 Clinical considerations 1

Failure of one or more muscle groups to work can result in quite dramatic gait abnormalities. Weakness of the hip abductor muscles is quite common and will make it difficult for the patient to keep their pelvis level during the stance phase. Under these circumstances the pelvis may drop to the unsupported side producing a 'Trendelenberg gait'. Such a gait is often quite uncomfortable for the patient; to avoid discomfort, they may laterally flex their trunk towards the stance side. This shifts their centre of gravity over the stance limb and the pelvis no longer drops painfully to the unsupported side. A very obvious lateral movement of the upper trunk is apparent when observing this sort of gait.

Paralysis of the dorsiflexors will lead to a high stepping gait. Dorsiflexion is normally needed to ensure that the toes will clear the floor in the swing phase and dorsiflexion is also needed to facilitate heel strike. When patients cannot dorsiflex, they have to increase the range of hip and knee flexion in the swing phase in order to avoid dragging their toes on the ground. Under these circumstances, heel strike is not possible and the patient's toes hit the ground first with the heel slapping down immediately afterwards. If you look at the shoes of a patient who has paralysis of the dorsiflexors, you will find that the toe area on the affected side is excessively scuffed and worn.

WALKING BACKWARDS

Although it is uncommon for anyone to walk backwards for any great distance, it is necessary in everyday life to be able to take one or two steps in that direction. The pattern of joint movement is very similar to normal gait but the step length is reduced and heel strike is replaced by toe strike (Vilensky et al 1987).

WALKING UP AND DOWN STAIRS

This is a modified walking activity using similar patterns of joint movement and muscle action. There is a stance phase, a swing phase and a period of double support (Fig. 10.5). The range of hip and knee joint movement is greater than in walking and there is considerable vertical translation of the centre of gravity (Andriacchi et al 1980, McFadyen & Winter 1988) making it an activity which requires high energy levels. Because stairs vary greatly in height, the range of movement and the vertical translation of the centre of gravity will vary according to the height of stair tread.

Stance phase on ascent

This phase is sometimes referred to as 'pull up' and it starts from the moment of foot contact on the step above. It normally requires a longer stance period than flat walking. Weight is initially taken on the anterior and middle third of the foot and then transferred to the remainder of the foot in readiness for full weight bearing.

On weight acceptance there is strong concentric contraction of the hip and knee extensors to extend the lead limb and raise the body up to and over the step. Gastrocnemius and soleus also work during 'pull up', moving the tibia posteriorly on the talus. As the single support phase is entered, the hip abductors on the stance limb work strongly to prevent the pelvis dropping to the unsupported side and to pull the trunk laterally over the supporting limb.

In the latter part of the stance phase, when body weight is fully on the stance limb and the knee extended, the quadriceps work isometrically to maintain joint position as the centre of mass passes in front of the stance foot. In some subjects the dorsiflexors undergo a low magnitude contraction at this stage to facilitate the movement of the centre of gravity forwards.

In the final stages of the stance phase there is plantar flexion produced by strong contraction of gastrocnemius and soleus to push the body forwards and upwards onto the new weight bearing limb. At this stage there is minimal activity in the knee and hip extensors (Andriacchi et al 1980, McFadyen & Winter 1988).

Swing phase on ascent

The swing limb must swing past the intermediate step and over the top step on which its stance will occur before the foot can be placed on that step. For this to occur there has to be a flexion of the whole lower limb involving concentric work of the hip and knee flexors and the dorsiflexors.

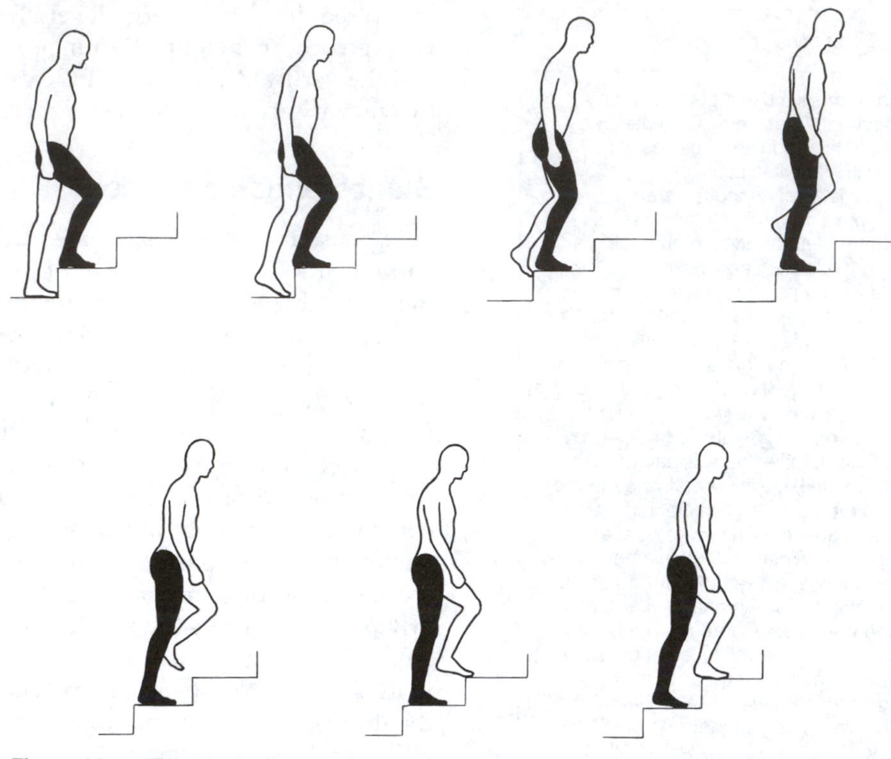

Figure 10.5 The pattern of movement in stair climbing.

In early swing the hip joint flexes and the hamstrings flex the knee joint to pull the leg and foot posteriorly to achieve intermediate step clearance.

By mid-swing the knee flexors are no longer contracting because the hip joint is sufficiently flexed to ensure intermediate step clearance. At this stage there may be some eccentric work of the quadriceps to control unwanted knee flexion.

In the later swing phase the hamstrings contract again to increase knee flexion so that the foot clears the top step, where it will eventually be placed. In order to avoid the foot catching on the top step, the amount of hip and knee flexion is quite extensive and this results in the foot being well above the step immediately before foot step contact. To gain step contact the foot has to be lowered onto the step and this is achieved by slight hip extension controlled by eccentric activity of the hip flexors.

Through most of the swing phase, tibialis anterior works isometrically to hold the ankle joint in dorsiflexion so that the toe will not stub on the steps. Immediately before foot contact the dorsiflexors work eccentrically to lower the forefoot onto the step ready for weight acceptance on the forefoot (Andriacchi et al 1980, McFadyen & Winter 1988).

Stance phase on descent

The patterns of movement which occur when going downstairs are illustrated in Figure 10.6 and for convenience they are subdivided into the weight acceptance phase and the lowering phase.

On weight acceptance, the initial foot contact is made with the anterior and lateral border of the foot. The ankle joint moves from the initial step-contact position of plantar flexion into a neutral or dorsiflexed position controlled by eccentric work of the calf muscles. The hip joints are in very slight flexion and the knee joint may flex up to 50° to cushion the instant of foot-step contact. This is controlled by eccentric work of the hip

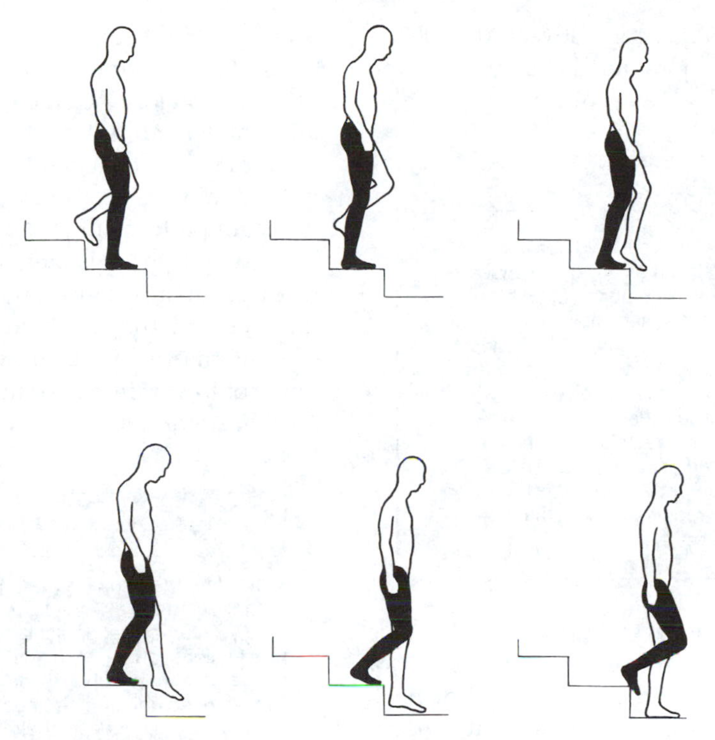

Figure 10.6 The pattern of movement in stair descent.

and knee extensors. The quadriceps then contract concentrically to extend the knee about 10° whilst the trunk moves horizontally to carry the centre of gravity over the stance limb. Tibialis anterior co-contracts with the calf muscles to control ankle position and to maintain weight bearing on the lateral border of the foot.

To lower the body weight (mid-stance) to the next step entails controlled hip and knee joint flexion and ankle joint dorsiflexion. This mainly involves eccentric action of the quadriceps and, to a lesser degree, the calf muscles and hip extensors. The stance ankle is in maximum dorsiflexion with the body weight tending to force the movement further. To prevent over dorsiflexion at the joint, the plantar flexors may need to contract. Throughout this phase the hip abductors on the stance side maintain the level of the pelvis and pull the trunk over the stance limb.

Swing phase on descent

In the swing phase, the limb has to be raised off the higher step and swung forwards and downwards clearing the intermediate step until it is in position to take weight at the start of the next cycle. The hip and knee flexors work concentrically to raise the foot off the top step and pull the limb forwards. Then the limb starts to extend ready for foot placement with eccentric work of the hip flexors controlling hip extension and the hamstrings working eccentrically to decelerate the extension of the knee joint. The ankle joint drops into plantar flexion, controlled by eccentric work of the anterior tibial muscles which also maintain the foot in inversion in preparation for weight to be taken on the lateral border of the foot. The hip abductors contract just before the end of the swing phase in preparation for maintaining pelvic levels on weight acceptance (McFadyen & Winter 1988).

The amount of joint range needed for ascent and descent of stairs depends on the depth of tread, but for standard sized tread (16.5 cm) the hip joint must be able to move between full extension and about 60° of flexion, the range required at the knee joint is between 0°–100° of flexion and the ankle joint needs full dorsiflexion (Tata et al 1983, McFadyen et al 1988).

Box 10.4 Clinical considerations 2

You will be aware from your own personal experience that stair climbing requires much more energy than walking on the flat. The vertical component of stair climbing causes the main problem, but if you are to get to the top of the stairs you have no option but to go up. In contrast, walking on the flat involves components specifically designed to avoid or reduce any vertical movement. The energy expenditure required of stair descent is less because the movement is in the direction of gravity and mainly requires control through eccentric muscle activity.

In addition to muscle strength, good balance ability is also needed for ascent and descent of stairs. When going up stairs the period of peak muscle torque and greatest instability occur simultaneously at the start of the swing phase. The hip and knee joints of the stance limb are in considerable flexion and substantial effort is needed from the extensor muscles to raise the body; at the same time effort must also be directed to the maintenance of balance (Tata et al 1983).

When these facts are taken into consideration, it is no wonder that elderly and frail people find stair climbing difficult. To be able to use stairs safely it is necessary to have a wide range of movement at hip, knee and ankle joints, muscles capable of generating considerable force through a wide range and a good sense of balance. Rehabilitation programmes for patients who have difficulties with stairs should always include activities which will increase joint range and the torque generating capacity of muscle and improve balance.

MOVING FROM SITTING TO STANDING

The ability to rise from sitting to standing is essential for the achievement of many everyday activities, yet interestingly this activity has not been studied in the same depth as walking and contradictory evidence has been produced.

For the majority of the population, getting out of a chair is an automatic activity requiring no thought unless the chair is particularly low or the individual is feeling tired or weak. As with gait, the main patterns of joint movement are usually stereotyped, though there can be significant differences, mainly caused by the initial position of the feet and the height of the chair. It is self-evident that a low chair will require a greater range of lower limb joint movement and a higher level of energy.

Box 10.5 Task 3

It is important to be aware of the different ways in which people get out of chairs.

Watch people, particularly in relaxed situations, and note the different ways in which they rise from a chair. Work out what abilities they need (joint range, muscle function, balance) in order to stand up in both a 'conventional' and a 'non-conventional' manner.

Are you able to notice any differences in the way older and younger people rise from chairs?

Whether the upper limbs are involved in the process of getting out of a chair depends on the strength of the individual, the height of the chair and whether there are armrests. Under normal circumstances the situation is similar to walking in that the upper limbs are not essential to the activity and can be used for carrying or manipulating objects during the activity of standing up or sitting down. However, if the individual is feeling weak or tired, they will choose to use their arms to push themselves up out of the chair, thus reducing the load on the hip and knee extensor muscles.

When considering the movement of sitting to standing in detail it is best to divide it into a seated phase and a stance phase. In the seated phase, the subject prepares for standing by adjusting the position of the limbs and trunk so that the centre of gravity moves forward until it is almost over the feet.

In the stance phase, the weight is taken

through the lower limbs and the centre of gravity is transferred forwards and upwards. This phase can be further subdivided. Initially, the centre of gravity continues to transfer forwards until it is slightly anterior to the lateral malleoli, then there is an extension phase as the subject rises from the chair and achieves an erect posture (Fig. 10.7).

Both the sitting and the stance phases occur whether or not the chair arms are used. The period of time needed to complete both phases is very variable but takes, on average, one to three seconds (Kerr et al 1991). The sitting phase consists of about 30% of the total movement time and the transfer and extension phases take approximately 20% and 50% respectively.

Seated phase

Usually when a person is sitting in a relaxed manner they lean against the back of the chair and their centre of gravity is well behind their feet. If the centre of gravity remains behind the feet, it is not possible to rise. The ideal preparatory position for standing up is with the knees flexed to at least 90° so that the feet lie directly under the knee joints or with the knees flexed up to 115° placing the feet more posterior (Ikeda et al 1991). This position of the lower limbs reduces the horizontal distance that the centre of gravity has to travel by the end of the transfer stage and thus reduces the energy expenditure. If the feet are not in this position, the initial movement usually consists of knee joint flexion to place the feet, at the very least, under the knee joints. Movement of the feet can occur either prior to movement of the trunk or simultaneously.

The trunk needs to angle forwards in order to reposition the centre of gravity about 2 cm anterior to the ankle joints. This occurs by hip flexion and initially there is a minor burst of activity in the hip flexors in order to initiate the movement. Once that has been achieved, the flexors stop acting and the force of gravity takes over the movement. Towards the end of the hip flexion phase there may be slight eccentric work of the hip extensors to control the forward movement

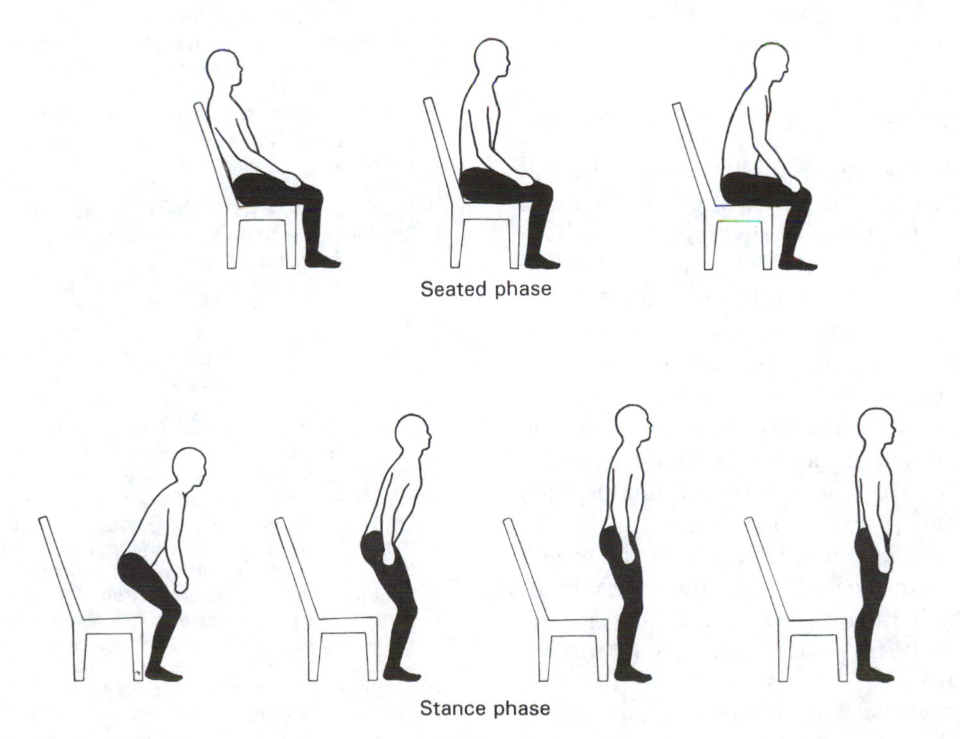

Figure 10.7 The pattern of movement when rising from a chair.

(Kelly et al 1976, Ikeda et al 1991). Knee flexion may increase slightly at this time and there is some dorsiflexion.

Very little motion is seen in the joints of the vertebral column except for the cervical spine, which extends to keep the vertex of the skull uppermost. If the upper limbs are not being used to push up from the chair arms, they usually flex between 11° to 53° at the shoulder joint. This combination of trunk and shoulder movement serves to move the centre of gravity forwards and also provides horizontal momentum which will contribute to the transfer phase and be translated into a vertical movement in the extension phase (Riley et al 1991).

The seated phase ends with the lift off from the chair. There is considerable variation in the degree flexion at the hip and shoulder joints during this phase and also in the exact position of the head. The velocity of the movement also varies between individuals and under different circumstances. Possible explanations for such variety include the height of the chair relative to the lower limb length of the subject, the age of the subject and the degree of their joint mobility and muscle strength (Kerr et al 1991).

Transfer phase

The transfer phase begins on lift off from the seat and continues until the centre of gravity is about 7 cm anterior to the ankle joints, where it will remain until the erect posture is achieved (Kelly et al 1976, Ikeda et al 1991). The transfer phase is short and is completed before much hip and knee joint extension has occurred. There is peak quadriceps activity at the instant of lift off when the knees extend slightly to raise the body off the seat and this is sometimes accompanied by slight hip extension. Dorsiflexion of the ankle joints reaches its maximum during this phase.

The trunk and upper limbs continue the movement begun in the seated phase and the momentum developed in the seated phase carries the centre of gravity forwards until it is anterior to the ankle joints.

The transfer phase is relatively short taking approximately 20% of the total stand up time but it is vital to the whole process as standing is not possible without transference of the centre of gravity over the feet. It is a time of peak muscle activity in the quadriceps and the hip extensors (gluteus maximus and the hamstrings) and also a time of instability because the body is no longer supported by the seat and the centre of gravity is moving forwards. Failure to complete the transfer phase is often responsible for patients being unable to stand up from a chair.

Box 10.6 Clinical considerations 3

Elderly and disabled people who have a limited range of joint movement and a significant degree of muscle atrophy frequently find getting up out of a chair a problem. This is partially because of the large range of lower limb movement required but also because a large muscle force is needed to overcome inertia and move the body in a direction opposed by the force of gravity. In addition, the extensor muscles of the lower limb have to generate peak torque when the hip and knee joints are in more than 90° of flexion; in this joint position the muscles are working in outer to middle range and are not at their strongest. A standard seat has a height of between 43 cm and 46 cm. To rise from such a chair the subject needs to have a range of knee flexion which is greater than 90°.

Raising the chair height, providing armrests to push on and keeping the feet under the knees if not posterior to them all facilitate the process of rising from sitting to standing (Kelly et al 1976, Kerr et al 1991). As the height of the seat increases, the maximum angle at the hip, knee and ankle joint decreases.

The force the lower limb muscles need to generate decreases as the height of the seat increases and the chair arms are used. The largest decrease in force would appear to be in the knee extensor muscles, though some decrease around the hip has been noted. Interestingly, no decrease in ankle moment is seen. Rising from a chair which has a motorised lift on the seat or is spring assisted requires significantly less muscle force until the subject's body loses contact with the seat. At this moment there is a sudden increase in force which has been found to be greater than any force generated by rising smoothly from a normal chair (Kerr et al 1991). Frail patients find the sudden requirements to generate high levels of muscle force very difficult and it is for this reason that spring or motor assisted chair seats are not as successful as might have been expected.

Extension phase

Once the centre of gravity is over the feet, horizontal movement is replaced by vertical movement. Extension of the hips and knees begins in earnest and there is also slight plantar flexion of the ankle joints. As the joints become progressively extended, the trunk also starts to extend and the cervical spine flexes slightly to keep the vertex of the skull uppermost. The upper limbs return, by eccentric muscle work, to their normal resting position.

Rising from a chair is an important functional activity which requires greater joint range and muscle torque than walking and most stair climbing (Kelly et al 1976). To be successful, it is necessary to have more than 90° of flexion at the hip and knee joints and virtually full range dorsiflexion. It is also necessary to have good balance because the centre of gravity is moving but the base of support is not. If the relationship of the centre of gravity to the feet is not correct, the individual is likely to lose their balance and fall. Flexibility of the vertebral column is particularly important because balance requires correct head positioning and, if the vertebral column is stiff, then the appropriate positions may not be achieved.

The individual must also have the strength and the balance ability to complete the transfer phase, which is arguably the most crucial part of the whole process.

REFERENCES

Andriacchi T, Andersson G, Fermier R, Stern D, Galante J 1980 A study of lower limb mechanics during stair climbing. Journal of Bone and Joint Surgery 62A: 749–757

Arakawa K 1993 Hypertension and exercise. Clinical & Experimental Hypertension 15: 1171–1179

Hardman A E, Hudson A 1994 Brisk walking and serum lipid and lipoprotein variables in previously sedentary women; effect of 12 weeks of regular brisk walking followed by 12 weeks of detraining. British Journal of Sports Medicine 28: 261–266

Ikeda E, Schenkman M, Riley P, Hodge W 1991 Influence of age on dynamics of rising from a chair. Physical Therapy 71: 473–481

Inman V, Ralston H, Todd F 1981 Human walking. Williams and Wilkins, Baltimore

Kelly D, Dainis A, Wood G 1976 Mechanics and muscular dynamics of rising from a seated position. In: Komi P (ed) Biomechanics. V B International Series on Biomechanics, University Park Press, Baltimore

Kerr K, White J, Mollan R, Baird H 1991 Rising from a chair: a review of the literature. Physiotherapy 77: 15–19

McFadyen B, Winter D 1988 An integrated biomechanical analysis of normal stair ascent and descent. Journal of Biomechanics 21: 733–744

Murray M P 1967 Gait as a total pattern of movement. American Journal of Physical Medicine 46: 290–333

Murray M P, Drought A B, Kory R C 1964 Walking patterns of normal men. Journal of Bone and Joint Surgery 46A: 335–359

Murray M P, Sepic S B, Barnard E J 1967 Patterns of sagittal rotation of the upper limbs in walking. Journal of the American Physical Therapy Association 47: 272–284

Riley P, Schenkman M, Mann R 1991 Mechanics of a constrained chair rise. Journal of Biomechanics 24: 77–85

Smidt G L 1990 Gait in rehabilitation. Churchill Livingstone, New York

Tata J, Peat M, Grahame R, Quanbury A 1983 The normal peak of electromyographic activity of the quadriceps femoris muscle in the stair cycle. Anatomischer Anzeiger (Jena) 153: 175–188

Vilensky J A, Ganiewicz E, Gehlsen G 1987 A kinematic comparison of backward and forward walking in humans. Journal of Human Movement Studies 13: 29–50

CHAPTER CONTENTS

11

Function of the spine

A. P. Moore N. J. Petty

OBJECTIVES

When you have completed this chapter you should be able to:

1. **Describe how the spine functions to give support to the body during movement and weight bearing activities**
2. **Discuss how the spine is able to give protection to soft tissues and vital organs during weight bearing and physiological movements**
3. **Describe how the spine gives attachment for the muscles of the abdomen, thorax and upper and lower limbs**
4. **Explain how the spine allows movement of the human body**
5. **Explain how the spine contributes to the enhancement of movement of the upper and lower extremities**
6. **Describe how the spine is able to act as a shock absorber**
7. **Describe how the vertebral column gives shape to the human body in static and dynamic postures**
8. **Describe how the spine contributes to changes from static to dynamic postures.**

INTRODUCTION

The spine is a complex, multisegmented structure which has many functions. It will be seen from this chapter that the spine is essential for weight bearing, protection and movement of the human body. A knowledge of the relationship of structure and function of the spine is of great importance to the clinician. When writing this chapter, the reader has been assumed to have a knowledge of the basic structure of the spine.

The spine as a whole functions in a variety of ways. It gives support to the head, upper limbs and thoracic cage during movement and weight bearing activities. It gives protection to the vital organs, such as the heart, lungs and to soft tissues such as the spinal cord, during physiological movements and weight bearing activities. It provides attachment for the muscles of the abdomen and thorax and for some muscles of the upper and lower limbs. It allows movement to occur throughout its length and enhances movement of the upper and lower extremities. It enhances the visual and hearing fields. In addition, the spine gives shape to the human body in static and dynamic postures and facilitates changes from static to dynamic postures. Finally, it acts as a shock absorber. Each of these functions will now be considered in turn.

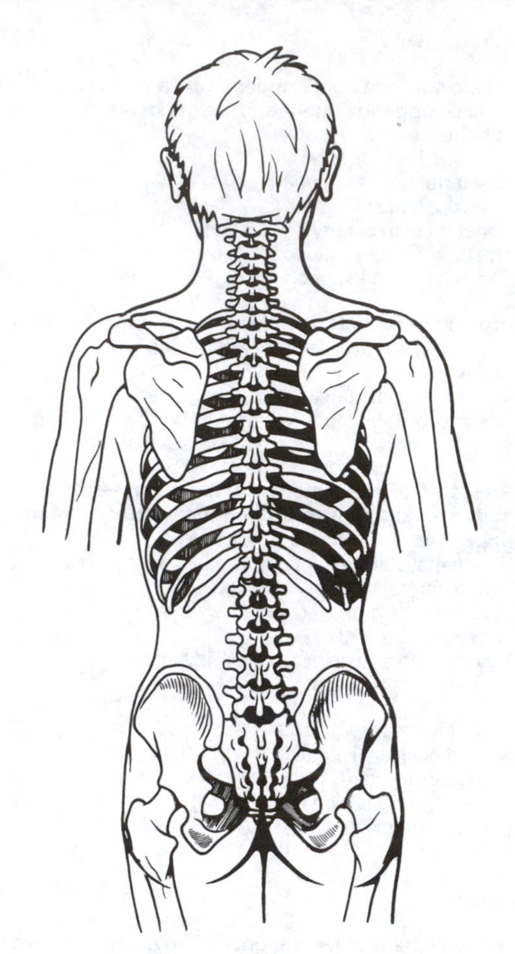

Figure 11.1 The vertebral column, posterior view (reproduced with permission from Kapandji 1974).

SUPPORT FOR THE HEAD, UPPER LIMBS AND THORACIC CAGE DURING MOVEMENT AND WEIGHT BEARING ACTIVITIES

Normal movement can only take place if adequate support is available for the head, upper limbs and thoracic cage (Fig. 11.1). Such support is offered by the vertebral column and is discussed in this section.

The spine is composed of 33 vertebrae (Fig. 11.2), most of which comprise a vertebral body and a vertebral arch (Fig. 11.3). The vertebral bodies are, in the main, separated from each other in life by an intervertebral disc. The exceptions to this are found at the atlanto-occipital joint and the atlantoaxial joint, where no such disc exists, and in the sacrum, where the five sacral vertebrae are fused and do not contribute to spinal movement. At the atlantoaxial joint there is a large range of rotation available, and the inclusion of an intervertebral disc at this level would severely limit range. It is also important that, in an area where the emerging brain stem is potentially vulnerable, this junction is well supported by ligamentous tissue. The union is, therefore, completed by the upward projecting dens fitting into the osseofibrous ring rather than through a discal union. Similarly, at the atlanto-occipital joint a large range of flexion extension movement is permitted since there is no intervertebral disc restricting range of movement. The sacral vertebrae are fused in order to contribute to a solid and stable osseous pelvic ring which has to support the forces generated in weight bearing and locomotion. The vertebral column is completed

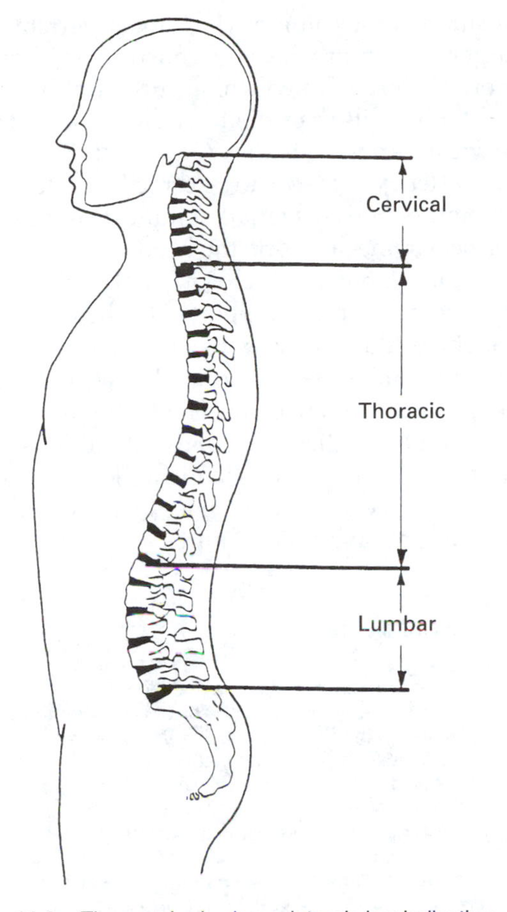

Figure 11.2 The vertebral column, lateral view indicating the spinal regions (reproduced with permission from Oliver & Middleditch 1991).

by the coccygeal region, which is composed of four vertebrae linked together by fibrous tissue: there is variable osteophytic union between either the first and second, second and third or third and fourth bones. The coccyx plays no supporting role in terms of spinal function but its movements are important in order to allow defaecation to occur.

Vertebral body support

The vertebral body consists of a cylinder of cancellous bone with trabeculae surrounded by a thin layer of cortical bone. The trabeculae act like struts strengthening the vertebral body: the vertical trabeculae resist compressive forces and horizontal trabeculae resist bowing of the struts and thus increase its strength (Bogduk & Twomey 1991). This arrangement of the trabeculae can be seen in Figure 11.4. The vertebrae are thus able to resist the compressive and torsional stresses during movements of the spine and the tension caused by contraction of muscles which attach to it. The structure resembles a cardboard box which is full of packing material which prevents its collapse under compression during everyday movement. The compact bone represents the cardboard box and the trabeculae the packaging inside.

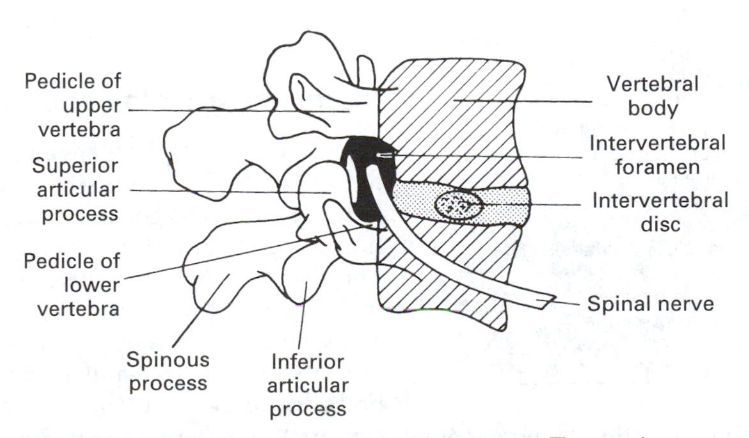

Figure 11.3 A lateral view of a spinal motion segment. The anterior weight bearing part is shaded.

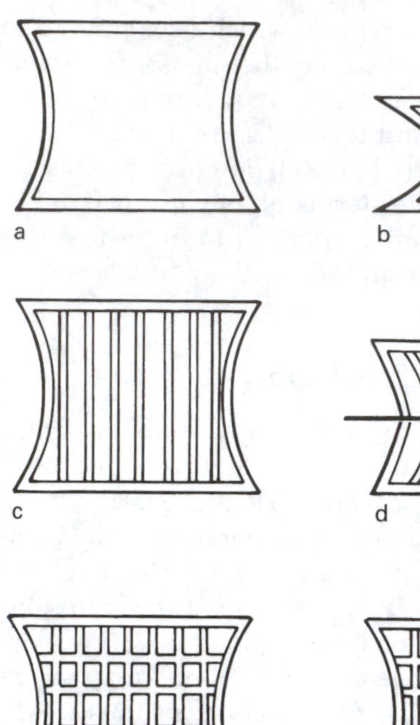

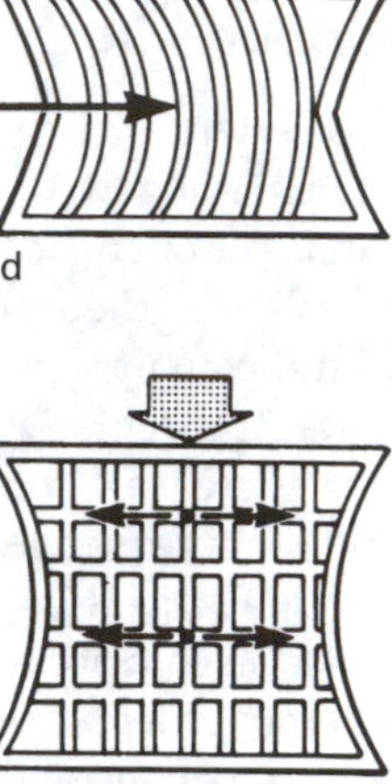

Figure 11.4 Reconstruction of the internal architecture of the vertebral body. a) with just a shell of cortical bone, a vertebral body is like a box, and collapses when a load is applied (b). c) internal vertical struts brace the box (d). e) transverse connections prevent the vertical struts from bowing, and increase the load-bearing capacity of the box. Loads are resisted by tension in the transverse connections (f) (reproduced with permission from Bogduk & Twomey 1991).

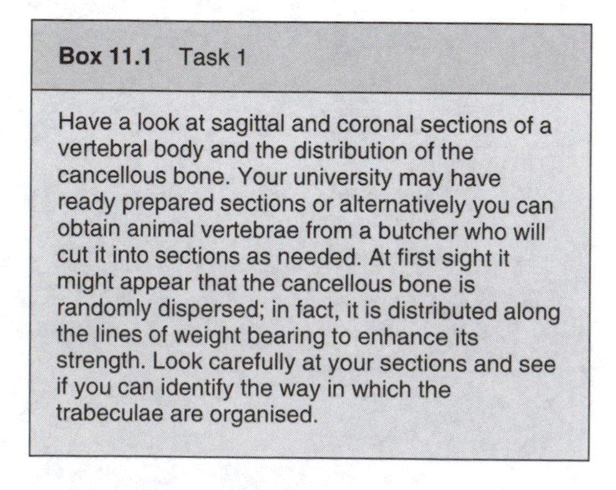

Box 11.1 Task 1

Have a look at sagittal and coronal sections of a vertebral body and the distribution of the cancellous bone. Your university may have ready prepared sections or alternatively you can obtain animal vertebrae from a butcher who will cut it into sections as needed. At first sight it might appear that the cancellous bone is randomly dispersed; in fact, it is distributed along the lines of weight bearing to enhance its strength. Look carefully at your sections and see if you can identify the way in which the trabeculae are organised.

Vertebral bodies throughout the vertebral column vary in size and shape depending on their position in the column. The lumbar vertebrae are larger and more heavily constructed than the thoracic vertebrae which, in turn, are more substantially built than the finer and more intricate cervical vertebral bodies (Fig. 11.2). The size of the vertebrae appears to be directly related to the amount of body weight that it must support. The cervical spine supports the head which accounts for approximately 10% of normal body weight, the thoracic spine supports the head and the weight of the upper limbs and thoracic organs, the lumbar spine supports the weight of the head, neck, thoracic cage and abdominal contents and functions to transmit all this body weight through the sacroiliac joints via the sacrum to the pelvis and hence to the lower limbs in static and dynamic postures.

Box 11.2 Task 2

Look at a vertebral column and identify differences in the size and shape of the vertebral bodies in the five regions. See if you can suggest reasons for the features that you can see, noticing in particular the upward projections at the lateral aspects of the upper surfaces of the cervical bodies and the reciprocal bevelled surfaces of the lateral aspects of the lower surfaces of the cervical vertebrae. The upward projections are called *uncinate processes* and articulate with the vertebrae below at what is known as the *uncovertebral joint* (joints of von Luschka). These joints influence the degree and direction of movement and are thought to have a protective function for the lateral aspects of the intervertebral joints.

Intervertebral disc support

The intervertebral discs are interspersed between the vertebrae from the second cervical vertebra to the sacrum and constitute approximately one-fifth of the total length of the vertebral column. The shape of each disc corresponds to the shape of its adjacent vertebral body. The discs both allow and restrict movement between the vertebral bodies and transmit loads from one vertebral body to the next. The discs vary in shape and size in the different regions of the spine. The discs are wedge shaped, thicker anteriorly than posteriorly in the cervical and lumbar spine, which contributes to lordosis in these regions. They are thinnest in the upper thoracic region and thickest in the lumbar region in order to bear a greater proportion of body weight. In proportion to the height of the vertebral body, the discs are thickest in the cervical region and this, in part, enables the cervical spine to have a greater range of physiological movement than in the other regions of the spine. The discs form the main connection between adjacent vertebral bodies and are held in place around the periphery by Sharpey's fibres (Jackson 1966) The discs serve to keep the vertebral bodies apart during the maintenance of static and dynamic postures and therefore are well placed to contribute to the support mechanisms provided by the vertebral column.

The disc is composed of three parts:

- end plate
- annulus fibrosus
- nucleus pulposus.

The end plate is permeable and lies between the disc and vertebral body (Fig. 11.5). It is composed of hyaline cartilage (Ghosh 1990a). Water and nutrients pass between the nucleus and the cancellous bone of the vertebral body through the end plate.

The annulus fibrosus is a ring-shaped structure composed of concentric layers (or lamellae) of collagen fibres bound together and prevented from buckling by a matrix of proteoglycan gel. Approximately 70% of the annulus is composed of water, although this amount varies according to the load on the disc and its age. The inner annulus is attached above and below to the vertebral end plate and the outer part of the annulus is attached to the periosteum and epiphyseal ring of the vertebral body and is strengthened by the anterior and posterior longitudinal ligaments.

The annular collagen fibres run parallel to each other and obliquely at between 40–70° to the horizontal, lying in opposite directions in adjacent layers giving the annulus a lattice-like appearance (Fig. 11.6). The annulus contains both Type I and Type II collagen fibres (Ghosh 1988). Type I is found in tissues that are designed to resist tensile and compressive forces, and Type II is found in tissues designed to resist compressive forces (Bogduk & Twomey 1991) The presence of Type I and II reflects the tensile and compressive loads applied to the annulus during static and dynamic postures. The lattice-like arrangement of the lamellae helps to limit movement between adjacent vertebrae. For example, during rotation half the fibres of the lamellae are

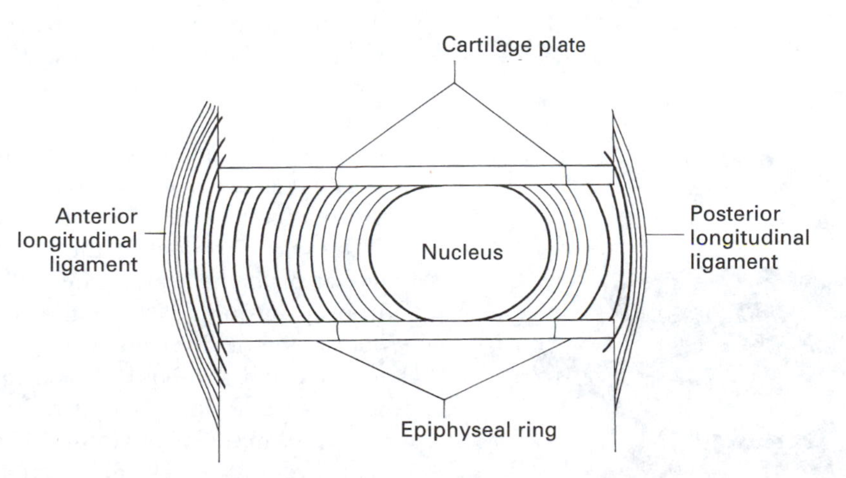

Figure 11.5 The cartilage end plate lies between the intervertebral disc and the vertebral body. The nucleus lies adjacent to the end plate. The anterior and posterior fibres of the annulus are attached to the anterior and posterior longitudinal ligaments respectively (reproduced from Macnab 1977).

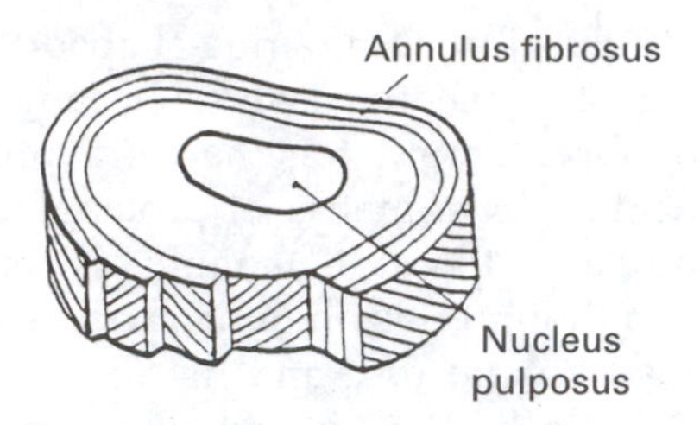

Figure 11.6 Horizontal section through a disc showing lattice-like arrangement of the annulus fibrosus (reproduced with permission from Oliver & Middleditch 1991).

stretched and the other half are relaxed (Fig. 11.7). On flexion, all the fibres posterior to the axis of movement are stretched and all the fibres anterior are relaxed. On extension, the anterior fibres are stretched and the posterior fibres are relaxed. On lateral flexion, the fibres ipsilateral to the direction of the motion are relaxed but on the contralateral side they are stretched.

The nucleus pulposus is a semifluid gel making up 40–60% of the disc and lies adjacent to the vertebral end plates of the vertebrae above and below. At its periphery the nucleus blends with the annulus in such a way that there is no distinct separation between the two components. The nucleus is made up of a loose network (not in layers like the annulus) of mainly Type II collagen fibres (to resist compression) and a proteoglycan/water gel with some elastic fibres. Approximately 70–90% of the nucleus is composed of water.

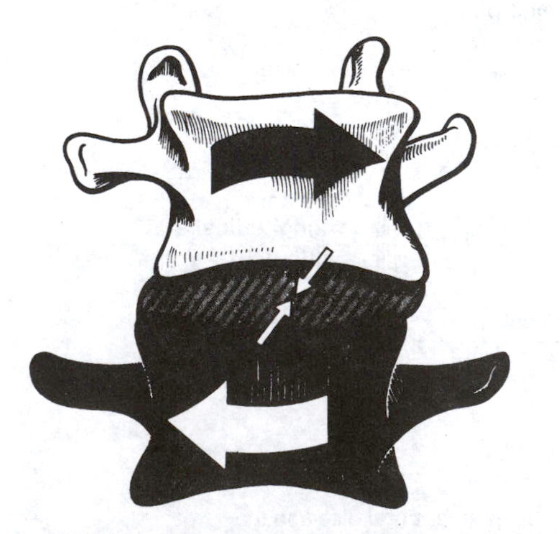

Figure 11.7 Anterior view of a spinal motion segment in the lumbar spine showing half the annular fibres stretched during rotation (reproduced with permission from Kapandji 1974).

The main property of proteoglycan gel is its water-imbibing capacity: by giving the disc a high osmotic pressure, water content is maintained even under high compressive loads generated in normal weight bearing postures and movement. The disc imbibes water and nutrients through the end plate from the vertebral body, the imbibition of water creates a hydrostatic pressure within the disc which keeps the annulus under tension and helps to keep the vertebral bodies separated. The hydrostatic pressure varies according to the level of the disc, the posture adopted, and any weights which are lifted. Nachemson (1976) measured intervertebral disc pressure in the lumbar spine during the maintenance of various postures and during various activities; the results can be seen in Figure 11.8.

A knowledge of how posture and load bearing can influence hydrostatic pressure within the disc is important to the clinician as it explains why, in some pathological states, some postures may be more painful and unachievable by the patient than others. For example, in a patient with an acute disc lesion, sitting is often very difficult to maintain for more than a minute: such a patient would prefer to lie in a weight relieving posture in order to minimise pain. Pain in this situation may be caused by a rise in hydrostatic pressure within the affected disc. Raised hydrostatic pressure may create or increase the symptoms arising from a disc lesion, the extruded disc material being brought into closer proximity with pain-sensitive structures and giving rise to local and referred pain.

The water content of the nucleus varies with age: at birth the water content is over 85% but drops to around 70% in the mature disc. The water content does not vary very much in the annulus and is about 70% throughout life. In humans, total body height reduces by approximately 19 mm (1% of stature) from the morning to the evening (Tyrell et al 1985) due to the loss of fluid from the discs on weight bearing and the imbibition of fluid in non-weight bearing positions, particularly when recumbent.

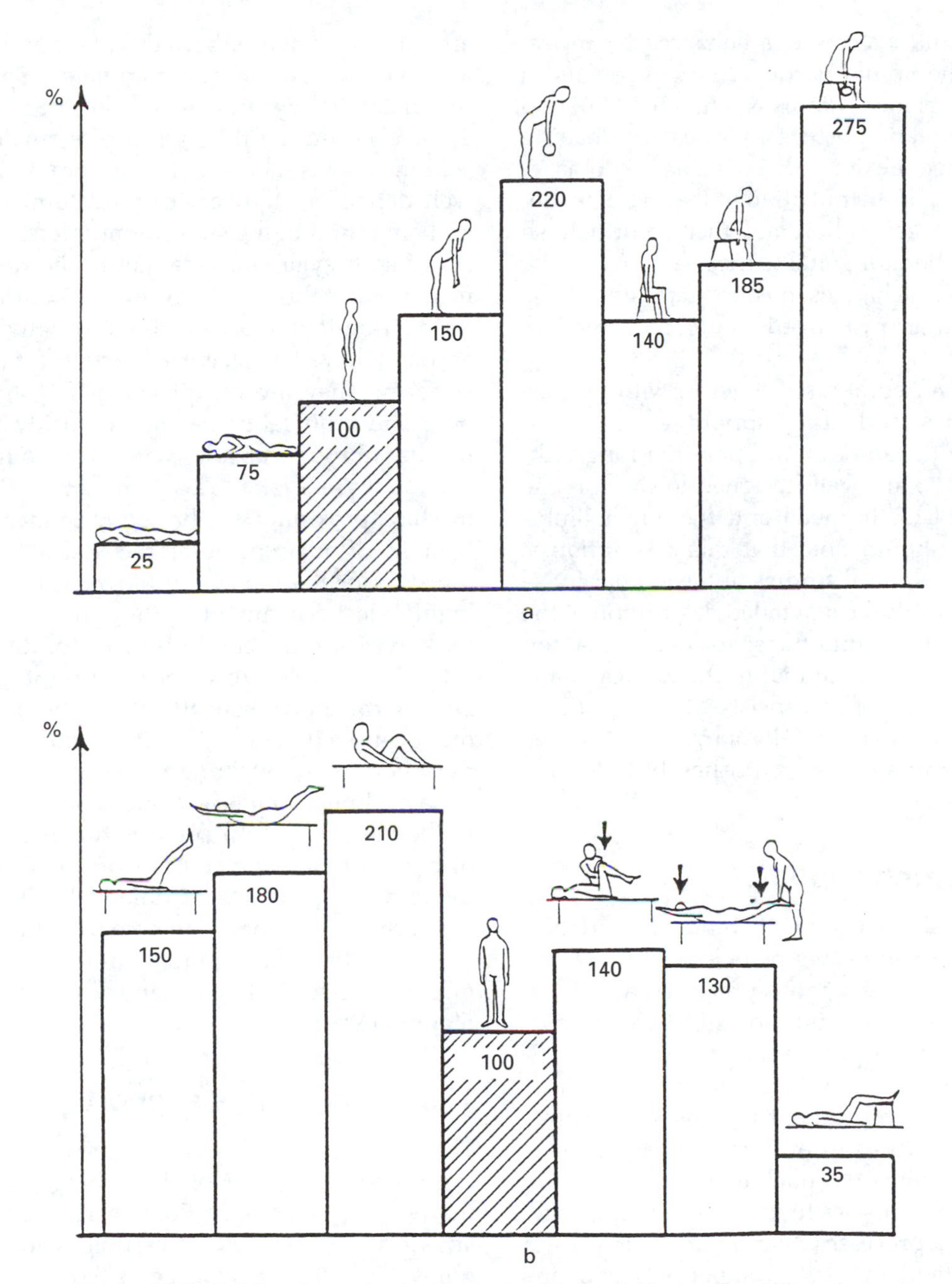

Figure 11.8 Relative change in pressure (or load) in the 3rd lumbar disc: a) in various positions; b) in various muscle strengthening exercises (reproduced by permission of Lippincott–Raven Publishers from Nachemson 1976).

Nutrition of the disc

The disc is avascular after the first decade of life so it then receives its nutrition mainly via tissue fluid exchange through the vertebral end plates and also from small blood vessels at the periphery of the annulus. The anterior annulus receives a better supply of nutrients than the posterior annulus. Fluid exchange is enhanced by movements of the spine, particularly movements in the sagittal plane (Adams & Hutton 1986), and reduced by static postures, particularly loading at end range flexion or extension (Adams & Hutton 1985). If the nutrition of the disc is inadequate, disc degeneration may then occur (Ghosh 1990b). It is therefore vital to the health of the disc that the spine changes posture throughout the day, and that prolonged static postures are avoided.

The intervertebral discs, together with the vertebral bodies and the supporting ligaments, particularly the anterior and posterior longitudinal ligaments, are well designed to support the body weight of the head and the upper limbs. The reader should note that this description is based on anatomical studies of the lumbar intervertebral disc. While a detailed description of the cervical disc is beyond the scope of this chapter, the reader should note that in the cervical spine the nucleus pulposus consists of fibrocartilage and the nucleus is not fully surrounded by the annulus fibrosus (Bland & Bushey 1990, Mercer 1995).

Vertebral arch support

The vertebral arch, lying behind the vertebral body, is made up of two pedicles and two laminae from which the spinous processes, two transverse processes, two inferior and two superior articular processes project (Fig. 11.3).

The nature, shape and direction of these processes vary in different regions of the spine. The spinous processes and transverse processes function as points of attachment for supporting ligaments and muscles to increase their leverage. The articular processes bear an articular surface called an articular facet. The superior facet of one vertebra articulates with the inferior facet of the vertebra above, forming a zygapophyseal (apophyseal or facet) joint. These zygapophyseal joints are synovial joints and therefore have a synovial membrane lying deep to the fibrous capsule. The capsules which surround the joints are fairly lax to allow movement to occur. Zygapophyseal joints are plane joints. When they are viewed in the dissected state, however, the articular surfaces are not completely flat—they undulate slightly and in life these undulations are evened out by the presence of small meniscoid inclusions. These inclusions are particularly well defined in the cervical and lumbar spines and are found in the superior and inferior recesses of each zygapophyseal joint. The role of the meniscoid inclusions is to increase surface area for distribution of loads acting through the joints and may have a protective function for articular surfaces. They are composed of fatty cartilaginous and synovial tissue and are firmly attached to the fibrous capsule which surrounds the zygapophyseal joint. The ligamentum flava lies in close proximity to the joint, connecting the laminae of adjacent vertebrae and attaching to the anterior margin of the fibrous capsule. This highly elastic ligament is thought to assist the back extensor muscles initiate restoration of the fully flexed spine to a more upright position. During this movement the elasticity of the ligament causes it to return to its normal length so that there is no buckling which could compromise the lumen of the vertebral canal. In addition, in the lumbar spine it protects the intervertebral disc by not allowing full flexion to be achieved too abruptly (Oliver & Middleditch 1991). Other supporting ligaments lie remote from these joints; they are the supraspinous, interspinous and intertransverse ligaments and can be seen in Figure 11.9.

The thoracic cage support

The thoracic cage forms a semi-rigid structure composed of 12 pairs of ribs which, apart from the two lower pairs of floating ribs, are united anteriorly by their costal cartilages to the sternum (Fig. 11.1). The cage is completed posteriorly at the junction of the ribs with the vertebrae, the upper 10 ribs articulating with the vertebral bodies and the transverse processes of each of the upper 10 thoracic vertebrae. The thoracic cage forms the basis of attachment for muscles which link and give support to the upper limbs, via the shoulder girdle.

Due to its position relative to the spine and its

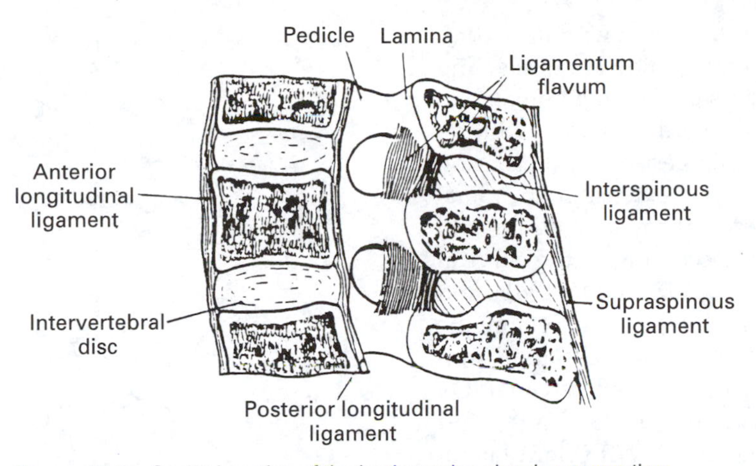

Figure 11.9 Sagittal section of the lumbar spine showing supporting ligamentous structures (reproduced with permission from Oliver & Middleditch 1991).

relative inflexibility, the thoracic cage serves to restrict the physiological movements of flexion, extension and lateral flexion of the spine.

Movement of the thoracic cage occurs during respiration. Small movements of the ribs at the costovertebral joints produce large movements anteriorly of the sternum and laterally of the rib shafts. Because of the long leverage of the rib shafts and the direction and shape of the costotransverse joint surfaces, these changes in anteroposterior and transverse diameters of the thoracic cage increase its volume, reduce intrathoracic pressure and enable inspiration to occur.

The pelvic girdle support

The pelvic girdle comprises the two pelvic bones, the sacrum and the articulations between them (Fig. 11.1). Posteriorly, the sacrum is wedged between the massive ilia of the pelvis and articulates with it via the two sacroiliac joints. The pelvic rim is completed anteriorly by the union of the two pubic bones at the symphysis pubis. The sacroiliac joints, which are synovial, are supported by some of the strongest ligaments in the body in order to maintain stability of the pelvic girdle. The symphysis pubis is a fibrous junction which is equally stable. Together, these articulations allow minimal movement during weight bearing and locomotion and maintain a solid base of support for the spine, head and upper extremities. The pelvic girdle is intimately linked with the lower extremities via the hip joints. Body weight is transmitted via the pelvis to the lower extremities, likewise impact from ground reaction forces during weight bearing and locomotion are transmitted via the lower extremities through the pelvis and on to the spine.

PROTECTION FOR SOFT TISSUE AND VITAL ORGANS DURING PHYSIOLOGICAL MOVEMENTS AND WEIGHT BEARING ACTIVITIES

Box 11.3 Task 3

List for yourself all the possible soft tissue structures which may be protected by the spine, including the structures which lie anterior to it. As you read the rest of this chapter you will be able to check the accuracy of your list.

Also have a look at a vertebral column and work out how it is adapted to fulfil a protective role in relation to soft tissue structures which pass through or lie near to it.

Vertebral canal structures

The vertebral canal serves to support and protect

the spinal cord and cauda equina with its accompanying spinal meninges, blood vessels and lymphatic drainage vessels. During movement of the spine, the spinal cord, together with its meninges, undergoes changes in length, tension and position. From spinal extension to spinal flexion there is between 5–9 cm elongation, with most of the movement occurring in the cervical and lumbar regions (Breig 1978, Louis 1981), and increase in tension (Butler l991a). On flexion, the spinal cord and meninges elongate, become thinner and move anteriorly in the spinal canal. On extension they become shorter and fatter and move posteriorly (Breig 1978). With right lateral flexion, the right hand side of the spinal cord shortens on the right and elongates on the left hand side (Breig 1978). Thus, with normal physiological movements the space taken up by the spinal cord and nerve roots in the spinal canal will vary. If pathology causes any encroachment into the spinal canal, the spinal cord or nerve roots may be compromised and this may be accentuated by certain physiological movements. It must be remembered that any blood vessels accompanying the spinal cord and its meninges or the nerve roots and their dural sleeves may also be compromised by pathology. Limb movements can also affect the neural tissue in the spinal canal, for example the straight leg raise (SLR) increases tension in the lumbar and sacral nerve roots and their meningeal covering.

The boundaries of the vertebral canal formed by the vertebral body anteriorly and the vertebral arch posteriorly are well adapted to protect the spinal cord and its meninges as they are made of compact bone which is very resistant to compressive forces.

The lumen of the vertebral canal varies in its shape in different parts of the spine depending upon the size of the neuromeningeal tissue passing through it. It is triangular and large in the cervical spine to allow for the enlarged spinal cord close to the brain stem, in the thoracic spine it is smaller and circular, and in the lumbar region the lumen widens and becomes more triangular in shape to accommodate the cauda equina (Fig. 11.10). Within the vertebral canal are small clusters of fat pads which fill in the recesses

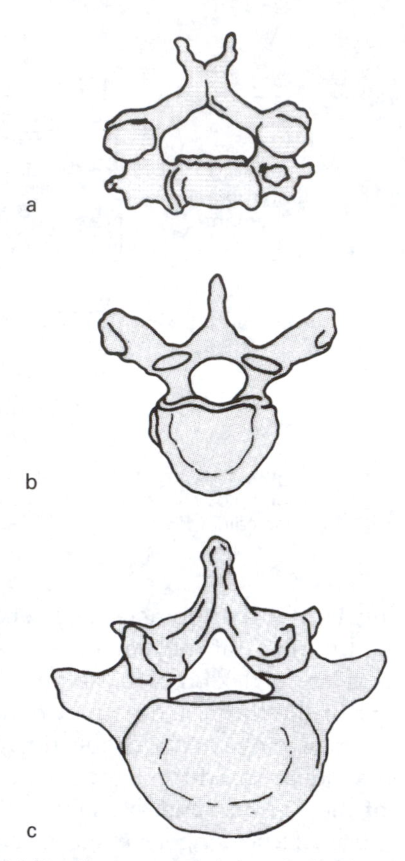

Figure 11.10 Segmental variations in the shape of the spinal canal: a) cervical; b) thoracic; c) lumbar (reproduced with permission from Butler 1991).

of the canal and act as cushions to the soft tissue structures during movements of the spine. This is a very important function during rapid spinal movement when the spinal cord moves forwards in the canal during flexion and backwards during extension and also when physiological movements are accompanied with compression, for example when jumping. The fat pads protect the sensitive neuromeningeal tissues from sudden impact.

The intervertebral foramen as a protective structure

The intervertebral foramina are gaps between the vertebrae and lie laterally (Fig. 11.3). The posterior wall of an intervertebral foramen is formed by the zygapophyseal joint, the anterior wall by the vertebral bodies and intervening disc, and the superior and inferior walls by the pedicles of the vertebrae above and below, respectively.

Within each intervertebral foramen lies the spinal nerve, sinuvertebral nerve, adipose tissue, blood and lymphatic vessels. The adipose (fatty) tissue together with the osseous fibrous ring serves to protect the structures in the intervertebral foramen during physiological movements when the diameter of the intervertebral foramen is altered. The cross-sectional area alters significantly during flexion and extension; in the lumbar spine flexion increases the area by 30% and extension decreases the area by 20%, whereas rotation and lateral flexion reduce the area (on the side to which the movement is directed) by 2–4% (Panjabi et al 1983).

Normally the spinal and sinuvertebral nerves occupy one-third to one-half of the cross-sectional area of the foramen, and in normal circumstances the alteration of the intervertebral foramen with movement does not adversely affect the enclosed tissues. However, certain individuals have transforaminal ligaments in the lumbar spine which are vestigial ligaments and have been described in detail by Golub and Silverman (1969). They span the intervertebral foramen and so reduce its vertical height and cross-sectional area. They occur quite naturally to a variable extent in some individuals but are absent in others. If they exist in association with minor pathology, the foraminal space, during physiological movements, may be significantly reduced and compression of the soft tissue structures may occur leading to clinical signs and symptoms of a large space-occupying lesion. In other words, the presence of these ligaments may give a false impression to the diagnostician in terms of the true nature and size of the lesion. Transforaminal ligaments may also be present in the thoracic and cervical spines, but the evidence for this is far from extensive at the present time.

In the thoracic cage, the vertebral column, which lies posteriorly, is perfectly positioned to offer protection to the vital organs and their protective membranes, i.e. lungs, pleura, heart and pericardium. In addition, the descending thoracic aorta is well protected from external trauma as it lies deep within the thorax on the anterior surface of the vertebral column.

In the cervical spine the transverse processes are punctuated by the foramen transversaria for the passage of the vertebral arteries. The vertebral arteries on the left and right join anterior to the brain stem to form the basilar artery and this feeds into the circle of Willis which supplies a large area of the brain (Fig. 11.11).

The foramen transversaria offer the two vertebral arteries protection from compression during physiological movements and external trauma. This protective device can, however, create some difficulty during physiological movements in the pathological state, when osteophytic growth from zygapophyseal joints can impinge on the vertebral artery impeding blood flow and producing vertebrobasilar insufficiency. The most common symptom is dizziness; other symptoms which depend on the area of the brain stem affected can include: 'drop attacks', visual disturbance, diplopia, nausea, disorientation, dysarthria, dysphagia, ataxia, impairment of trigeminal sensation, sympathoplegia, hemianaesthesia and hemiplegia (Bogduk 1994).

PROVISION OF ATTACHMENTS FOR THE MUSCLES OF THE ABDOMEN AND THORAX AND FOR SOME MUSCLES OF THE UPPER AND LOWER LIMBS

The spine via its many bony processes offers direct or indirect attachment to muscle structures which have one or more of the following functions:

- segmental stabilisation of the spine during movement and normal posture
- production of gross movement over a large number of segments
- stabilisation and physiological movement of the limbs relative to the trunk.

Segmental stabilisation of the spine during movement and normal posture

Muscles whose function relates to segmental sta-

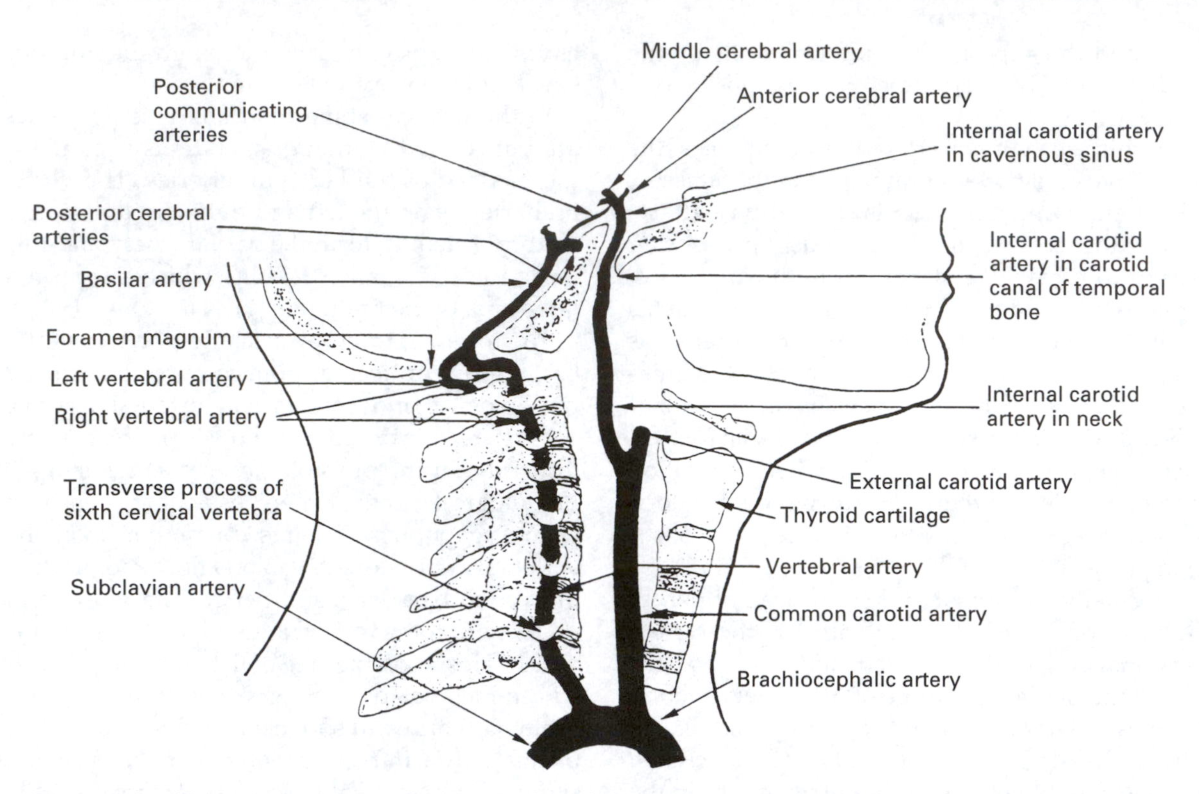

Figure 11.11 The pathway of the vertebral arteries through the cervical vertebrae to the brain stem (reproduced from Palastanga et al 1994).

bilisation and posture lie much closer to the vertebral column than muscles which produce gross movement. By virtue of the length of the vertebral processes, muscles increase their mechanical advantage because of the greater leverage that is available.

A spinal motion segment consists of two adjacent vertebrae with their intervening disc as shown previously (Fig. 11.3). Each individual motion segment is potentially unstable without its supporting ligaments and muscles. The muscles which span the motion segment are very important in stabilising adjacent vertebrae during gross movements of the spine which are produced by larger muscle groups. The deep stabilising muscles include multifidus, rotatores, interspinales and intertransversarii. Multifidus spans between one to three or four vertebrae and the latter two muscles link adjacent vertebrae. The main action of multifidus is to produce posterior sagittal rotation of the vertebrae which occurs during extension, and to control this movement during flexion (Macintosh & Bogduk 1994). During rotation of the trunk the contraction of the prime movers, the oblique abdominal muscles, would tend to produce trunk flexion. Multifidus in the lumbar spine acts with erector spinae to oppose the flexion pull of the obliques ensuring pure axial rotation (Macintosh & Bogduk 1986).

The rotatores, interspinales and intertransversarii muscles are thought to act as stabilisers. One suggestion is that they act as large proprioceptive transducers since they have been found to contain two to six times the density of muscle spindles found in the longer muscles (Bastide et al 1989, Nitz & Peck 1986, Peck et al 1984). They would thus provide feedback on spinal position and movement.

In standing, the spine is well stabilised by its joints and ligaments so that there is little back muscle activity; individuals vary, however, and there may be slight continuous activity, intermittent activity or no activity (Valencia & Munro 1985). Back muscle activity in sitting is similar to that in standing (Andersson et al 1975) but with the arms supported or with the backrest reclined there is reduced back muscle activity (Andersson et al 1974).

The degree of lumbar lordosis and the position of the pelvis are interdependent and are to some degree controlled by the surrounding muscles. Contraction of the back extensors and hip flexors will tend to increase lumbar lordosis and cause an anterior pelvic tilt; contraction of the abdominals and hip extensors, on the other hand, will produce a flattening of the lumbar lordosis and posterior pelvic tilt. The balance of contraction of these muscle groups can be influenced by pathological processes of the spine (Aspinall 1993, Cooper et al 1993, Hides et al 1994, Jull & Janda 1987) and, therefore, assessment of the muscle function is important in the examination of a patient with spinal pain.

Production of gross movement over a large number of segments

Muscles which produce this type of gross movement tend to be more remote from the vertebrae, for example it is the more superficial members of the erector spinae group which are concerned with the gross movement of the trunk. This is because these superficial members of the erector spinae group span up to five or six vertebral segments and are thus able to produce more gross segmental movements. In the upright position the trunk muscles, notably the abdominals and back extensors, initiate movements into flexion, extension and lateral flexion. Once the centre of gravity is displaced, the contralateral group will contract eccentrically to control the movement against gravity. For example, on spinal flexion, the trunk flexors will initially contract to displace the centre of gravity forwards, then the movement will be controlled by eccentric work of the back extensors which increases with increasing angles of flexion (Shultz et al 1982). It should be noted that because of the direction of the back extensor muscle fibres (being parallel to the spine), activity in these muscles causes a proportional increase in intradiscal pressure.

The abdominal muscles which lie some distance from the spine achieve the movements of physiological flexion, side flexion and rotation of the trunk in combination with other muscles of the trunk and in association with the deeper muscles of the back. None of the abdominal muscles is attached directly to the spine: however, transversus abdominis and the internal abdominal oblique muscles have attachments to the lower thoracic and lumbar spines via the thoracolumbar fascia. Together these three abdominal muscles and their fascial attachments provide a complex bracing mechanism to protect the lumbar spine during flexion movements and lifting activities. The exact mechanism of lifting still remains unclear, despite much research in this area. A detailed discussion of the various mechanisms put forward is beyond the scope of this chapter.

Stabilisation and physiological movement of the limbs relative to the trunk

The spine gives attachment for levator scapulae, serratus anterior, latissimus dorsi, trapezius and rhomboids minor and major which are all important muscles for the production of shoulder girdle movement and for stabilisation of the scapula in order to facilitate movements of the upper limb. In addition, the spine affords attachment for psoas major and piriformis, muscles which have direct influence on the lower extremity during gait.

SPINAL MOVEMENT

Functionally, the spine is considered to consist of a large number of spine motion segments which contribute to overall spinal movement (Fig. 11.3). The motion segment consists of the interbody joint which allows movement to occur under compression and the two zygapophyseal joints which are concerned with guiding the direction of the movement which takes place. We have considered these two joints earlier in the chapter. The motion segment is well developed to allow movement to occur between adjacent segments since the collagenous fibres of the annulus are compressible to a small extent, are capable of being torsioned and are also capable of being stretched longitudinally.

In addition, the nucleus pulposus acts rather like a water cushion stabilised by the surrounding annulus and adjacent vertebral segments and is capable of deforming in response to changes of both static and dynamic postures (Fig. 11.12).

The size of the intervertebral disc varies according to vertebral level. The discs are thickest in the most mobile segments of the vertebral column, i.e. in the lumbar and cervical spine, and thinnest in the thoracic spine.

The two synovial zygapophyseal joints together complete the triad motion segment. Their structure has been described earlier in this chapter.

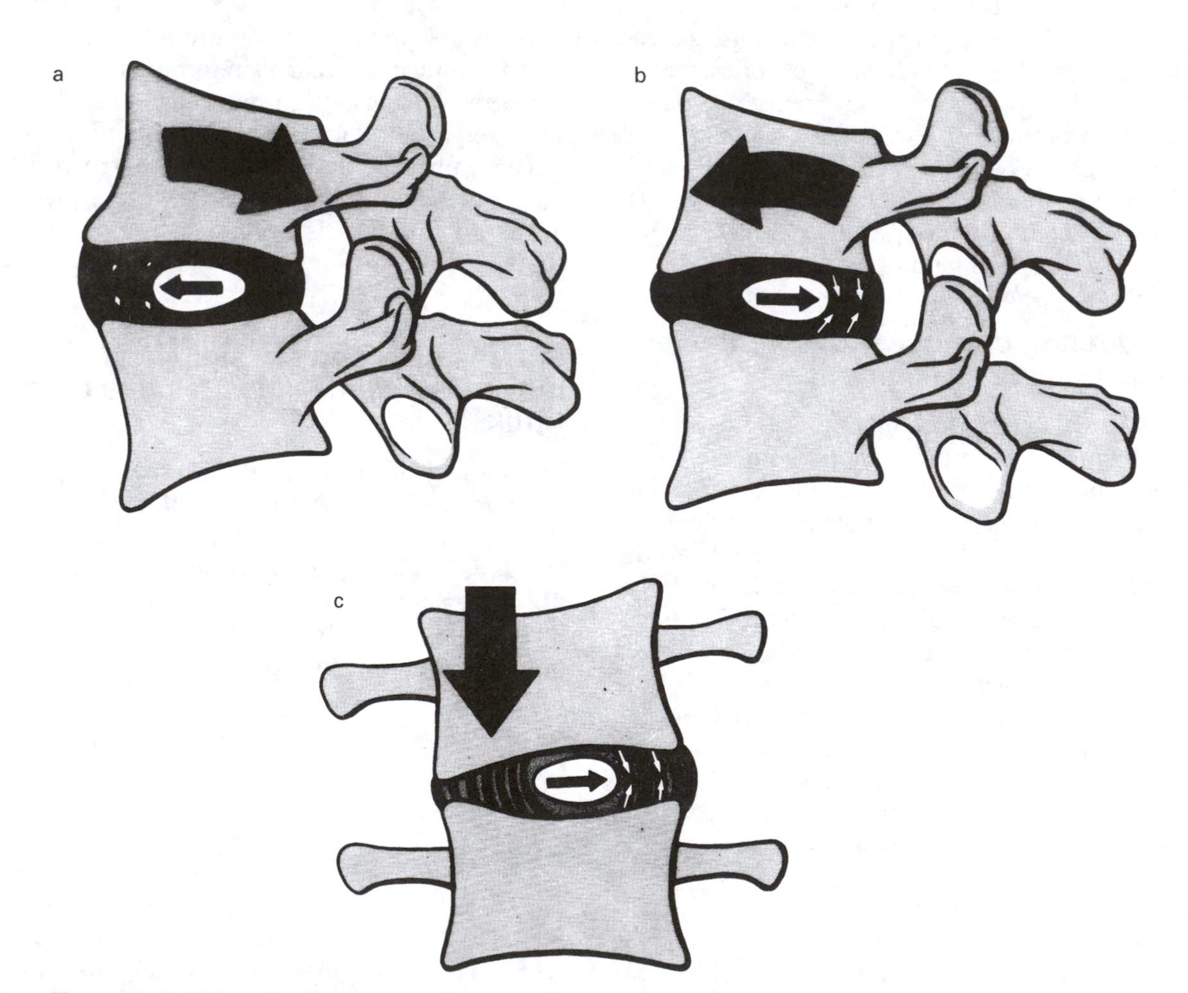

Figure 11.12 Effect of movement on deformation of the intervertebral disc: a) extension, the upper vertebra moves posteriorly and the nucleus increases the tension in the anterior part of the annulus; b) flexion, the upper vertebra moves anteriorly and the nucleus increases tension in the posterior part of the annulus; c) lateral flexion, the upper vertebra tilts towards the side of flexion and the nucleus moves in the opposite direction, increasing tension in that part of the annulus (reproduced with permission from Kapandji 1974).

Box 11.4 Task 4

Look at the vertebral column and notice the changes in the direction of the articular surfaces in each region of the spine.

It is to be noted that in the cervical region (apart from C0/1 and C1/2) the inferior articular facets face downwards and forwards at an angle of approximately 45 degrees (Fig. 11.13). The superior articular facets of the vertebra below lie in a complementary position, facing upward and backward. In the thoracic spine the zygapophyseal joints are orientated so that the inferior facets of the vertebra above face forwards and slightly medially, lying almost in a coronal plane and therefore the superior facets of the vertebra below face backwards and laterally. By contrast, the facets of the lumbar spine are curved so that the inferior articular facets face both laterally and forwards, the superior facets face medially and backwards.

As segments are viewed progressively from C2 to the sacrum it will be noted that the change in the direction of the articular facets is a gradual process. The inclination of zygapophyseal joints will affect the range and direction of the motion available at each segmental level as can be seen below (Figs 11.14 and 11.15).

The upper cervical joints, the atlanto-occipital

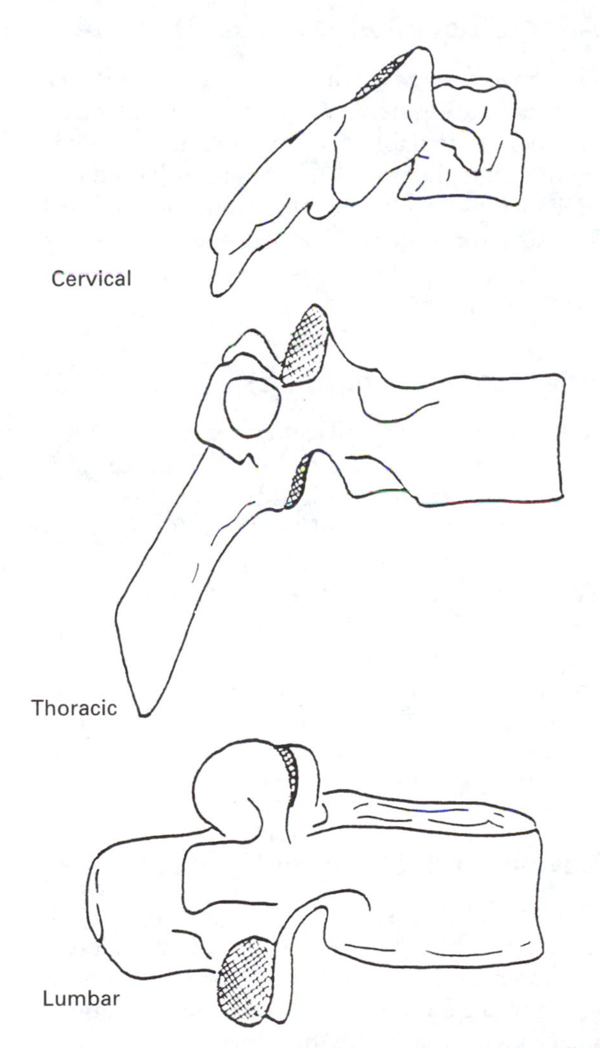

Figure 11.13 Orientation of the articular facets (shaded) in the cervical, thoracic and lumbar regions (reproduced from Palastanga et al 1994).

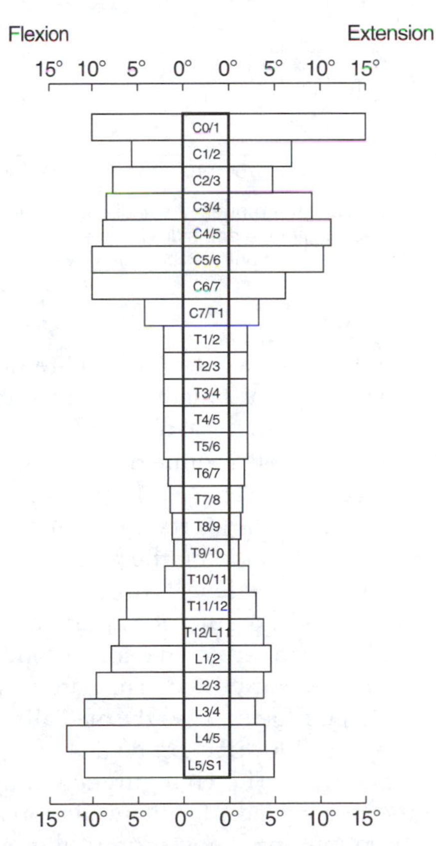

Figure 11.14 Average ranges of spinal segmental movement (flexion and extension) (reproduced with permission from Oliver & Middleditch 1991).

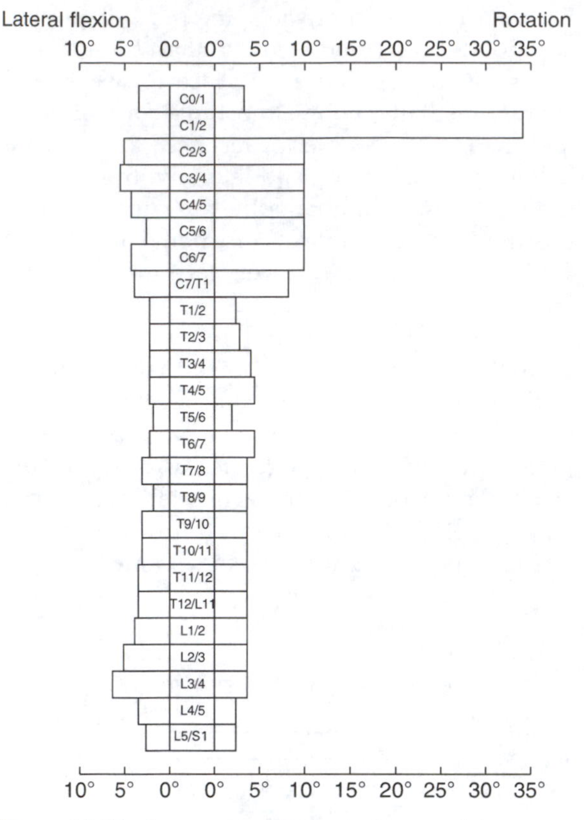

Figure 11.15 Average ranges of spinal segmental movement (values given to one side) for lateral flexion and rotation (reproduced with permission from Oliver & Middleditch 1991).

(C0/1) and atlantoaxial (C1/2), are atypical motion segments since there is no intervertebral disc present between these two junctions and the direction of the facets is quite different to the rest of the cervical spine. The C0/1 articulation is the only joint within the vertebral column which does not have a triad joint, there being only two articulations at this level which are synovial condylar joints. The superior facets of C1 are significantly expanded to enable the condyles of the occiput to be supported. They are elongated and cup shaped facing slightly medially and in the anteroposterior direction lie at 45 degrees to the sagittal plane. The configuration and direction of the joint surfaces facilitate anteroposterior sagittal rotation and translation of the occipital condyles on the superior articular facets of C1 allowing a large range of flexion and extension at this level. The C1/2 joint is formed by a pivot joint between the odontoid peg of the axis and the anterior arch of the atlas. This junction creates a very mobile joint in terms of physiological ranges of rotation as can be seen from Figure 11.15.

There are 6 degrees of freedom at each spine motion segment: sagittal rotation and translation, coronal rotation and translation, horizontal rotation and translation (Fig. 11.16). Flexion consists of anterior sagittal rotation and anterior translation, extension consists of posterior sagittal rotation and posterior translation.

Atlanto-occipital joint (C0/1)

Flexion and extension are the largest ranges available at this segment, with slight lateral flexion also being available. During flexion of the head on the neck, the occipital condyles roll on the lateral masses of C1 and also translate forwards. The atlas translates backwards and tilts upwards and posteriorly. In extension the reverse movements occur.

Atlantoaxial joint (C1/2)

Rotation is the largest range available. During rotation to the left, the right inferior facet of C1 moves forwards and slightly upwards on the superior facet of C2 and the left inferior facet of C1 moves backwards and slightly downwards. The forward and backward movements constitute rotation and the upward and downward movements constitute lateral flexion, therefore rotation is accompanied by some lateral flexion movement. Some flexion, extension and lateral flexion movements are also available at this level (Figs 11.14 and 11.15).

Lower cervical region (C2–C6)

Flexion and extension, lateral flexion and rotation are all possible at these levels. During flexion the vertebrae undergo anterior sagittal rotation and anterior translation. The intervertebral foramen increases in size during this movement.

During extension the vertebrae undergo posterior sagittal rotation and posterior translation.

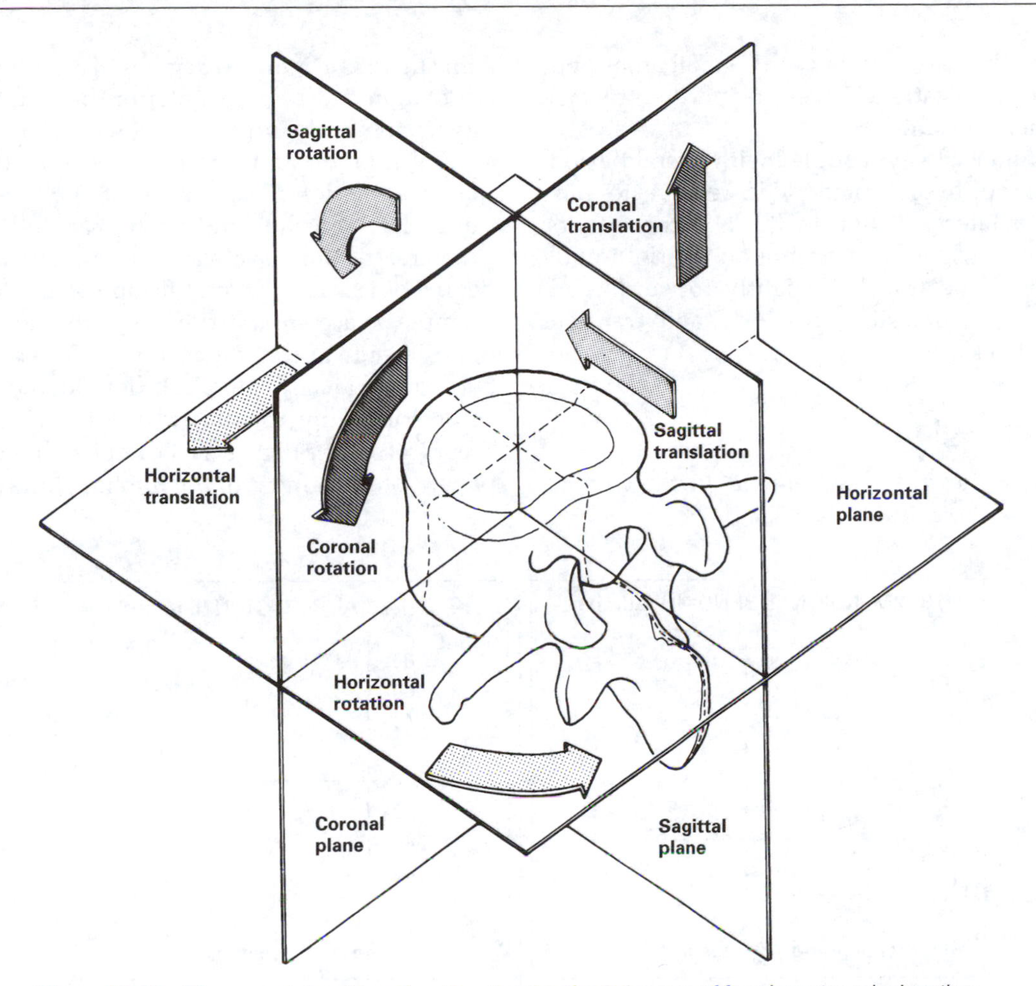

Figure 11.16 Planes and directions of motion showing the 6 degrees of freedom at a spinal motion segment (reproduced with permission from Bogduk & Twomey 1991).

The intervertebral foramina decrease in size during this movement.

Rotation is coupled with lateral flexion to the same side. For example, with rotation to the right, the left inferior articular facet of the upper vertebra glides superiorly, anteriorly and laterally on the superior articular facet of the vertebra below; the right inferior articular facet of the upper vertebra glides inferiorly, posteriorly and medially on the superior articular facet of the vertebra below. The anterior, posterior, medial and lateral movements constitute rotation movements and the inferior and superior movements constitute lateral flexion.

In the same way lateral flexion is accompanied by rotation. With lateral flexion to the right, the left and right inferior articular facets of the upper vertebrae glide in a similar way as in right rotation. The inferior and superior glides produce the lateral flexion, the anterior, posterior, medial and lateral movements produce the rotation.

Cervicothoracic junction (C6–T2)

Similar movements occur in this region as above.

Thoracic region (T3–T10)

This is the least mobile area of the spine. During flexion the inferior articular facets of the superior vertebra slide superiorly with a small amount of forward translation. In extension the reverse movements occur.

Rotation is always coupled with lateral flexion, the zygapophyseal joints slide relative to each other. In lateral flexion to the left, the inferior facets of the superior vertebra on the right glide superiorly and translate slightly forwards and, on the left side, slide inferiorly and translate slightly backwards.

Lumbar region

Flexion is the freest movement. In the lumbar spine there is around 10 degrees of anterior sagittal rotation and 2 mm of anterior translation during flexion and around 3 degrees of posterior sagittal rotation and 1 mm of posterior translation on extension (Pearcy et al 1984). Movements during lateral flexion and rotation are less clear.

Lateral flexion is always accompanied by a degree of rotation. Lateral flexion to the left, for example, is accompanied by axial rotation to the opposite side at the upper lumbar levels. At the two lower levels, lateral flexion to the left is accompanied by rotation to the left (Pearcy & Trebewal 1984). Figure 11.17 depicts the overall pattern of movement of the lumbar spine during

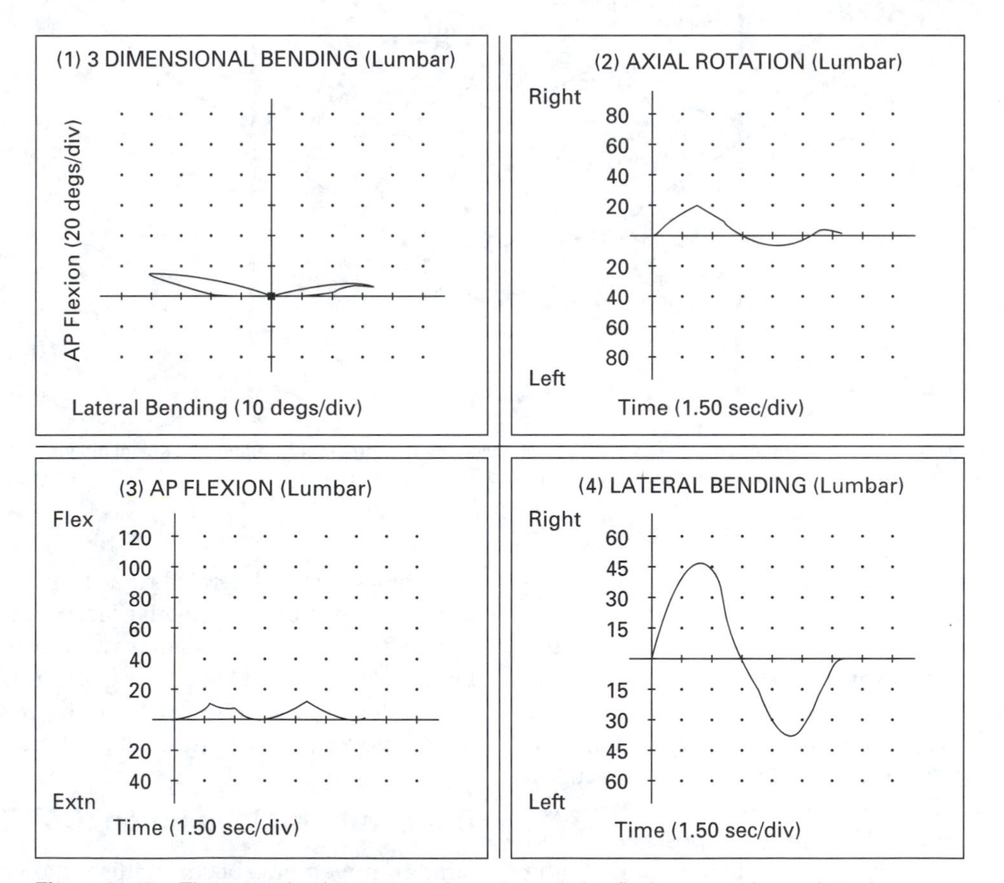

Figure 11.17 These graphs demonstrate the accompanying flexion, extension and rotation movements which occur during lateral flexion movement of an asymptomatic subject. The bottom right graph shows lateral flexion (the primary movement). Top left graph shows the overall movement as if looking from above; bottom left graph shows accompanying flexion and extension movement; top right shows accompanying rotation to the left and right. The CA-6000 Spine Motion Analyser is able to measure three-dimensional spinal movement in real time (reproduced with permission from Orthopaedic Systems Incorporated, Hayward, California, 1993).

the active physiological movement of lateral flexion in standing of a young asymptomatic male subject. It can be seen that lateral flexion to the left is accompanied by rotation to the left and with lateral flexion to the right, there is rotation to the right. The lateral flexion movement in this case is accompanied by flexion; however, in other subjects lateral flexion may be found to be accompanied by extension.

It should be noted that range of movement is not static, there is a reduction in range with increasing age. The lumbar spine, for example, has a reduced range of movement, in both males and females, with increasing age (Leighton 1966). This is due to an increase in stiffness of the intervertebral disc (Twomey & Taylor 1983). Range of spinal movement also varies between males and females, although there is conflicting evidence from the literature. One study found that up to the age of 65, men had a greater sagittal mobility than females, but the reverse was true after 65 (Sturrock et al 1973).

Sacroiliac joints

Movement of the sacroiliac joints occurs during flexion and extension movements of the trunk. Anterior rotation of the base of the sacrum with posterior rotation of the apex is termed *nutation*. The reverse movement is known as counternutation. There is approximately 1° of nutation during flexion and 1° of counternutation during extension of the lumbar spine in standing (Jacob & Kissling 1995). The direction of the movement varies between individuals, in some flexion is accompanied by nutation and in others by counternutation.

This movement does not occur as a result of muscle activity but as a result of mechanical forces placed upon the base of the sacrum during load bearing through the lumbar spine. The movement is restricted by the sacrotuberous and sacrospinous ligaments and also by the interosseous ligaments which bind the sacrum together with the ilia. These nutation and counternutation movements occur readily during gait and weight bearing activities.

During stance phase, the upward pressure from the supporting limb causes a reaction force from the ground to be transmitted via the femur through the hip to the ipsilateral pelvic bone. This causes a tendency for shear (sliding) to take place at the symphysis pubis and the SI joint, (Fig. 11.18). This shearing force is enhanced by the weight of the dependent leg on the contralateral side. Also during gait, anterior and posterior rotation of the pelvis relative to the sacrum occurs. It should be noted, however, that due to the very strong ligaments supporting the sacroiliac joints the range of movement is extremely small.

ENHANCEMENT OF MOVEMENT OF THE UPPER AND LOWER EXTREMITIES AND ENHANCEMENT OF VISUAL AND HEARING FIELDS

The spine serves to enhance movement of both

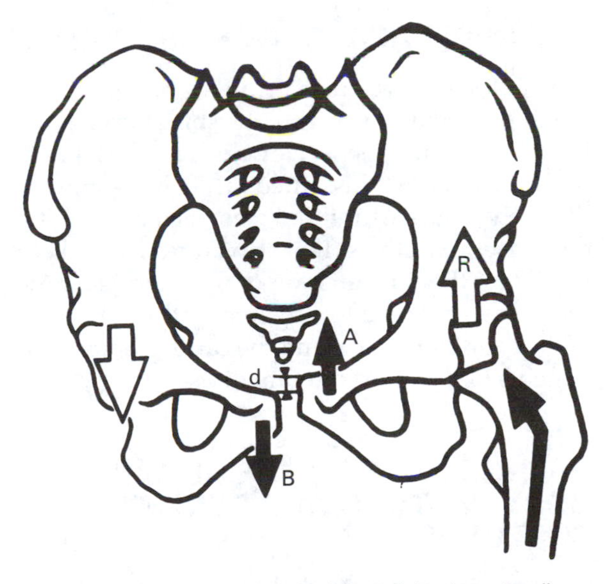

Figure 11.18 The forces around the pelvis when standing on the left leg (stance phase) and taking the right leg forward in walking. The ground reaction force (arrow R) elevates the left hip while the right hip is pulled down by the weight of the free leg. This causes a shearing force at the pubic symphysis tending to raise on the left (A) and lower on the right (B). The forces will be in the opposite direction at the sacroiliac joints, the left ilia will tend to lower and the right ilia will tend to be raised (reproduced with permission from Kapandji 1974).

the upper and lower extremities: for example, in reaching activities the range of motion of the upper limb can be significantly enhanced by rotation and side flexion of the trunk, and an example relating to the lower limb can be seen in hurdling activities where the trunk side flexes above the flexed hip and knee in order to gain clearance of the hurdle.

The trunk also serves to increase the field of vision. This is accomplished by rotatory movements of the head on the neck or of the trunk as a whole, allowing the eyes to be brought into a more optimum position for viewing the targeted object. Trunk movements can also enhance hearing fields in the same way.

Movements of the spine generally aid maintenance of balance by allowing the subject's centre of gravity to be brought over the base of support; for example, in one leg standing the trunk will side flex, rotate, flex and extend in whatever sequence is necessary to maintain the upright posture.

THE SPINE AS A SHOCK ABSORBER

Impact forces, e.g. during running and jumping, are transmitted upwards to the spine through the lower limbs and pelvis. Downward forces due to body weight are transmitted through the spine to the pelvis; these are reduced significantly by the presence of the spinal curvatures which help to stagger the transmission of these forces. These forces are absorbed to a degree by the trabeculae of the cancellous bone and the cartilaginous components of the intervertebral discs.

GIVING SHAPE TO THE HUMAN BODY IN STATIC AND DYNAMIC POSTURES

The spine in normal subjects takes on a characteristic appearance in both dynamic and static postures. If spinal contours are enhanced or lost, it can be a manifestation of poor postural control, muscle weakness or imbalance, bony deformity or bony and/or ligamentous pathology. It can also relate to habitual postural stances which can be related to work or leisure pursuits, or these habitual postures can be a manifestation of a psychological disturbance. For example, in depression the upper cervical spine is often extended so that the chin is poked forward, the lower cervical spine and thoracic spine is flexed, the lumbar spine is flexed producing a flattened lordosis, and there is some flexion of the hips and knees and the patient walks with a shuffling gait. A change in the contours of one region will often be accompanied by compensatory changes in other regions.

The spine also moves in a characteristic way. Analysis of spinal movement is important for the clinician in order to assess the presence of pathology or spinal dysfunction. It is important that contributions by all segments of the spine are monitored in terms of a total regional/spinal movement. If one spine motion segment is blocked (hypomobile), movement of the whole spinal region may be affected causing limited or abnormal movement and this is sometimes compensated for by the development of an area of hypermobility in adjacent segments.

THE SPINE FACILITATING CHANGES FROM STATIC TO DYNAMIC POSTURES

The spine serves a useful function when the body requires movement from a static to a dynamic posture or from one static posture to another static posture. The mobility in the spine allows the subject's centre of gravity to be brought over the base of support, for example when getting up from a chair the thorax and trunk are flexed forwards over the pelvis, towards the knees, so that the lower limbs can raise the trunk from the chair, whilst the trunk remains in a stable posture.

It is likely that the trunk can be used to produce deceleration of a moving body in conjunction with the lower limbs, for example, in running the trunk flexes slightly over the lower limbs. When the runner decelerates, the trunk is brought into a more upright, slightly extended position which increases air resistance by increasing turbulence and also brings the centre of gravity more posteriorly. This in itself retards motion.

In most circumstances the trunk, head and neck can initiate gross movements of the body, e.g. in rolling where either the head and neck or the pelvis can initiate the movement followed by the upper or lower extremity.

REFERENCES

Adams M A, Hutton W C 1985 Gradual disc prolapse. Spine 10: 524–531

Adams M A, Hutton W C 1986 The effects of posture on diffusion into the lumbar intervertebral discs. Journal of Anatomy 147: 121–134

Andersson B J G, Jonsson B, Ortengren R 1974 Myoelectric activity in individual lumbar erector spinae muscles in sitting: a study with surface and wire electrodes. Scandinavian Journal of Rehabilitation Medicine (suppl) 3: 91–108

Andersson B J G, Ortengren R, Nachemson A L et al 1975 The sitting posture: an electromyographic and discometric study. Orthopaedic Clinics of North America 6: 105–120

Aspinall W 1993 Clinical implications of iliopsoas dysfunction. Journal of Manual and Manipulative Therapy 1: 41–46

Bastide G, Zadeh J, Lefebvre D 1989 Are the 'little muscles' what we think they are? Surgical and Radiological Anatomy 11: 255–256

Bland J, Bushey D R 1990 Anatomy and physiology of the cervical spine. Seminars in Arthritis and Rheumatism 20: 1–20

Bogduk N 1994 Cervical causes of headaches and dizziness. In: Boyling J D, Palastanga N (eds) Grieve's Modern Manual Therapy, 2nd edn. Churchill Livingstone, Edinburgh

Bogduk N, Twomey L T 1991 Clinical anatomy of the lumbar spine, 2nd edn. Churchill Livingstone, Edinburgh

Breig A 1978 Adverse mechanical tension in the central nervous system. Almqvist & Wiksell, Stockholm

Butler D S 1991 Mobilisation of the nervous system. Churchill Livingstone, Edinburgh

Cooper R G, Stokes M J, Sweet C, Taylor R J, Jayson M I V 1993 Increased central drive during fatiguing contractions of the paraspinal muscles in patients with chronic low back pain. Spine 18: 610–616

Ghosh P 1988 The biology of the intervertebral disc, Vol 1. CRC Press, Boca Raton, Florida

Ghosh P 1990a Basic biochemistry of the intervertebral disc and its variation with ageing and degeneration. Journal of Manual Medicine 5: 48–51

Ghosh P 1990b The role of mechanical and genetic factors in degeneration of the disc. Journal of Manual Medicine 5: 62–65

Golub B S, Silverman B 1969 Transforaminal ligaments of the lumbar spine. Journal of Bone and Joint Surgery 51A: 947–956

Hides J A, Stokes M J, Saide M, Jull G A, Cooper D H 1994 Evidence of lumbar multifidus muscle wasting ipsilateral to symptoms in patients with acute/subacute low back pain. Spine 19: 165–172

Jackson R 1966 The cervical syndrome. Thomas, Springfield, USA

Jacob H A C, Kissling R O 1995 The mobility of the sacroiliac joints in healthy volunteers between 20 and 50 years of age. Clinical Biomechanics 10: 352–361

Jull G A, Janda V 1987 Muscles and motor control in low back pain: assessment and management. In: Twomey L T, Taylor J R (eds) Physical therapy of the low back. Churchill Livingstone, Edinburgh pp 253–278

Kapandji I A 1974 The Physiology of the joints. Vol 3: The trunk and the vertebral column. Churchill Livingstone, Edinburgh

Leighton J R 1966 The Leighton flexometer and flexibility test. Journal of the Association for Physical and Mental Rehabilitation 20: 86-93

Louis R 1981 Vertebroradicular and vertebromedullar dynamics. Anatomica Clinica 3: 1–11

Macintosh J E, Bogduk N 1986 The biomechanics of the lumbar multifidus. Clinical Biomechanics 1: 205–213

Macintosh J E, Bogduk N 1994 In: Boyling J D, Palastanga N (eds) Grieve's Modern Manual Therapy, 2nd edn. Churchill Livingstone, Edinburgh

Macnab I 1977 Backache. Williams and Wilkins, London

Mercer S R 1995 Clinical anatomy of cervical disc instability. Proceedings of the Manipulative Physiotherapists Association of Australia 9th Biennial Conference, November. Gold Coast, Australia, 101–103

Nachemson A L 1976 The lumbar spine: an orthopaedic challenge. Spine 1: 59–71

Nitz A J, Peck D 1986 Comparison of muscle spindle concentrations in large and small human epaxial muscles acting in parallel combinations. American Surgeon 52: 273–277

Oliver J, Middleditch A 1991 Functional Anatomy of the Spine. Butterworth-Heinemann, Oxford

Palastanga N, Field D, Soames R 1994 Anatomy and human movement, structure and function, 2nd edn. Butterworth-Heinemann, Oxford

Panjabi M M, Takata K, Goel V K 1983 Kinematics of lumbar intervertebral foramen. Spine 8: 348–357

Pearcy M, Portek I, Shepherd J 1984 Three dimensional X-ray analysis of normal movement in the lumbar spine. Spine 9: 294–297

Pearcy M J, Trebewal S B 1984 Axial rotation and lateral bending in the normal lumbar spine measured by three-dimensional radiography. Spine 9: 582–587

Peck D, Buxton D F, Nitz A 1984 A comparison of spindle concentrations in large and small muscles acting in parallel combinations. Journal of Morphology 180: 243–252

Shultz A, Andersson G B J, Ortengren R et al 1982 Analysis and quantitative myoelectric measurements of loads on the lumbar spine when holding weights in standing postures. Spine 7: 390–397

Sturrock R D, Wojtulewski J A, Dudley Hart F 1973

Spondylometry in a normal population and in ankylosing spondylitis. Rheumatology and Rehabilitation 12: 135–142
Twomey L T, Taylor J R 1983 Sagittal movements of the human lumbar vertebral column: a quantitative study of the role of the posterior vertebral elements. Archives of Physical Medicine and Rehabilitation 64: 322–325
Tyrell A R, Reilly T, Troup J D G 1985 Circadian variation stature and the effects of spinal loading. Spine 10: 161–164
Valencia F P, Munro R R 1985 An electromyographic study of the lumbar multifidus in man. Electromyography and Clinical Neurophysiology 25: 205–221

CHAPTER CONTENTS

12

Function of the upper limb

A. Hinde

OBJECTIVES

At the end of this chapter you should be able to:

1. **Outline the development of the upper limb**
2. **Explain the classification of functional activities of the upper limb**
3. **Describe the structure of the upper limb**
4. **Discuss how this structure and the function of the upper limb are interrelated**
5. **Discuss the functional aspects of the individual segments of the upper limb.**

DEVELOPMENT OF THE UPPER LIMB

The upper limb is usually considered to be the bones distal to the glenohumeral joint, the clavicle and the scapula with the muscle groups controlling those bones, and the periarticular structures. The only joint between the limb and the axial skeleton, as Moffat (1994) points out, is the sternoclavicular joint, the scapula being sandwiched between layers of muscle against the upper part of the posterior surface of the thorax.

At approximately 4 weeks, limb buds appear on the fetus and growth occurs until the eighth week when the limb shape is fully formed (Williams et al 1989). From then until birth the tissues differentiate to form the bone, the periarticular and the musculotendinous structures of the limb. There is some migration of the musculotendinous structures of the axial skeleton to become attached to the limb, e.g. latissimus dorsi.

Radiographs of the upper limb at birth show ossification in the long bones but not in the short bones, these still being cartilaginous. Comparison of the proportions of the segments, either from radiographs or photographs at birth, shows the hands to be much larger in comparison with other segments (upper arm, forearm) than during childhood, adolescence or adulthood.

Box 12.1 Task 1

Look at a series of 'family' photographs of a developing child (yourself or a relative) and by estimation plot a histogram of the proportions of the upper limb segments over time. Then compare the result with the results of colleagues.

DEVELOPMENT OF THE USE OF THE UPPER LIMB

The work done by Illingworth (1987) to identify the sequences of behaviour and the change in ability over time has produced the concept of 'milestones of development' as a method of identifying progress in an individual's development by comparing actual with expected performance. From this work it is clear that there are two avenues of development to be followed:

- the development of activities of the upper limb as a discrete unit
- the development of the whole individual as a result of the use of the upper limb.

To illustrate the first point: the grasp reflex is present until approximately the eighth week and it is only when the reflex is disappearing that intentional activity can start — including exploration of 'things' (teething rings, cubes, paper) or environment (splashing water in the bath). These unrefined activities lead onto play and the gradual refining of movement. To illustrate the second: once the upper limb is useful, the rest of the body benefits, e.g. by supporting the body on the forearms when placed prone (at approximately 12 weeks) or by taking weight on the hands when crawling or pulling the body into standing (at approximately 40 weeks). During this period and succeeding weeks the upper limb is also used in play and to communicate needs by gesture and begins to develop self-help abilities during dressing.

Box 12.2 Task 2

Try to identify, then list, the uses to which you put your upper limb. This might seem initially to be infinite but if you look for 'common denominators' you should be able to produce a smaller list of 'generalities'.

An interesting point about the developmental sequence and the upper limb is that it often contributes heavily to the acquisition of abilities by other body segments and then reduces its contribution to those segments, i.e. it has an enabling function. The limb then moves on to another task. Its contribution is not only in these concrete actions but also in its contribution to the cognitive ability, socialisation and habituation of the individual. The importance of these contributions is highlighted when a specific upper limb function does not itself develop leaving either a void to be filled by compensatory techniques (equipment or trick movements) or halting progress in some specific aspect of the individual's development. McClenaghan (1989) describes a situation where poor development of sitting ability has been compensated for by the use of the upper limb as a support (i.e. as a direct replacement for an undeveloped postural ability) and the consequences of that adaptive behaviour for the accurate performance of tasks later in life.

FUNCTION

. . . function
Is smother'd in surmise, and nothing is
But what is not.

William Shakespeare, Macbeth I. iii. 141

Structure and function are inextricably linked, each determining the other: although the structure allows the function, the function in its own right demands the structure. However, this chapter will work from structure to function.

In order for accurate movements to take place there must normally be a stable base about which the limb segments (or links in a mechanical sense) may move. At the other end of the chain of links will be the 'active' segment, i.e. that which is 'doing something'. While undertaking Task 2 you should have become aware that some of the uses involved the hand being moved freely in space and that others involved the hand becoming the fixed point for the movement of the rest of the body. If you did not appreciate this point look back through the list again.

The appropriate way to describe these movements is to use the term 'open chain' for free movements of the hand and 'closed chain' for those movements where the hand is the fixed base. The open chain movements enable some manipulative activity: the hand is the focus of the movement being positioned by the other segments of the upper limb, for example when using tools, cutlery or washing hands. The closed chain movements enable the body to be moved by allowing the segments of the upper limb to act to compensate for underactivity of some other body segment.

Examples of closed chain activity include pushing on chair armrests to assist rising from a chair and holding stair handrails to help maintain balance. Crutch walking and wheelchair–toilet/bed transfers are examples of closed chain compensatory use in dysfunction.

Box 12.3 Task 3

Repeat Task 2 but this time note the type of activity (open or closed chain) and the base of support and moving segments. Divide the upper limb into three components: hand, forearm/arm and pectoral girdle. Tabulate the results, e.g.

Activity	Type	Hand	Forearm/ arm	Pectoral girdle
Biscuit to mouth	Open chain	Grip on biscuit	Freely moving	Fixed base

You should by now be more aware that upper limb function goes beyond being the support and mover of the hand in the way that the springs, joints and arms of an 'Anglepoise' desk lamp serve only to position the light.

Structures permitting function

There are three types of structure in the upper limb which permit function to occur:

- levers, i.e. bones
- joints about which levers move (including ligaments which restrict range or direction)
- muscles causing or controlling movements of the levers.

Each type of structure has its own specific properties and limitations and the student is directed to the References for more detail of these. It is the combination of these structures that permits the whole limb to be more than the sum of the parts, i.e. that gives the limb its ability to undertake functional use.

Hand function

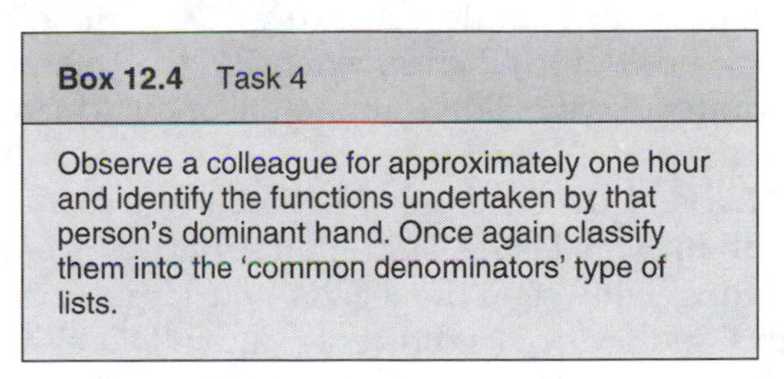
Box 12.4 Task 4

Observe a colleague for approximately one hour and identify the functions undertaken by that person's dominant hand. Once again classify them into the 'common denominators' type of lists.

The carpals are short irregular bones which lock together to form a stable block. During most actions involving the hand and needing long flexor and long extensor muscle activity, the carpals are in compression proximal to distal. Transverse section through the carpals shows the arching form of the group of bones which protects structures anteriorly from damage by items pressed into the palm and helps to maintain arterial supply and nerve conduction when gripping strongly. During movement the carpals work as a disc so that movements at the wrist joint have their axes through the carpals, with the scaphoid and lunate sliding laterally during ulnar deviation and medially during radial deviation. During flexion the articular surface of the scaphoid and lunate glide posteriorly and during extension, anteriorly. Cailliet (1984) explains how the many facets on the distal surfaces of the trapezoid and capitate lock the second and third metacarpals into a fixed unit, with the first metacarpal, and the fourth and fifth mobile to either side.

A transverse section through the metacarpals also shows an arched formation, concave anteriorly. This arch is more mobile than the carpal arches and will allow flattening or cupping of the hand with the outer units moving about the second and third metacarpals as stated above. The metacarpal heads have condylar shapes with an articular area on the distal and anterior aspects giving a sweep area or surface arc for the proximal phalanx of approximately 180°. The surface arc of the base of the proximal phalanx is approximately 20° and this bone glides across the surface of the metacarpal head to give a range of movement between 90° of flexion and 20° hyperextension. The joint is very stable at 90° of flexion because the tension in the collateral ligaments is at its maximum (i.e. in the close pack position). When in extension the collateral ligament laxity allows abduction and adduction to occur.

Each of the phalanges is similarly structured with proximal bases and small surface arcs articulating with condylar heads and large surface arcs. In these joints, however, the collateral ligaments are in tension in extension at the joint making this the close pack position. Salter (1987) emphasises the importance of these details when splinting the hands and fingers. Flexion and extension are the only movements produced by direct muscle action but rotation does occur as an accessory movement when the metacarpophalangeal (MCP) and interphalangeal (IP) joints are not in their close pack position.

The static function of the hand is usually called 'grip' and the grips are classified by different authors in different ways, for examples see Backhouse & Hutchings (1989) and Salter (1987). All have three features:

- a stable foundation to locate the levers
- the ability to arrange the levers appropriately
- the ability to apply forces through the levers.

The stable foundation will normally be provided by the carpals providing a closely packed foundation upon which the metacarpal bases will be anchored. When arranging the levers, the arch of the metacarpals is actively or passively adjusted. Intrinsic muscles of the hand actively pull the bones into a transverse arch. This positions the phalanges so that their tips converge towards a central point for gripping between thumb and finger pads, or gripping larger, roughly spherical objects between the palm and phalanges. The arch is passively flattened during gripping so that the phalanges can move along parallel pathways to grip cylindrical or cuboidal shapes. Bray et al (1989) and Wirhed (1988) indicate that the force of contraction of a muscle depends directly on its physiological cross-sectional area. The forces that can be generated during gripping can be very high but there is no space in the palm of the hand to hold an object and have large muscle bellies. The upper limb reconciles these requirements by placing the source of the forces (long flexor muscles) outside the hand and using tendons (with very much smaller volumes) and pulley-like arrangements to transmit the forces to the levers (phalanges). Cailliet (1984) has some very clear illustrations of normal and dysfunctional arrangements. Backhouse & Hutchings (1989) illustrate and describe the actions.

The long (extrinsic) muscles controlling the movements of the fingers are each described anatomically by their actions. The intrinsic muscles are named by their location (interossei), action (opponens pollicis) or appearance (lumbricals). Each group has its own function: to provide force, or to position the bones for the efficient application of that force. However, the true functions of the muscles of the hand do not become apparent until the combinations of actions are studied, i.e. it is the interplay of intrinsic and extrinsic actions that give functional ability to the hand rather than merely anatomical movement. The work of Cailliet (1984) and Salter

(1987) provides details of these interplays and their results.

It is worth examining the combination of the concepts of types of grip and the arches of the hand. The *power grip*, for example, places the tool handle across the supports of the index–little finger arch and then clamps between these points with the third support (from the thumb). The *lateral pinch grip* and the *tip grip* use only the index–thumb arch supports, while the *span grip* uses all three arch supports in a centripetal motion. The *dynamic tripod grip* uses the index–thumb arch to hold the tool (pen); Salter (1987) emphasises the role of the little finger as a support for the 'working' arch (Fig. 12.1).

Box 12.5 Task 5

Place a ruler on a piece of paper and draw a line; look at the arches formed by the 'supporting' hand and compare them with the arches formed by the 'drawing' hand.

An understanding of these arches and the bones and muscle actions which cause and maintain them is essential when splinting or otherwise treating a dysfunctional hand, whatever the cause of the dysfunction.

As a means of communication the hand and fingers may parody the spoken word, replace the spoken word or express (as body language) otherwise concealed communication. These communication gestures may take the form of pointing out features, e.g. on a poster, chalkboard or in a landscape. They might involve the hands being used to indicate the shape of, or relationship between, objects. A formalised sign language has been developed and is in use to compensate for dysfunction of speech or hearing by the Association of Sign Language Interpreters.

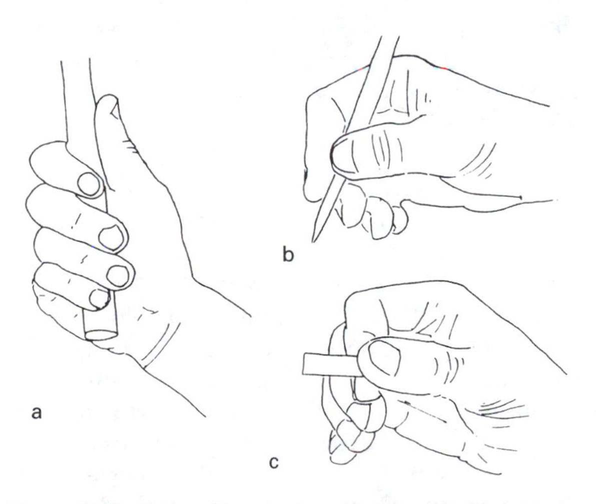

Figure 12.1 Illustrations of grips: a) power grip; b) dynamic tripod grip; c) pinch grip.

Activities which are intentional movements take their origins in the cerebral cortex. There are many unconscious movement patterns which express an individual's values and subconscious judgement. Moods may be indicated by hand activity, but mood may also modify hand activity by the influences on the anterior horn cells through the medial pathways and limbic system. Pain and anger may inhibit movement by increasing muscle tone, lassitude or relaxation having the opposite effect.

Sensory functions

Review of the sensory homunculus will show the relatively large area of sensory cortex 'allocated' to the hand. This offers a clue about the importance of the hand as a source of exteroception and proprioception.

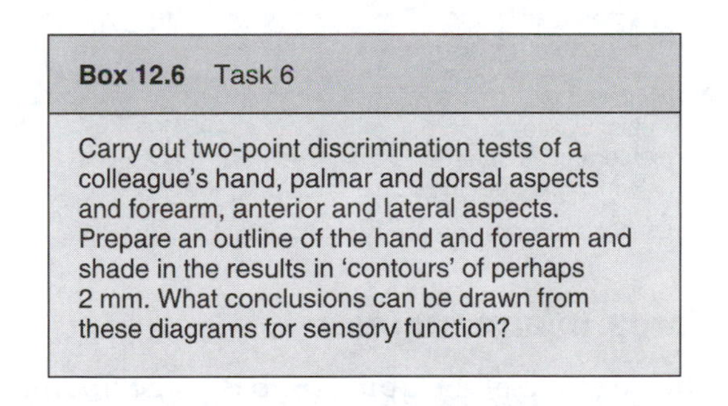

Box 12.6 Task 6

Carry out two-point discrimination tests of a colleague's hand, palmar and dorsal aspects and forearm, anterior and lateral aspects. Prepare an outline of the hand and forearm and shade in the results in 'contours' of perhaps 2 mm. What conclusions can be drawn from these diagrams for sensory function?

The types and function of sensory end organs are described more formally elsewhere (Williams et al (1989) for example), but their functional uses are worth consideration in more detail. Whether surfaces are hotter or colder than skin temperature can be identified by exteroceptors, surface texture can similarly be identified. The proprioceptive function of the lumbrical muscles is essential for accurate functional activity from the hand and fingers. Beyond this, the sensory data is generated by the combination of extero- and proprioception; shape, dimensions and weight all require information from proprioceptive nerve endings in joint capsule, muscle and tendon as well as receptors in the skin. Most importantly, experience and practice are necessary for the data to be correctly interpreted — this differentiates the skilled craftsman or Braille reader from the beginner, for example. It is worth reflecting at this point about Illingworth's (1987) indication of the importance of maximal sensory experience during development and the consequences for children with congenital dysfunctions compared to those with acquired dysfunctions.

Combinations of motor and sensory activity applied within specific contexts can then be seen within a wider definition of function. Maslow (1954) classified the needs of individuals into a hierarchy of five levels:

- physiological (to satisfy the needs of homeostasis)
- safety or security (shelter, protection)
- belonging (social grouping or position)
- self-esteem (having a sense of 'value')
- self-actualisation (achieving full potential).

Box 12.7 Task 7

Using each subgroup of Maslow's (1954) hierarchy of needs, identify two hand functions which may be undertaken to satisfy that need.

Forearm/arm function

This segment of the upper limb is made up from the three long bones and their intermediate joints; further to this there is a synovial joint with the adjacent segment at each extremity, the wrist joint distally and the glenohumeral joint proximally. The segment has two main functions:

- lever systems for movement or positioning
- the location of muscle attachments.

The wrist joint connects the hand to the forearm and, because only two degrees of freedom are available at the joint, the extensors and flexors carpi assist in the alignment of the hand in readiness for use, or stabilise the carpals for activities of the fingers. Where the problems of muscle insufficiency (active or passive) occur, the mobility of the wrist increases the range of the movements possible at the fingers. The third degree of freedom, sacrificed by the wrist joint, is transferred to the radioulnar joints. This arrangement increases the stability of the carpal/forearm joint without the loss of potential positions for the hand. The major tension forces applied to outside loads at the hand are applied through the long flexor or extensor muscles and the phalanges. The loads on the carpals and metacarpals are relatively low and are close to the joint axis. The mechanical arrangement is further assisted by the pisiform bone acting as a sesamoid bone to alter the alignment of the flexor carpi ulnaris tendon.

The elbow joint is a complex of three articulations where two have given up freedom of movement for stability and the third is a compromise between movement and stability. The humerus/ulna articulation has a single degree of freedom (flexion and extension) as does the superior radioulnar joint (rotation). These are both very stable joints. The humerus/radius articulation offers both rotation, and flexion and extension. The complex serves to assist the wrist and inferior radioulnar joints to position the hand in a required position and to move the hand and any load to a new position.

The muscles acting on the elbow joint have an attachment close to the axis, cross the joint and attach the length of the forearm or arm away. This makes the lever system very disadvantageous in terms of effort applied, but this is compensated for by the advantageous velocity ratio giving large distance and rapid movement for the extremities of the segments in return for small length changes in the muscle during contraction. The need to produce high forces in contraction is satisfied by having the muscles work in two groups; those in the arm attached close to the elbow on the radius and ulna, and those in the forearm attached close to the elbow on the humerus. Each group will cross the joint and attach to the further extremity of the opposite long bone. The combined action can be represented by drawing triangles of force to estimate their cumulative effect.

When the hand is used in open chain functional movements, elbow extension will usually be caused by gravity under the control of the flexor muscles in eccentric contraction. For closed chain actions, elbow extension is usually produced against resistance and so the elbow extensor comes into action. The effect of the division of the triceps muscle into three heads can be estimated by drawing polygrams of force, but more importantly the division allows an increase in the physiological cross-sectional area and therefore the capacity to generate high forces increases. When the hand is fixed and the triceps muscle is moving the body (as when pushing on chair arms to assist rising to standing), the lever system is of the second order and triceps has a slight mechanical advantage. From a practical point of view, the distance between the olecranon process and the trochlear notch is small when expressed as a proportion of the length of the forearm, so mechanical advantage is minimal. The velocity ratio still maintains its advantage and for a physically small range of contraction of triceps the body is raised considerably.

The glenohumeral joint has three degrees of freedom permitting the hand to describe almost the whole surface of a sphere (the axial skeleton occupying the missing part). In order to achieve this range of movement the joint sacrifices some stability and relies on muscle action to maintain congruity of the surfaces when under load.

Box 12.8 Task 8
Return once again to your notes for Task 2 (or carry it out again). This time identify the axis and plane of the movements of the upper limb.

For many of the activities involving the upper limb the actions have two components at the glenohumeral joint:

- the primary purpose of the movement, e.g. writing, stirring a cup of coffee
- a synergistic positioning of the segments to allow more efficient performance of the primary purpose.

The muscles acting for the synergistic purpose are the deltoid and the rotator cuff muscles, which typically need only to work against gravity's action on the mass of the upper limb. Primary activity is usually generated by the large muscles attached to humerus and thorax or humerus and scapula. This arrangement allows large forces to be generated in either flexion/adduction or extension/adduction at this joint. The importance of these relationships is well illustrated in the case histories chosen for publication by Bullock (1990).

For many activities of the upper limb, the axis and plane of movement between the glenoid fossa and humeral head remains the same even when the direction moved by the limb changes. This apparent contradiction is explained by the adjustment in the plane of the glenoid fossa when the scapula is moved in relation to the thorax. Although the limb may be swinging in flexion/extension movements at the glenohumeral joint, the movement seen in relation to the axial skeleton may be across the front, parallel to the sagittal plane, or forwards and outwards depending on the degree of protraction/retraction of the pectoral girdle.

The pectoral girdle is the remaining structure permitting function in the upper limb. It has only one synovial joint with the thorax, at the sternoclavicular joint, and the scapula lies sandwiched between layers of muscle. The girdle acts like the jib of a crane (especially when viewed from above) to position the glenohumeral joint in preparation for activity in the arm, forearm and hand. The glenohumeral joint cannot slide around the thorax in a horizontal plane because the clavicle acts like the spoke of a wheel (unless fractured) and props the acromion outwards. When the scapula does slide forward the glenoid is moved forwards in the movement called protraction, retraction pulls the scapula towards the vertebral column and the glenoid moves backwards still a clavicle-length from the sternum.

The muscles joining the scapula to the thorax may act together to produce rotation of the scapula about a sagittal axis, raising or lowering the glenoid in elevation or depression, respectively. These four movements have the advantage of increasing the volume of space falling within the reach of the hand in the way a telescopic arrangement would in a machine. The number, size and alignment of these muscle groups would seem at first to be excessive for open chain movements of the upper limb. It is in closed chain movements when the axial skeleton is being moved in relation to the upper limb that their purpose becomes apparent.

Anthropologically, a major function of the upper limb was to transport the body, as it still is in primates. When hanging from an overhead support the animal needs to move 93.5% of its body mass (Wirhed (1988) suggests the upper limb is 6.5% of body mass) and therefore needs muscles of large physiological cross-sectional area and mechanically advantageous arrangements in terms of force diagrams and lever systems. Although man no longer relies on this form of transport, the potential remains useful when compensation for dysfunction is required, for example when using walking aids to relieve lower limb weight bearing for people with spinal cord lesions when the axial skeleton becomes slung between the bilateral props of the upper limbs and crutches or parallel bars.

REFERENCES

Backhouse K M, Hutchings R T 1989 A colour atlas of surface anatomy; clinical and applied. Wolfe Medical Publications Ltd, London

Bray J J, Cragg P A, Macknight A D C, Mills R G, Taylor D W 1989 Lecture notes on human physiology, 2nd edn. Blackwell Scientific Publications, London

Bullock M I (ed) 1990 Ergonomics: the physiotherapist in the workplace. Churchill Livingstone, London

Cailliet R 1984 Hand Pain and Impairment, 3rd edn. FA Davis Co., Philadelphia

Illingworth R S 1987 The development of the infant and young child, 9th edn. Churchill Livingstone, London

Maslow A H 1954 Motivation and personality. Harper and Row, New York

McClenaghan B A 1989 Sitting stability of selected subjects with cerebral palsy. Clinical Biomechanics 4: 213–216

Moffat D B 1994 Lecture notes on anatomy, 2nd edn. Blackwell Scientific Publications, London

Salter M I 1987 Hand injuries; a therapeutic approach. Churchill Livingstone, London

Williams P L, Warwick R, Dyson M, Bannister L H 1989 Gray's anatomy, 37th edn. Churchill Livingstone, London

Wirhed R 1988 Athletic ability and the anatomy of motion. Wolfe Medical Publishing, London

CHAPTER CONTENTS

13

Strength, power and endurance

D. J. Newham

OBJECTIVES

At the end of this chapter you should be able to:

1. **Define and explain the concepts of work, strength, force and power**
2. **Discuss the effects of various factors on the above concepts**
3. **Discuss the measurement of the above**
4. **Describe fatigue and the other limits to the above concepts**
5. **Explain the factors involved in muscle training for strength, power and endurance**
6. **Discuss the parameters involved in muscle training with the different aspects of muscle functioning**
7. **Discuss the physiological effects of the training on the three aspects of muscle function**
8. **Describe the factors involved in therapeutic exercise**
9. **Develop and justify an exercise programme for the three concepts of muscle function.**

INTRODUCTION

The aim of this chapter is to give an understanding of the factors affecting force generation and how they are utilised in training programmes for both clinical and athletic purposes. Factors influencing strength, power and endurance are discussed, along with how they may be affected by training. The chapter builds on existing knowledge of the structure and function of skeletal muscle.

We require our muscles for two main purposes: to maintain a given posture (when they generally adopt a static role) and to move our bodies in a coordinated fashion and enable effective functional or athletic performance.

The concepts for force, strength, work and power are fundamental to an understanding of muscular action but are often used incorrectly. As they each have precise and different meanings it is important to clarify them.

Force cannot be seen but its effects can be measured. It can be defined as that which changes, or tends to change, the state of rest or motion of matter. For example, when a critical amount of force is applied to a stationary object, it will start to move. The application of an increased force will result in faster movement in the absence of any opposing resistive force such as friction. Application of a force less than the critical amount will not result in movement, but can still be measured with the appropriate technology. The unit of measurement of force is the newton (N) which is the force required to accelerate a mass of one kg at one metre/sec/sec.

Strength is the ability to generate force and is obviously central to muscle performance. A critical amount of strength is required to perform any function: if the strength of an individual is below this level, that function cannot be performed. Strength is measured in newtons or torque. The latter is the effectiveness of a force to produce rotation about an axis, e.g. a joint, and is measured in newton meters (N.m).

Work is the product of the force exerted and the distance through which it acts. The unit of measurement is the joule (J). It is important to note that work is only done when movement occurs; therefore, someone exerting considerable muscle strength and force by pushing against an immovable object is not doing any work, since there is no external movement.

Power is the rate of doing work, i.e. work divided by time, and is measured in watts (W). As velocity is distance divided by time, power can also be calculated by multiplying force by velocity.

The interactions between these terms is illustrated by considering two individuals who each weigh 70 kg, but who have muscles of very different strengths. When standing still, no work is being done as there is no movement but each is exerting the same force of about 700 N (70 kg × force of gravity (9.81 m/s/s)). This force will represent a lower proportion of the maximal strength for the stronger person. If they are unable to exert a force of 700 N they will be unable to stand.

If each one covers a distance of 10 m, they will have moved and therefore work will have been done. The work done, 7000 J (70 kg × 10 m), will be the same for each. In life there is a requirement for speed and therefore the rate of doing work, power, is important. If the 7000 J of work is performed by one person in 2 s, the power output (7000 J/2 s) will be 3500 W. However, if the other person performs the same work in more time, their power output will be less. Covering the same distance in 4 s will give a power output of 1750 W.

It can therefore be seen that the key components of muscle function are strength and velocity. Endurance, the ability to continue activity and withstand fatigue, is also essential. These factors will be considered in further detail along with the causes and effects of improved performance with training. Strength and endurance must be considered separately as either primary or secondary muscle problems may affect either one alone or both together.

Box 13.1 Task 1

Make sure you understand the definitions of force, work and power. Think of some examples of normal activities and work out which definitions do, and do not, apply.

STRENGTH

Intuitively we think that bigger muscles are

stronger than smaller ones. It has also been shown experimentally that the strength of a muscle is proportional to its cross-sectional area (CSA) and independent of its length (Jones & Round 1990). In other words, the number of sarcomeres in parallel determines the strength of a muscle.

Contraction force

Most activities do not require a person of normal strength to make maximal contractions. However, each action requires a certain amount of absolute strength, i.e. lifting a bag of shopping or raising the body weight from sitting to standing. For a weak person these activities will require contractions at a greater proportion of their maximal strength than for a stronger person. The weaker person will therefore fatigue more rapidly.

Therefore, a given amount of strength is necessary for normal function and the greater the difference between the required force and the maximal force of the active muscle, the easier it is to perform an activity. Strength below the critical value will result in an inability to perform a particular function and this may have such profound functional consequences as being unable to rise from a chair or toilet seat.

For activities involving moving the body, the body weight is of great importance, but this may easily be overlooked. A person with weak thigh muscles is further disadvantaged by being overweight; losing a few kilograms of body weight may be the difference between being able or unable to stand from sitting, even in the absence of any increase in strength.

Power

As previously described, we frequently want our muscles to produce power and bring about movement rather than simply generate force. The determinants of power are force and velocity and so an alteration in either will affect power output.

Muscle velocity (speed of contraction and relaxation) is largely influenced by the distribution of fibre types within a muscle. It is little affected by reduced activity or muscle injuries but may be changed by some neurological and primary muscle diseases. Fibre typing is also relatively constant whilst the muscle remains normally innervated and is only marginally influenced by training.

Therefore, power output is most likely to be affected by changes in muscle strength.

Development, ageing and gender

Reliable measurements of strength can be made in children from about the age of 5 years. From this age until puberty, when a growth spurt occurs, there is a steady increase in size and muscle strength which is similar in boys and girls (Jones & Round 1990).

In adolescent girls and women, the relationship between muscle strength, height and weight is similar to that of younger children. In adolescent boys, there is an additional hypertrophy which particularly affects the upper torso and limbs and which is probably an effect of testosterone.

Peak muscle strength is reached in approximately the early 20s and in the fifth decade there is a progressive atrophy (reduction in size and strength). This affects both genders equally until the menopause, when women show a greater loss of muscle size and strength which can to some extent be prevented by hormone replacement therapy (HRT).

The progressive loss of muscle mass can be as much as 30% by 90 years. The reduction in the size of a whole muscle is said to be greater than the atrophy of individual muscle fibres (Grimby & Saltin 1983) and it may be that the loss of neurones in the brain is accompanied by a loss of anterior horn cells in the spinal cord resulting in a reduction of motor units.

There is no evidence that habitual exercise can prevent the muscular effects of ageing. However, the preservation of cardiac and respiratory function brought about by regular exercise will help the elderly to make optimal use of the remaining muscle. Muscle performance can still improve in response to strength training, although there is some indication that the improvements are brought about more by learning and less by hypertrophy than in the young (Moritani & DeVries 1979). This learning adaptation to exercise implies that the central nervous system 'forgets' how to perform with disuse and is an additional reason why the elderly should be encouraged to remain active.

Body types (somatotypes)

As well as differing in gender and height, people vary in their somatotype (Fig. 13.1) and this is another congenital determinant of musculature and, therefore, physical strength. The majority of people contain elements of more than one type, but there are some who show a clear domination of one or another.

There are correlations between somatotype, lifestyle, disease and sporting ability. Certain somatotypes dominate particular athletic events and endomorphs are unlikely to excel in any type. The different ratios between body weight and muscle mass and also the biomechanical consequences of varying the length of body segments provides a genetic advantage, or disadvantage, in different physical activities (Tanner 1964).

Alterations in body weight with diet and exercise do not overcome the basic somatotype characteristics and the basic body proportions are retained despite the most rigorous attempts to change them.

In the absence of injury, disease or training, each individual has a natural position on a spectrum of strength and physical performance (Fig. 13.2). Incidence of acute pathology and also disuse may shift them temporarily towards the weak end of the spectrum. Primary diseases of

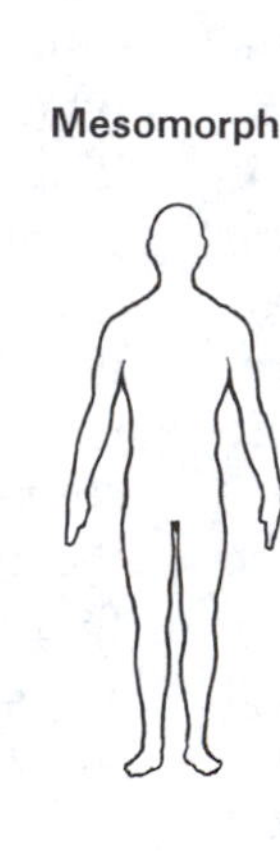
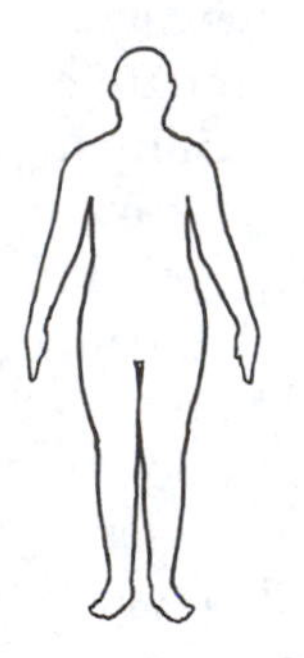

Figure 13.1 Illustration of the three basic body types and their physical characteristics. Elite athletes are usually ectomorphs or mesomorphs. Endomorphs are unlikely to excel at any athletic activity, despite rigorous training.

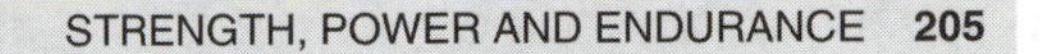

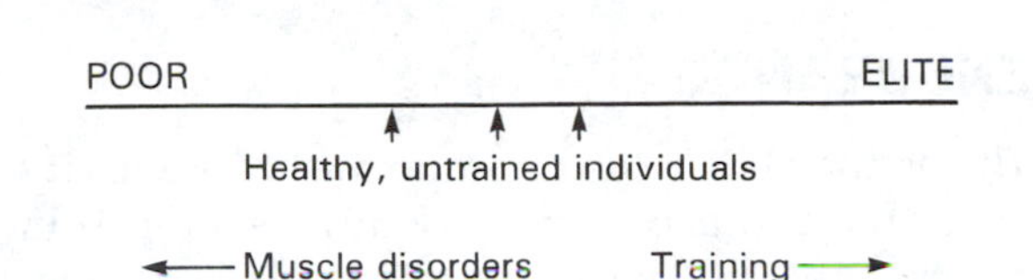

Figure 13.2 The spectrum of human performance. Healthy, untrained individuals can be pushed to the left by injury, disease and disuse or to the right by training.

nerve and muscle may have long-lasting, permanent or progressive effects. Training can usually improve strength, although, even in the absence of pathology, each of us has a predetermined ceiling which cannot be exceeded. Any increases brought about by training will disappear once training ceases and the healthy individual will revert to their original place on the spectrum.

Measurements of strength

Objective

Theoretically, either size or force can be measured to indicate muscle strength. However, the only measurements of size that are useful are those of the physiological cross-sectional area (PCSA) (Fig. 13.3). As many muscles have a pennate structure and the fibres are not arranged parallel to the line of pull of the muscle, the anatomical cross-sectional area (ACSA) is not usually the same as the PCSA. Therefore, measurements of limb circumference are of very little use as an increase in the PCSA does not necessarily result in a proportional increase in the ACSA (Jones & Round 1990).

A number of factors, including disease, disuse and training, can alter the thickness of subcutaneous fat and also muscle bulk, but circumference measurements are unable to discriminate between the two (Stokes & Young 1986). Furthermore, the difference in measurements of limb circumference made by different people and even by the same person on different occasions is considerable (Stokes 1985). Accurate measurements of muscle CSA can only be made using ultrasound or computerised axial tomography (CAT) scanning techniques, which are normally unavailable to the therapist.

There are a number of relatively inexpensive hand-held devices which measure either force or pressure. These give accurate and reliable results if testing position is standardised. Their only limitation is that they rely on manual resistance by the

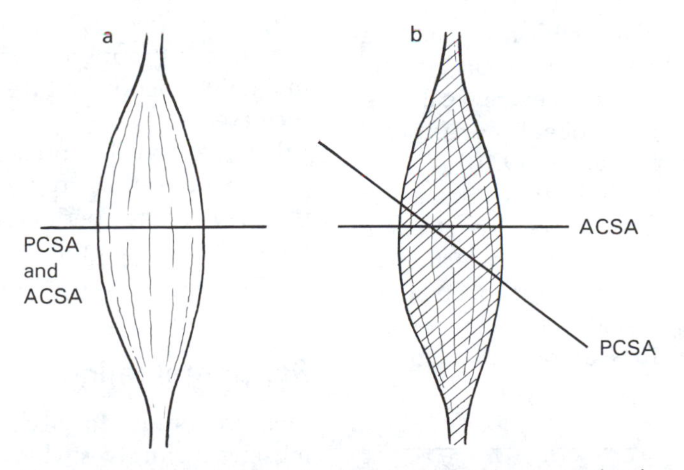

Figure 13.3 Force is determined by physiological cross-sectional area (PCSA) but measurements of limb circumference only identify the anatomical cross-sectional area (ACSA). In a muscle whose fibres lie parallel to the line of pull (a) the PCSA and ACSA are the same. Most muscles do not have this arrangement of fibres, which often lie at an angle to the line of pull and where PCSA and ACSA are very different (b). In this situation, measurements of ACSA will not detect large changes in PCSA.

therapist; the large muscle groups of adults cannot therefore be measured unless they are particularly weak. Some of these devices overcome this problem by having attachments which can be connected to immovable objects, such as a table leg.

An objective measurement of strength which can be made with the minimum of equipment is the 1 RM, which may be performed under isometric or dynamic conditions using free weights. In an isometric contraction, a 1 RM is the maximum weight that the patient can hold in a particular part of the range. In a dynamic contraction, it is the maximum weight that they can move through the available range. Care must be taken to standardise the joint position during isometric testing and the velocity of movement during a dynamic one as both muscle length and angular velocity will influence the force a given muscle is able to generate. In practice, it is easier to monitor joint position accurately than angular velocity and therefore isometric testing is probably more reliable.

Subjective

Manual muscle testing scales usually range from 0 (no contraction) to 5 (normal strength) and are based on a method devised to measure muscle strength in infants (Martin & Lovett 1915). However, it has been well established that their accuracy and sensitivity are so poor that they are only of any use in cases of very severe weakness. Even then their use should probably be restricted to smaller muscle groups (Sapega 1990).

Manual techniques may indicate similar scores despite significant increases in objective strength measurements (Krebbs 1989), not detecting as much as a 50% increase in strength (Watkins et al 1984) and rate as 'good' a muscle strength which is only about 10% of normal (Agre & Rodriquez 1989).

Box 13.2 Task 2

What reliable and valid techniques for measuring muscle strength are available to the therapist? Which techniques are unacceptable due to either technical reasons or lack of accuracy?

ENDURANCE

This is the ability to continue muscular activity over time and is obviously highly developed in athletic sports such as distance running. Endurance activities predominantly rely on aerobic metabolism and therefore depend on mechanisms which transport oxygen to, and metabolic waste products from, the working muscles as well as those involved in aerobic metabolism within the muscles and intramuscular energy stores. Therefore, both the muscles themselves and the cardiovascular and respiratory systems are involved in endurance activities.

The importance of endurance capacity is that it reduces the effects of muscular fatigue. This is an inevitable consequence of activity which results in a decrease in strength or performance over time. In healthy subjects it is a temporary event, although the rate of both fatigue and recovery varies between individuals and is influenced by training. Fatigue is an inevitable consequence of activity and cannot be eliminated, but the resistance to fatigue, i.e. endurance, can be improved.

The extent of fatigue varies between a sensation of tiredness and complete exhaustion. As fatigue progresses, additional motor units and also muscle groups are recruited in an attempt to maintain function and the energy cost of activity increases.

Fatigue can be broadly divided into two groups according to whether the underlying mechanisms are peripheral, i.e. in the muscles themselves or central, i.e. within the central nervous system.

Peripheral fatigue

This occurs due to mechanisms in the working muscles themselves which can affect contraction, excitation or both (MacLaren et al 1989, Jones & Round 1990, Kukulka 1992). Working muscle is a highly active tissue in biochemical terms and the substances released (e.g. acetylcholine, potassium, phosphate and lactic acid) act alone or in combination to decrease force and power output.

Active muscle often achieves internal pressures which are sufficient to occlude their own blood supply and therefore exacerbate fatigue.

It was thought that the release of lactic acid and consequent decrease in intramuscular pH was the sole cause of metabolic fatigue, but it is now clear that this is not the case (Jones & Round 1990).

Peripheral fatigue impairs the muscles' ability to generate force, even though it is maximally activated either through the nervous system or by external electrical or magnetic stimulation.

Central fatigue

This term describes mechanisms within the central nervous system which result in decreased voluntary activation, irrespective of the muscles' ability to produce force. Prolonged activity may be very boring and uncomfortable and this alone tends to increase the sense of effort and decrease motivation.

In addition, reflex mechanisms resulting in either decreased excitation, increased inhibition or an increased threshold to stimulation of the anterior horn cells can also cause central fatigue in the presence of full motivation. Both pain and joint pathology (Hurley et al 1992) also result in a reflex determined failure of voluntary activation.

The presence of central fatigue is confirmed when the superimposition of external (electrical or magnetic) stimulation upon a maximal voluntary contraction generates additional force (Rutherford et al 1986a).

Limitations and ageing

There are physiological limits to the amount of oxygen that can be taken up by the body. This is called the maximal oxygen consumption or VO_2 max and is influenced by a number of factors including genetics, age and gender in addition to training (Åstrand & Rodahl 1988, McArdle et al 1991).

After the mid-twenties, the VO_2 max steadily declines and by the age of 70 may be in the order of 50% of that at 20 years. Maximal heart rate also declines with age; a rough guide for a particular age is 220 minus the age in years (Åstrand & Rodahl 1988, McArdle et al 1991). It is interesting that the decline in endurance ability with age is less than that for muscle strength.

VO_2 max also decreases with inactivity and particularly with bedrest (Saltin et al 1968). This is separate from any direct effects of pathology and should be remembered during treatment.

Measurement of endurance

Objective

Objective methods include the performance of maximal or submaximal activity using cycle ergometers, treadmills or step tests (Åstrand & Rodahl 1988, McArdle et al 1991). The measurements made may include measurement and analysis of expired air, heart rate and blood pressure. Appropriate clinical measurements can be made using these techniques, or by measuring the number of repetitions possible with a given load or resistance, or timing a specified activity such as walking a fixed distance or climbing a flight of stairs.

All these measurements and tests involve strength as well as endurance to some degree and this is unavoidable. Without adequate muscle strength, no movement is possible and weak muscles result in slow movement. However, by using a relatively low force activity and using a high repetition number over time, a good indication of endurance capacity can be obtained and any changes observed.

An increase in endurance capacity is indicated by a reduced heart rate or sense of effort (Borg 1982) for a given amount of activity, an extension of the duration for which an activity can be carried out or an increase in the number of repetitions possible in a given time.

Subjective

Reported symptoms of fatigue are very difficult to interpret. Most patients will complain of weakness, rather than fatigue, and find it difficult to distinguish between the two.

Box 13.3 Task 3

What factors should be considered when a patient presents with fatigue as the main symptom? What assessments should be made?

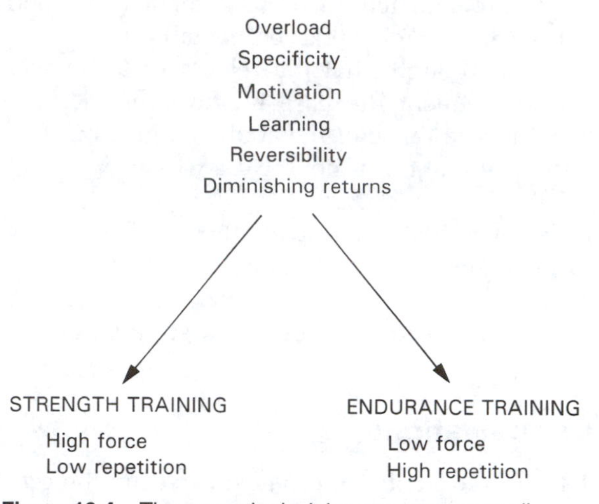

Figure 13.4 The general principles are common to all forms of training, but the type of exercise required to improve strength or endurance is very different.

MUSCLE TRAINING

Training may be defined as the preparation of an individual for purposeful, skilled and effective specific movements which may range from the general activities of daily living and function to specific and highly skilled activities. All activities comprise elements of:

- strength
- endurance
- flexibility
- coordination
- skill.

Training programmes need to be worked out for each individual according to their current condition and the goals and aims of training.

General principles of training

There are some principles which are common to all types of training (Fig. 13.4).

Overload

This means that a muscle, or physiological system such as the cardiovascular, is required to work harder than it is accustomed to. For strength training this applies to the force of contraction; for endurance training it means a repetition number of relatively low force contractions.

For a given stimulus to remain an overload, it must constantly be increased as muscle performance improves. The training programme needs to be kept at the same relative intensity and so must change in proportion to the changes in the muscle itself and, in the case of endurance training, the cardiovascular system.

Specificity

There is now considerable evidence that after training the greatest improvement is seen in the activity which was used for training (Jones et al 1989, Jones & Round 1990) and this is known as *specificity*. It also means that an improvement in one type of activity is not automatically reflected in other activities. There is also little or no overlap between strength and endurance training: if both are to be improved, they must be individually addressed in the training programme.

Motivation

It takes considerable mental and physical effort to carry out an effective training programme and only well motivated individuals will be able to do this. One of the key roles of the therapist is to inform and motivate the patient so that they can see the reason and relevance of what they are being asked to do.

Learning

There are two aspects of learning which relate to training. One is the motor learning which occurs with the practice of a motor event and is discussed later with strength. The other is that the patient must know exactly what they are required to do and what not to do. This puts the therapist in the role of educator, not only for the patient's knowledge about their own condition, but what they can do about it.

If the learning component is not addressed satisfactorily, the patient may well develop inappropriate movements and exercise programmes, but is unlikely to be sufficiently motivated to persevere with training.

Reversibility

After an acute or reversible pathological event a patient should regain their original position on the performance spectrum (Fig. 13.2). In the absence of ongoing disease or further incidents, they should be able to maintain that position without further training.

In some cases, and particularly with athletic training, they are in a position on the spectrum which is further to the right of their natural one. Once training ceases, the tendency will be for them to revert to their original, congenitally determined position. This is why athletes need to continue training to maintain their enhanced performance.

Diminishing returns

The response to training is roughly inverse to the physical condition at the start of training. Thus it is easier to bring about a 10% improvement in someone who is grossly weak or unfit than in an individual who is already at a high level of performance. All people have their own ceiling for physical performance and the closer one is to it, the harder it becomes to improve. This is probably one of the factors which persuades athletes to resort to illegal and often dangerous means of attempting to improve performance.

Despite these common principles, the physiological stimulus for increasing strength and endurance are very different. Strength training (Jones et al 1989) involves the generation of relatively high forces and therefore only low repetition is possible before fatigue sets in. In contrast, endurance training requires a high repetition of low force contractions.

Warm-up

Therapists and athletes commonly use a brief period of light exercise before embarking on the training or competitive event. This will certainly have the effect of increasing muscle blood flow and temperature (Åstrand & Rodahl 1988) which enhance performance. It also may allow a practice for the desired movement or activity and may possibly increase arousal and alertness.

It is also believed that a warm-up decreases the probability of strains and injury in athletes, although there is little direct evidence for this claim.

Warm-up regimes may take the form of a generalised, whole body activity designed to increase muscle blood flow and soft tissue flexibility (DeVries 1986). An alternative approach is to practise the specific movement to be used, but at low intensity (Shellock & Prentice 1985). There is little conclusive evidence on the relative merits of these two approaches.

STRENGTH TRAINING

The most usual method of providing resistance is by means of free weights, pulleys and springs or elasticated bands. In recent years there has been an increase in the number and type of equipment available for muscle testing (dynamometers). These tend to be isotonic or isokinetic systems which can be used to perform concentric or eccentric, isometric and dynamic contractions. Most are computerised systems which give measurements of strength and power in addition to joint angle and velocity. The display of force or power output acts as visual feedback which can aid and encourage the patient. Dynamometers are both large and expensive and will not be available to many therapists, nor are they a possibility for home use by the patient.

The terms *isokinetic* and *isotonic* are often used incorrectly and it is necessary that they are understood in order to avoid confusion and inaccuracy (Table 13.1).

Table 13.1 The definition and key features of different types of exercise. A single contraction may be purely of one type or contain elements of more than one

Type of contraction	Definition	Features
Isometric	Constant muscle length	No movement May be isotonic or force may vary
Isotonic	Constant force output	May be isometric or dynamic Velocity may vary
Isokinetic	Constant velocity	Usually not isotonic

Types of muscle activity

Isometric exercise (Fig. 13.5a)

During these static contractions the muscle remains the same length, no movement occurs and therefore no external work or power is performed. The contractions may be maximal or submaximal and the force generated may vary during the contraction or it may remain constant, i.e. isotonic.

If a maximal isometric contraction is maintained for more than a few seconds, the force will start to decline as fatigue sets in.

Isometric contractions are all that is available to a muscle which is immobilised. They are also useful when activity in a certain joint position or muscle length is to be avoided. They are also relatively simple procedures and the amount of learning and practice needed to perform them is minimal. It is important that the rest of the body is well stabilised so that effort can be directed to the required activity rather than to maintaining posture.

Isometric exercises can be performed with relatively simple and inexpensive equipment and are ideal for home use when a fixed object, such as a wall or table, can be used as the resistive force.

Isometric force can be measured objectively and reliably using inexpensive equipment such as the 1 RM or hand-held dynamometers and strain gauge systems in addition to the large and expensive isotonic or isokinetic dynamometers.

High force isometric contractions, particularly

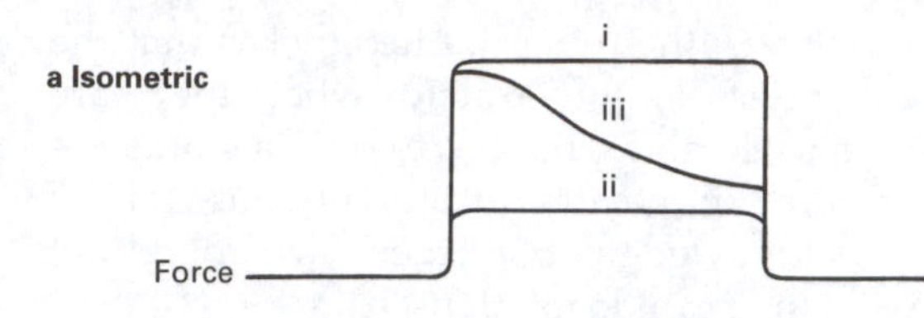

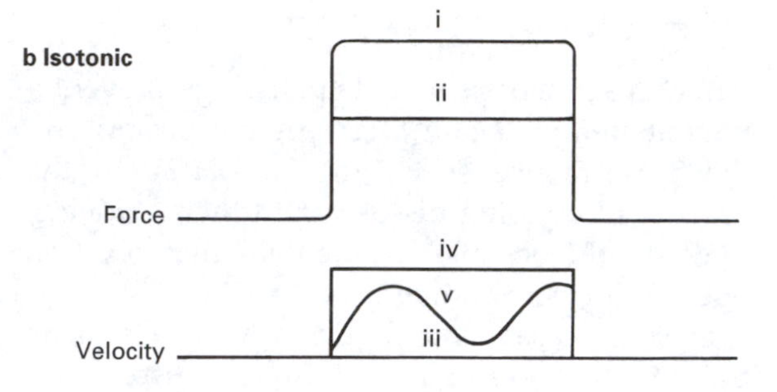

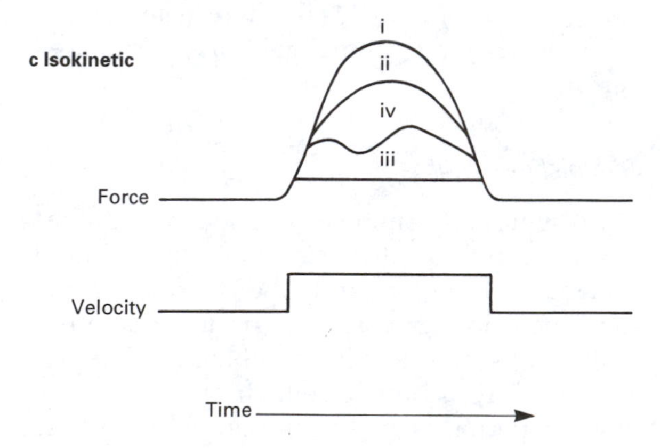

Figure 13.5 Possibilities for force and velocity in different contraction types. a) During an isometric contraction there is no movement. The force may be isotonic at either maximal (i) or submaximal (ii) levels, but it may also vary (iii). b) In an isotonic contraction the force remains constant at either maximal (i) or submaximal (ii) levels while the movement velocity may be zero, i.e. isometric (iii), isokinetic (iv) or variable and therefore neither isometric or isokinetic (v). c) In an isokinetic contraction the movement velocity remains constant but the force may be maximal (i) or submaximal (ii). It may be isotonic throughout most of the contraction (iii) or may fluctuate (iv).

of the upper limbs, cause large increases in blood pressure. They should therefore be avoided with individuals in whom this might be dangerous.

Isotonic exercise (Fig. 13.5b)

It is often thought that these are always dynamic contractions, but in fact they may be either dynamic or isometric (static). The key feature is that the force generation remains constant throughout the contraction. This is what happens when lifting or holding a given weight. If movement occurs, it may be isokinetic or the velocity of movement may vary.

If isotonic contractions are dynamic, it should be remembered that they cannot be maximal throughout the entire range of movement. Due to the length:tension relationship and also the biomechanical changes occurring during movement, the greatest force is usually generated in mid-range and is less at the extremes of range. Therefore, a force which is maximal at either the shortest or longest muscle length will be submaximal at mid-range. This is illustrated by the difference between force lines i and iii in Figure 13.5c.

Isokinetic exercise (Fig. 13.5c)

The movement velocity is chosen by the therapist and entered into the isokinetic dynamometer (a dynamometer is something which measures (metron) work (dynamics)) which then controls movement so that the preset velocities cannot be exceeded, irrespective of the patient's strength or force generated. However, it is possible for the patient to move at velocities lower than that.

The range of movement is controlled by the therapist and may be either all or part of the available or anatomical range.

It is important to realise that there is an acceleration phase at the start of movement and a deceleration phase at the end. Therefore, the patient is only moving at the preset velocity for part of the time. The proportion of the movement which occurs at the preset velocity decreases as velocity increases.

An important point is that the term 'velocity' refers to the angular velocity, i.e. that of the moving body parts, and not the velocity at which the muscles themselves are changing length. The relationship between angular and muscle velocity is a complex one and, in the absence of any direct measurements, no assumptions can be made about the velocity of changes in muscle length.

The maximal angular velocity available with even the most sophisticated isokinetic equipment is 300 deg/s. Whilst this feels very fast, it is relatively slow compared to the maximal physiological angular velocity which can exceed 1000 deg/s.

What is the best type of exercise to increase strength and performance?

For years this has been a subject of controversy and discussion. There are still a number of unknown issues, but a reasonable degree of clarity has emerged in some areas (McDonagh & Davies 1984, Jones et al 1989, Jones & Round 1990, McArdle et al 1991) and these are outlined below. It is worth remembering that there is considerable individual variation in the response to training even in normal, healthy people and that individual variation is probably even greater in those with pathology.

It is generally true that both static and dynamic high force contractions increase strength and that there is no inherent advantage with either type. The training method used should be based on the needs of a particular patient and will obviously be governed by the clinical condition and available equipment.

Manufacturers of dynamometers and advocates of particular exercise regimes claim that one form of resisted exercise, for example isokinetics, is superior to others, but there is very little evidence for this. One claim often made for both isotonics and isokinetics is that they are the most physiological form of exercise and therefore their effects are most likely to carry over to functional activities. These claims are rather spurious in that normal movement is composed of all types of activity; it is rarely isokinetic and is often not isotonic. Postural muscles are most likely to perform isometric contractions, but there are often small amounts of movement. There is no evidence that one form of exercise is inherently better than another.

There has also been debate about whether eccentric exercise is best for strength training. The theoretical basis for this is that the greatest force is generated by a muscle under eccentric conditions and if high force generation is the stimulus for hypertrophy, then they should be the most effective. However, the available evidence does not support this theory (McDonagh & Davies 1984, Jones & Round 1990). Another factor to consider is that unfamiliar high force eccentric exercise will cause muscle pain and long-lasting fatigue and damage (Newham 1988). There is no reason to think that this improves the training response, in fact it seems more likely to hinder it.

Specificity

This takes the form of training and task specificity (Jones et al 1989, Jones & Round 1990).

Training specificity

There are some reports of training specificity in both muscle length and velocity. In other words, if training is carried out at a particular muscle length or velocity, the benefits are only, or mainly, seen at that particular speed or length. Theoretically, this seems unlikely as it implies an alteration in the length:tension or force:velocity relationship, but these are fundamental principles of the myofilaments and it is difficult to see how a bump could appear in either of these curves.

The solution may be that the claims for length specificity have been made after studies on whole muscle groups which comprise a number of individual muscles. The observed length:tension relationship is the product of all these muscles, which may show peak tension at different lengths. If one of these muscles hypertrophies to a greater extent, or they all show different amounts of hypertrophy, then the length:tension relationship of the whole muscle could be changed (Jones & Round 1990).

The evidence for velocity specificity is not convincing. Measurements of the force:velocity relationship are very difficult to make in large muscle groups. This is particularly so when voluntary contractions, rather than electrically stimulated ones, are used.

Task specificity

It is undoubtedly the case that being able to lift heavier weights does not necessarily mean improved performance during activities and that there is no simple relationship between strength and performance. People can become better at lifting weights without becoming stronger simply by practising the movements and becoming more skilled at them. Equally, people can become stronger by training with isometric contractions, but not perform better in another activity such as cycling (Rutherford et al 1986b, Rutherford 1988). Therefore, task specificity is a very important part of strength training which is brought about by a learning process in which the correct sequence of movements is laid down as a motor pattern in the central nervous system. This means that the training task must be identical, or as similar as possible, to the functional task or tasks which it is hoped to improve.

Intensity, repetition and frequency

Despite the vast number of training studies which have investigated different training regimes, the optimal programme has not emerged. However, it seems that a load which exceeds 60% of maximum will produce an increase in strength if as few as 10 repetitions are carried out at least three times a week. Higher force contractions will bring about greater increases in strength. It is, therefore, essential that the resistance is increased as strength improves, otherwise it will cease to act as an adequate stimulus.

Types of strength training programmes

A number of training programmes are well established and all have their advocates. The two main categories are progressive resistance exercises (PRE) and progressive rate training (Hellebrandt & Houtz 1958). In the former the resistance is increased proportionally with muscle strength, while in the latter the resistance remains constant and the speed at which a set number of repetitions is performed is increased. This is a rather complicated technique involving the use of a metronome and, theoretically, is unlikely to be effective as the resistance remains constant, irrespective of the extent of strength increases and may cease to act as an overload stimulus.

Common examples of PRE methods are as follows.

The Delorme programme (Delorme & Watkins 1948) is based on the 10 RM; the maximal weight which can be lifted throughout the available range 10 times. Resistance is applied by means of a weighted boot or barbell. The programme consists of 1 set of 10 lifts at 0.5, 0.75 and 100% RM with a rest between each one. This is performed 5 times weekly and the 1 RM is used as a test of progress at weekly intervals.

The Oxford programme (Zinovieff 1951) uses a system whereby the resistance is progressively reduced in each training session. It consists of 10 sets of 10 repetitions. The first is at 10 RM and thereafter reduced by 0.5 kg in each set. A rest is given between each set. This is recommended for 5 days per week.

The MacQueen programmes (MacQueen 1954, 1956) consist of two programmes, one each for hypertrophy and power. The hypertrophy programme consists of three sets of the 10 RM with a rest between each set. The power programme consists of one set of 10 RM followed by a rest period and further sets in which the weight is increased and the repetition number decreased, e.g. second set of 8 lifts at 8–6 RM.

It seems that none of these programmes is inherently more effective in increasing strength than the others and perhaps the most pragmatic solution is to use the MacQueen hypertrophy programme as it is the simplest.

Changes during strength training

Strength training appears to result in a sequence of events (Jones & Round 1990). The first phase is that of motor learning when performance improves but strength remains virtually constant — this may continue for 6–8 weeks. The second phase is an increase in the strength of the muscles which appears to occur without a parallel increase in muscle size. This might be because of an increased synchronisation of motor unit firing (Komi 1986), an increased ability to recruit all available motor units or changes in fibre architecture or increased packing (density) of contractile material. The final phase starts at about 10–12 weeks and thereafter there is a slow but steady increase in both muscle size and strength. The stimulus for this phase remains unclear but certainly requires the generation of high forces.

Hypertrophy or hyperplasia?

One possibility for increased muscle size and strength with training are either that the total number of fibres remains constant, but they all increase in size (hypertrophy), or that the fibre diameter remains constant but there is a growth of new fibres and therefore an increase in fibre number (hyperplasia). Whilst there is some evidence of fibre splitting leading to hyperplasia in animal studies, there is little human data to support this theory and it appears that hypertrophy is the main mechanism (McDougall et al 1984).

TRAINING FOR POWER

In many cases, the desired outcome of strength training is an increase in power rather than strength *per se*. Since power is the product of force and velocity, these are the only variables which can be manipulated to improve power. The evidence is that the maximal velocity is unlikely to be affected by training and, therefore, the only practical way to improve power is by increasing strength (Jones & Round 1990).

The importance of task specificity has already been discussed. This means that the greatest increases in power output will be seen in the task used for training and not transferred equally across all activities. Similarly, strength can increase substantially when isometric exercise is used, but with little or no improvement in activities. Therefore, the training task should be the same as that which is required to improve, or should be as close as possible in terms of movement.

TRAINING FOR ENDURANCE

The common principles of training apply to endurance activities. In contrast to strength training, it is based on a high repetition number of relatively low force contractions and the stress (overload) is the increased oxygen demand of the working muscles. If high force contractions are used, anaerobic metabolism is utilised, a form of strength training is implemented and endurance capacity will not increase.

As a general rule, exercise should not cause heart rate values greater than those shown in Table 13.2. There may be clinical reasons to train at even lower heart rates than this, particularly for those with cardiovascular pathology.

The muscles need to contract at about 30–50% of their maximal force generation and exercise needs to be continued for about 20–30 minutes three times weekly (McArdle et al 1991) for endurance to improve.

During the sessions the individual should be slightly breathless if they are performing activities such as treadmill walking or cycling on an ergometer, but they should not be distressed or exhausted at the end.

Individual muscle groups may be trained by the performance of either isometric or dynamic contractions, but the cardiovascular and respiratory systems are best influenced by whole body exercise.

Table 13.2 The maximal heart rates recommended during training at different ages. Recommendations of the World Health Organization (Anderson et al 1971)

Age (years)	Upper limit (beats/min)
20–29	170
30–39	160
40–49	150
50–59	140
60 and over	130

The American College of Sports Medicine (1978) has issued the following guidelines for training to develop and maintain cardiorespiratory fitness in healthy adults. Lower levels of training may be necessary for some patients, particularly at the start of treatment:

- Frequency of 3–5 sessions each week.
- Intensity at 60–90% of maximal heart rate (measured or estimated for age) or 50–80% of VO_2 max.
- Duration of 15–60 minutes for each session. Lower activity should be continued for a longer time. Low to moderate intensity for relatively longer duration is recommended for non-athletic training.
- Type of activity: any that uses large muscle groups, that can be maintained for the desired time and that uses aerobic metabolism. Exercise such as walking, running, swimming and cycling are effective.

In general, endurance training which occurs less than twice weekly, at less than 50% VO_2 max and for less than 10 minutes per session is inadequate for this purpose.

Changes during endurance training

Endurance training affects the muscles themselves and also the cardiovascular and respiratory systems so that the delivery and utilisation of oxygen is improved (Åstrand & Rodahl 1988, Jones & Round 1990, McArdle et al 1991). Capillary density increases in the trained muscles so that the diffusion distance for oxygen decreases. The quantity and quality of mitochondria changes so that they become better able to metabolise oxygen.

The cardiac muscle hypertrophies so that the cardiac output increases and the resting heart rate decreases. The body switches to fat metabolism earlier, prolonging the time for which exercise can continue.

CIRCUIT TRAINING

This comprises a series of four to six activities

which are performed in a single session with a rest period between each one (Åstrand & Rodahl 1988). This has the advantage that general and task specific exercises can be selected and combined for each individual. Activities planned to increase strength, endurance, flexibility, coordination and general fitness can be combined.

There is some evidence that strength training is less effective when combined with endurance training (McArdle et al 1991) but this is probably more relevant to athletic training than clinical treatment.

As with all types of treatment, the purpose of each activity must be considered by the therapist and explained to the patient. Intensity and duration must also be increased with performance.

The programme should be designed so that different types of activities and muscle groups are interspersed to avoid fatigue and boredom.

Box 13.4 Task 4

What are the similarities and differences of training regimes for increasing strength and endurance?

THERAPEUTIC EXERCISE

This forms the major part of most physical therapy programmes and can be defined as the use of body movement in the treatment of a patient. It may include movement and use of individual joints or muscles, as well as graduated training for the activities of daily life. Exercise is often used in conjunction with other therapeutic modalities.

Assessment and measurement

Before commencing treatment, an accurate assessment must be thoroughly carried out. This should include the following:

- Diagnosis/pathology. Coexisting medical, psychological or social issues which may have an effect on their management, treatment, prognosis or outcome should be determined.
- The extent of handicap, impairment and disability should be established along with the limiting factors and their underlying causes.
- Sensory, perceptual, cognitive or behavioural problems. These may affect the treatment itself or the potential for the patient to become self-managing.
- Objective measures of muscle strength, joint range, functional activity relevant to the individual, their problems and lifestyle. Examples of functional activities are measurements of the time taken to walk a fixed distance, climb a flight of stairs, move from sitting to standing or the maximal weight that can be overcome.

Objective and subjective assessments are both essential so that a complete picture of the patient as an individual is built up, aims and goals for treatment can be established and progress monitored regularly and accurately.

It is essential that the therapist involves the patient in the assessment. Time must be taken to identify what the patient perceives to be a problem and what their expectations of therapy are. Aims and goals for treatment must be agreed between patient and therapist, otherwise it is unreasonable to expect the patient to comply and become sufficiently involved and motivated about their treatment. It is also essential that realistic goals are set to avoid disappointment and loss of motivation by the patient.

Both diagnosis and prognosis must be taken into account in addition to the patient's current state and their previous level of ability or disability. These will be revealed through a thorough examination and assessment and are the basis for treatment planning. The central questions to be asked are:

- What can the patient do?
- What can they not do and why?
- What do they want and need to do?

The assessment should include a detailed analysis of movement and any related problems. The present, absent and incorrect movements need to be identified along with the potential for improvement.

The principles of therapeutic exercise and

training are the same as those used in athletic training as described previously. The difference is that both the initial starting point and the desired outcome are likely to be at a lower level of performance. The training programme (and assessment) should take all the components of physical ability into account, i.e. strength, power, endurance, flexibility, coordination and skill.

Selecting and using an exercise programme

An exercise programme should take a holistic approach and consider the whole patient, not simply a clinical condition affecting one or more parts of the body. The exercise programme may also be part of a treatment programme by one therapy profession, or it may form part of a multidisciplinary rehabilitation plan. The latter must be carefully coordinated and effective communication with other health care professionals will avoid uneconomical use of time and unnecessary overlap in treatment.

An initial priority is to relieve pain and swelling as much as possible since both of these will negatively affect any exercise programme.

It is important to remember that whilst the goals may be similar for different patients, the best way to achieve these will vary from one individual to another and will be influenced by their present and past condition in addition to the functional tasks involved. Therefore, each programme must be individually tailored to each patient and the therapist must be vigilant in monitoring and making changes where necessary. Change may be necessary for clinical reasons or may be used simply to avoid boredom.

Where possible, activities should be used which the patient can carry on independently in their home environment. Any carers involved with a patient should also be involved in their treatment. They should be aware of the aims and goals of treatment, its expected outcome and time course. Carers can help in continued exercise programmes when a therapist is not present and may be important in keeping up motivation and ensuring that treatment continues and progresses.

Planning a treatment programme must involve the patient directly and closely. The setting of priorities, short- and long-term goals must be agreed between patient and therapist if the patient is to be expected to become sufficiently involved. Any restrictions to functional ability must be identified and taken into account. These may include the physical condition (motor and sensory) of the patient, their environment, psychological factors and also their ability to communicate, participate and learn.

A therapy programme should be developed on the basis of the individual patient's current level of performance and the desired outcome. Clear goals and aims must be set and adhered to. However, if it becomes apparent that they are no longer appropriate or realistic, they must be changed accordingly. An effective programme requires close interaction between patient and therapist so that problems are identified and a rational and mutually agreed programme is embarked upon.

The patient is required to make a considerable effort and it is unreasonable for the therapist to expect this input if the patient is not aware of, and in agreement with, the expected results. The therapist must involve the patient at every stage of the treatment and take care to find out from the patient what are their main problems and expectations. Any unrealistic expectations by the patient must be pointed out so that they do not become disappointed and lose motivation. Care must also be taken that the programme is varied and interesting, otherwise it may become boring for all concerned.

The setting of long- and short-term goals will help both patient and therapist to plan and monitor treatment. Regular and accurate assessment and measurements are essential for this. A problem-solving approach is the key to the optimal planning and execution of a successful treatment programme involving a number of factors and steps which are illustrated in Table 13.3. Those involved with caring for the patient should also be involved in all stages of this process.

Table 13.3 The clinical problem-solving approach necessary for successful and effective treatment

The therapist		The patient
Examines and assesses effectively		Communicates needs, desires and fears
Establishes long- and short-term goals	→	Gains knowledge of condition
Selects appropriate, interesting and challenging procedures	Feedback ←	Acquires understanding of what (and what not) to do and why
Designs effective independent home programmes		Becomes motivated to work
Gives correction and encouragement		Receives encouragement
Communicates effectively with patient and team members		Gains confidence in self and therapist(s)
Cooperates with patient and team members		Cooperates with those treating and managing
Replans when necessary		
Aims for patient takeover and independence		

Treatment outcomes

Relief of pain and oedema

Prevention of further damage and disability

Improvements of:
- strength and power
- endurance
- balance and stability
- flexibility
- coordination
- sensory and proprioceptive awareness
- cardiorespiratory function

Improved function, morale and satisfaction for the patient

Preparation for treatment

If there is severe loss of movement, the circulation to the affected parts may be reduced and there may also be some pain and stiffness. Both these would impinge on treatment and may be relieved by gentle warmth, massage and passive movements.

Body positioning

The patient's kinaesthetic sense may be reduced by sensory loss or prolonged disuse and the patient may have a poor sense of their posture. Correct body and limb alignment is important in gaining accurate and efficient movement and minimising the development of secondary postural problems. The therapist should explain this clearly to the patient. Continuous monitoring and correction of posture may be necessary.

The choice of activity

This should be governed by the findings of assessment, examination and measurement. Additional factors will also play a role in determining the choice of activity for a particular individual.

Psychological factors

The patient must be motivated to make the effort

necessary for exercise and training. They should have a clear idea of what is required, and why, as well as how it will affect their function or clinical condition.

There may be several possible reasons for a patient not being sufficiently motivated to comply with an exercise programme, including personal or social reasons for wishing to remain in their current state. Patient behaviour patterns can be adopted (Pilowsky 1994) either consciously or unconsciously. They are not easy to break out of and may be reinforced by carers, friends and relatives (Rowat et al 1994).

A patient may be genuinely convinced that improvement will not take place, or that if it does it will be so small as to make little difference to their lives. They may be frightened of failing to improve. If legal proceedings are proposed or pending, there may also be financial reasons for poor motivation. The therapist should explore these possibilities and deal with them as best they can, and care should be taken not to assume that poor motivation automatically indicates a difficult or malingering patient.

Anatomical and physiological factors

The patient needs to have the following in order for the desired activity to occur:

- necessary range of movement
- adequate muscle strength to maintain posture and simultaneously perform the movement. External stability or assistance is necessary if the patient is unable to provide this themselves
- adequate endurance and skill
- the sensory and motor components for movement control
- the cognitive and perceptual ability to understand and perform what is required.

Mechanical factors

- Gravity, friction and momentum may be utilised to assist or resist movement.
- The size of the base of support and type of supporting surface may help or hinder the desired movement.

Environmental factors

Concentration, motivation and the sense of effort can all be affected by a number of environmental influences. The most common disadvantages are:

- cramped space
- dim or over bright lighting
- excessive heat or cold
- noise and air pollution.

Patterns of movement

The technique of proprioceptive neuromuscular facilitation (PNF) utilises movement patterns relevant to normal functional activities which have diagonal and rotatory components (Knott & Voss 1968, Waddington 1976). It emphasises proprioceptive stimulation, visual, auditory, tactile and stretch, in an attempt to gain a maximal muscle response.

Irradiation, overflow and cross transfer

An attempt to make a maximal voluntary contraction of an individual muscle will quickly demonstrate the fact that it is virtually impossible to isolate the effort to only one muscle. This may be due in part to the simultaneous need for increased stabilising forces, but it is probably also due to the fact that the central nervous system tends to activate groups of muscles, rather than individual muscles. Furthermore, the excitatory influences descending from the higher centres are unlikely to be restricted to only specific anterior horn cells forming motor units within a single muscle (Rothwell 1994). Therefore, strong voluntary efforts directed at one muscle may irradiate, or overflow, to other muscles within the same functional group or to those in a group performing a synergistic or stabilising role. For example, it is difficult to imagine being able to voluntarily activate only one muscle of the quadriceps femoris group, or to activate these muscles without including the abdominals.

Similarly, it is difficult to limit strong muscle activity to a unilateral activity. It is highly likely that a maximal effort of the right quadriceps femoris muscles takes place with the muscles of the left limb being completely inactive. A number of unilateral training studies have reported a strength increase in the apparently 'untrained' limb.

It should be remembered that such techniques are only appropriate when a patient has a problem with the voluntary activation of a particular muscle either through disease, injury or disuse. The force of contraction elicited by such means is unlikely to be sufficient to bring about hypertrophy in all but the very weakest muscles. Therefore, their role is restricted to learning or relearning to use inactive muscle. Once voluntary activation is possible, it is more effective to use direct resisted exercises of the muscle, or muscle group, in question.

In a low force contraction it is easier to restrict muscle activity to the prime movers. As force increases, synergist muscles are progressively activated and with strong activity, particularly at extremes of range, there may be co-contraction of antagonists (Rothwell 1994).

Trick movements

If the prime mover for a particular action is too weak to bring about the movement, an instinctive attempt is made to carry out the movement using other muscle groups and body movements. A common example of this is the use of the lateral abdominal muscle to lift the pelvis if the hip abductors alone are inadequate for this purpose.

Such trick movements should be identified by the therapist and pointed out to the patient. Where possible they should be avoided as they can become habitual and may lead to deformity and musculoskeletal problems. However, they can be useful in very early stage treatment by utilising the principles of irradiation and overflow. Furthermore, it is clearly unreasonable to expect a patient to restrict their activities to those which can be performed correctly. In some cases this would markedly increase the level of disability and have severe functional consequences, particularly in chronic and progressive conditions.

Muscle imbalance

The importance of the relative strength of agonist and antagonist muscle pairs is the subject of much interest (Dirix et al 1988). This is particularly so for the knee and spine where there are claims that a disturbance of the normal balance of strength between flexors and extensors results in an increased susceptibility to musculoskeletal problems.

It remains unclear how important the balance of strength between muscles with an opposing action is, but it would not seem unreasonable to seek to correct any large discrepancies.

There is some evidence that resisted exercises of one muscle group cause greater strength increases if resisted contractions of the agonist are alternated with resisted reciprocal contractions of the antagonist, rather than with a rest period between contractions of the agonist (McArdle et al 1991). This is thought to be due to neuromuscular facilitation and the subsequent recruitment of more motor units.

Starting and finishing positions

All activity has a starting position which is the position of readiness from which activity is initiated. The finishing position may be a return to the initial starting position or might be the beginning of a new phase in a sequence of movement.

The starting position

For movement to be accurate, it is necessary that the patient understands the need for good postural alignment, is conscious of the position of their head and body and is able to correct when necessary. This process requires constant minor adjustments unless the patient is fully stabilised by external means.

The starting position may be used:

- as a foundation for an activity
- to provide fixation for one part of the body so that localised movement can be achieved
- to train posture and balance
- as an exercise to improve the stability and safety of the required posture.

An activity can initially use a very stable starting position and then be made more difficult by performing it in progressively more unstable positions. The four basic postures, from the most (easiest) to least (hardest) stable, are lying, sitting, kneeling and standing.

The factors necessary for selecting a starting position are:

- Appropriateness to the activity itself and relevance to the needs of the patient and acceptable to them.
- The supporting surface is stable, the base of support is of an adequate size and weight is distributed correctly.
- The position can be taken up, with or without help, held in good alignment and is capable of any necessary adjustment.
- The line of gravity falls within the base.
- Resistance to external forces can be withstood.

Progression may take the form of changing the starting position or by modifying the current one. An activity will be made more difficult if:

- the size of the base is reduced
- the centre of gravity is raised
- the speed of movement is increased
- the supporting surface is made more soft or mobile.

A position may fail for a number of reasons:

- an inappropriate choice
- muscle weakness or fatigue
- inadequate joint range
- joint or muscle pain
- lack of coordination or balance
- disturbed proprioception
- abnormal reflex activity.

Additional support may be provided, i.e. by wall or parallel bars, straps or splints, to help ensure correct posture and movement and to aid safety and confidence.

Teaching and learning

When helping a patient to learn a movement, whether one of an isolated joint or a sequence of movement patterns, it is beneficial to use as many teaching techniques as possible.

Explanation is used to describe what is required, why it is necessary and what the benefits will be.

Demonstration by the therapist will also act as a learning aid in combination with other techniques. However, even for trained observers, a demonstration is difficult to follow in any detail (Carlsöö 1972). In unilateral conditions it may be practised on the unaffected side and should be shown, with assistance and clear guidance, on the affected side.

This procedure will clarify the proposed movement to the patient and gives them the opportunity to practise, which is important for motor learning.

Illustrations are an important aid to learning for those able to use them. They may take the form of videotapes or photographs as well as written illustrations and instructions. It is helpful to give the patient some form of illustrated instruction that they may keep with them.

Review and progression

Regular reviews should assess and measure the patient's progress. Treatment should be modified according to the answers to the following questions:

- What has been achieved?
- Have the aims altered?
- What features should be added to, or removed from, the programme?

Once a contraction or movement is possible, the aim of training is to increase the intensity and complexity of the muscle contractions. Assistance should be phased out as quickly as possible and replaced by a resistance (overload) which increases with ability.

Resource management issues for therapists make it necessary for manual resistance to be replaced by other resistive forces as soon as possible in strength training. Lever length can be manipulated so that, initially, short lever arms are used and the patient progresses to longer ones. Resistance can be increased by changing the starting position so that first gravity and then mechanical resistance is used.

Movement patterns should be increased in complexity from simple movements of individual joints to those involving multiple joints and patterns which mimic those of functional activity and aid control of movement. In this way, independent work by the patient is encouraged.

If an increase in strength is the sole aim of treatment, there is little point in increasing the number of repetitions to more than approximately three sets of ten contractions performed three times a week. An increase in the number of repetitions above this in a single session will have little effect due to the onset of fatigue, which will temporarily reduce the strength of the muscle and not induce hypertrophy.

Most training programmes will aim to increase both strength and endurance, but it is important to remember that different types of exercise will need to be performed for these two purposes.

Early re-education of movement

In the early stages of treatment, a patient may be unable to perform the desired activity without assistance from the therapist, but such assistance should only be provided when absolutely necessary. Independent activity, albeit assisted by gravity or the patient themselves, should be implemented as early as possible. The patient should take responsibility for their own treatment so that they may continue it when the therapist is unavailable and in their own environment. When possible, the therapist should ensure that the patient has a good understanding of the exercises which they can do without assistance and how to progress them in order for the treatment to continue without the therapist.

Active movement may be impossible because of injury, disease, disuse or immobilisation. However, weakness and atrophy need to be unusually severe for any movement at all to be impossible. Denervation of a muscle of normal strength will result in the affected motor units being unable to generate any voluntary force, although either external electrical or magnetic stimulation will generate the available force.

When active movement is not possible, it is necessary to perform passive or assisted active movement throughout the available range in order to prevent or reverse any shortening of the soft tissues (muscle fibres, tendons and ligaments) causing a loss of range of movement and the flexibility that is necessary for optimal function.

For hypertrophy to occur, the muscles must generate a force which is close to their maximal capacity; passive movements will not, therefore, increase strength. In addition to preserving or increasing joint range, passive movements in which the patient attempts to participate may act as an aid to the central nervous system in learning, remembering and practising motor patterns. It may be beneficial to perform passive and assisted active movements in patterns which involve a number of joints in patterns of activity used in daily living. Thus, movements involving whole limbs, e.g. in flexion and extension patterns, may be preferable to moving each joint individually.

The starting position

A starting position should be selected which encourages and allows the desired movement to occur freely. In the case of very weak patients, the segment to be exercised may need to be positioned so that the effects of gravity are minimised if movement is to occur. An example of this is someone with very weak hip flexors who may be unable to produce movement when lying supine. They may, however, be able to do so if they are placed on their side, lying with the affected leg uppermost and its weight supported by the therapist.

Where possible, clothing should be adjusted so that the relevant parts of the body are uncovered and visual feedback can be used by both patient and therapist.

The patient should be comfortable and well supported so that they are free to concentrate on the desired movement and unnecessary strains and stresses are avoided.

Body stability and support are necessary to hold the body steady and in good alignment so that nearly maximal muscle contractions can be made with the minimum of stress and strain. For example, in quadriceps strengthening exercises, an ideal position is sitting with the thigh and back well supported.

Assisting and resisting forces

Manual assistance provided by the therapist is the most common form of assisting force. This may also be provided by the patient in some cases. Assistance may also be provided by water, smooth surfaces or supporting slings.

Resistive forces may include manual resistance by the therapist. Limb weight, friction or unwanted contractions or spasm in antagonist muscle groups will also increase the resistance to movement.

The ability to develop maximum force depends on adequate proprioception. The patient will be helped by the sight of movement and correct placement of the therapist's hands indicating the direction of movement or providing resistance. Clear instruction, encouragement and feedback are necessary.

Manual assistance

Ideally, the patient will gain confidence and reassurance in working closely with the therapist and it is important that manual contact is comfortable and pleasant. The therapist's hands should be warm, gentle and supportive and should help the patient understand what they are required to do. One of the therapist's hands should be used to control the proximal area and support the working part while the other is used for providing stretch, direction and assistance or resistance as appropriate.

The therapist should ensure that they move with the patient and that their own body position is maintained in a good position which will reduce the incidence of musculoskeletal injury.

Manual resistances

The role of manual resistance is restricted to situations where other forms of resistance are not possible. This might be due to extreme weakness or cases where the force generated in different parts of the range varies more than usual. It is also useful in the very early stages of treatment when the patient needs assistance from the therapist on how to perform the movement.

In practical terms, it is very difficult, or even impossible, for a therapist to offer sufficient manual resistance to adequately challenge the large muscles of adults with anything other than extremely weak muscles. To attempt to do so is putting the therapist at unnecessary risk of physical injury.

Manual resistance has the advantage that it is quick and easy to apply. However, it is not accurately measurable and can be influenced as easily by any changes in the physical and psychological state of the therapist as by real strength changes in the patient.

Stretch and vibration

A patient may have difficulty in activating a muscle which is extremely weak or which is not normally innervated with either motor or sensory impairment. In these situations, the muscle may be facilitated by a rapid stretch or mechanical vibration. The stretch should be applied to the contracting muscle in a lengthened position, although care must be taken not to damage the weakened muscles or the joints which they act over.

Both stretch and vibration activate the extrafusal muscle fibres through the muscle spindles. They also have the additional effect of inhibiting activity in antagonist muscle groups (Rothwell 1994).

Later stage treatment

Once the patient is able to perform movement against gravity, the emphasis of training and rehabilitation should be on independent exercise. Input from the therapist will still be necessary to correct any postural errors and to ensure that the training overload is increased appropriately. The performance of exercise which does not stress the system will not act as a stimulus for a training response.

The importance of task specificity must constantly be remembered and the training task should reflect as closely as possible the function requirements of each individual. In this way the effects of increased strength can be combined with motor learning processes which will act together to optimise function.

General maintenance

During all stages of treatment the strength and function of the whole body should be borne in mind and maintained as much as possible. The prognosis may be unknown: there may be improvement for some patients whilst others will deteriorate or remain static.

The situation must be reviewed regularly using both subjective and objective assessments and treatment goals, with aims and programmes revised accordingly. In addition to the peripheral effects of exercise on the muscles themselves, the effects of exercise on the heart and circulation are important and should always be remembered.

Box 13.5 Task 5

What factors should be taken into account when developing an exercise programme? Consider two individuals with an acute ankle injury: one is a competitive athlete and the other is an elderly person who has had a stroke previously.

An unsatisfactory training programme

If difficulties arise which the therapist cannot correct easily, the reason for the patient's difficulties must be analysed. Analysis should include how, why and where the movement fails. The cause may include errors such as:

- planning faults
  - — an overambitious or insufficient progression
  - — the movement sequence may be too complicated
  - — inadequate assistance, teaching and explanation
- psychological and emotional factors
  - — lack of ability of the patient to understand what is required of them
  - — poor memory
  - — poor awareness by the patient of their body positioning
  - — boredom, lack of interest and effort
  - — fear of pain and exacerbating the clinical condition
  - — fear of inadequacy in the outside world
  - — dislike or lack of rapport with the therapist
- teaching faults
  - — poor explanation and demonstration
  - — lack of suitable stimuli
  - — incorrect or inadequate help and correction
- other complicating factors
  - — poor vision and hearing
  - — poor general health resulting in lack of energy and motivation
  - — coexisting pathology
  - — sensory loss
  - — impending litigation
  - — the patient having become used to dependency or encouraged to be dependent by others.

Box 13.6 Task 6

A training programme is apparently not having the desired effect. What factors might contribute towards this? Consider this situation for a number of different individuals.

REFERENCES

Agre J C, Rodriquez A A 1989 Validity of manual muscle testing in post-polio subjects with good or normal strength. Archives of Physical Medicine and Rehabilitation 70(Suppl): A17–18

American College of Sports Medicine 1978 Position statement on the recommended quantity and quality of exercise for developing and maintaining fitness in healthy adults. Medicine and Science in Sports 10: vii–x

Anderson K L, Shephard R J, Denolin H, Vannauskas E, Masironi R 1971 Fundamentals of exercise testing. World Health Organization, Geneva
Åstrand P E, Rodahl K 1988 Textbook of work physiology. Physiological basis of exercise. McGraw-Hill, Singapore
Borg G A V 1982 Physiological base of perceived exertion. Medicine and Science in Sports and Exercise 14: 377–381
Carlsöö S 1972 How man moves. Heinemann, London
Delorme T L, Watkins A L 1948 Techniques of progressive resistance exercise. Archives of Physical Medicine 29: 263–273
DeVries H A 1986 Physiology of exercise for physical education and athletics. William C Brown, Dubuque
Dirix A, Knuttgen H G, Tittel K (eds) 1988 The Olympic book of sports medicine. Blackwell Scientific, Oxford
Grimby G, Saltin B 1983 The ageing muscle: a mini review. Clinical Physiology 3: 209–218
Hellebrandt F A, Houtz S J 1958 Methods of muscle training: the influence of pacing. Physical Therapy Review 38: 319–322
Hurley M V, Jones D W, Wilson D, Newham D J 1992 Rehabilitation of quadriceps due to isolated rupture of the anterior cruciate ligament. Journal of Orthopaedic Rheumatology 5: 145–154
Jones D A, Round J M 1990 Skeletal muscle in health and disease. Manchester University Press, Manchester
Jones D A, Rutherford O M, Parker D F 1989 Physiological changes in skeletal muscle as a result of strength training. Quarterly Journal of Experimental Physiology 74: 233–256
Knott M, Voss D E 1968 Proprioceptive Neuromuscular Facilitation. Harper & Row, New York
Komi P V 1986 How important is neural drive for strength and power development in human skeletal muscle? In: Saltin B (ed) Biochemistry of exercise VI. International Series on Sports Sciences Vol 16. Human Kinetics, Champaign, Illinois
Krebbs D E 1989 Isokinetics, electrophysiological and clinical functions relationships following tourniquet-aided knee arthrotomy. Physical Therapy 69: 803–815
Kukulka C G 1992 Human skeletal muscle fatigue. In: Currier D P, Nelson R M (eds) Dynamics of human biological tissues. F A Davies, Philadelphia
McArdle W D, Katch F I, Katch V L 1991 Exercise physiology: energy, nutrition and human performance. Lea & Fabinger, Philadelphia
McDonagh M J N, Davies C T M 1984 Adaptive response of mammalian skeletal muscle to exercise with high loads. European Journal of Applied Physiology 52: 139–155
McDougall J D, Sale D G, Alway S E, Sutton J R 1984 Muscle fibre number in biceps brachii in bodybuilders and control subjects. Journal of Applied Physiology 57: 399–403
MacLaren D P, Gibson H, Parry-Billings M, Edwards R H T 1989 A review of metabolic and physiological factors in fatigue. In: Pandolf K B (ed) Exercise and sport sciences reviews 17: 29–66
MacQueen I J 1954 Recent advances in the technique of progressive resistance exercise. British Medical Journal 2: 1193–1198
MacQueen I J 1956 The application of progressive resistance exercise in physiotherapy. Physiotherapy 40: 83–93
Martin E G, Lovett R W 1915 A method of testing muscular strength in infantile paralysis. Journal of the American Medical Association 65: 512–513
Moritani T, DeVries H A 1979 Neural factors versus hypertrophy in the time course of muscle strength gain. American Journal of Physical Medicine 58: 115–130
Newham D J 1988 The consequences of eccentric contractions and their relation to delayed onset muscle pain. European Journal of Applied Physiology 57: 353–359
Pilowsky I 1994 Pain and illness behaviour: assessment and management. In: Wall P D, Melzack R (eds) Textbook of pain, 3rd edn. Churchill Livingstone, Edinburgh
Rothwell J 1994 Control of human voluntary movement, 2nd edn. Chapman and Hall, London
Rowat K M, Jeans M E, LeFort S M 1994 A collaborative model of care: patient, family and health professionals. In: Wall P D, Melzack R (eds) Textbook of pain, 3rd edn. Churchill Livingstone, Edinburgh
Rutherford O M 1988 Muscle coordination and strength training: implications for injury rehabilitation. Sports Medicine 5: 196–202
Rutherford O M, Greig C A, Sargeant A J, Jones D A 1986a Strength training and power output: transference effects in the human quadriceps muscle. Journal of Sports Science 4: 101–107
Rutherford O M, Jones D A, Newham D J 1986b Clinical and experimental application of the percutaneous twitch superimposition technique for the study of human muscle activation. Journal of Neurology, Neurosurgery and Psychiatry 49: 1288–1291
Saltin B, Blomquist J H, Mitchell R L, Johnson J, Wildenthal K, Chapman C B 1968 Response to submaximal and maximal exercise after bedrest and training. Circulation 38 (Suppl 7): 1–78
Sapega A A 1990 Muscle performance evaluation in orthopaedic practice. Journal of Bone and Joint Surgery 72A: 1562–1574
Shellock F G, Prentice W E 1985 Warming-up and stretching for improved physical performance and prevention of sports-related injuries. Sports Medicine 2: 267–278
Stokes M 1985 Reliability and repeatability of methods for measuring muscle in physiotherapy. Physiotherapy Practice 1: 71–76
Stokes M, Young A 1986 Measurement of quadriceps cross-sectional area by ultrasonography: a description of the technique and its application in physiotherapy. Physiotherapy Practice 2: 31–36
Tanner J M 1964 The physique of the Olympic athlete. George Allen & Unwin, London
Waddington P J 1976 In: Hollis M (ed): Practical exercise therapy. Blackwell Scientific, London. Chaps 21–25
Watkins M P, Harris B A, Kozlowski B A 1984 Isokinetic testing in patients with hemiparesis. A pilot study. Physical Therapy 64: 184–189
Zinovieff A N 1951 Heavy resistance exercises. British Journal of Physical Medicine June: 129–133

CHAPTER CONTENTS

14

Evaluating and measuring human movement

M. Trew T. Everett

OBJECTIVES

When you have completed this chapter you should be able to:

1. **Understand why the measurement of human movement is difficult**
2. **Discuss the range of measurement techniques available**
3. **Be able to choose the appropriate tool for evaluating and measuring movement**
4. **Be able to evaluate the value of the different measurement techniques.**

INTRODUCTION

At the moment there is no single, satisfactory solution to the problem of how to measure human movement and, in particular, human movement when functional activities are being undertaken. Whilst it is relatively easy to measure static parameters such as limb length or the position of a stationary joint, it is considerably more difficult to measure and analyse the performance of functional activities. The problem lies in the multifactorial, multidimensional nature of human movement which results in the need to measure not only factors such as joint movement or muscle activity but also to take into account the control and the quality of the movement. Currently, it is not possible to measure the control and quality of movement, mainly because these terms are difficult to define and, until there is a consensus as to the exact meaning of quality, it is unlikely that satisfactory measures will be designed. Whether it will ever be possible to produce a global definition of quality of movement is open to discussion. Quality encompasses many factors such as precision, smoothness, velocity, range and appropriateness; the importance of each of these factors differs from movement to movement so that the likelihood of producing a universal measure of quality would seem remote.

Previous chapters have described the normal patterns of movement in some common functional activities and the information given was often obtained by collating the work of many scientists who were looking at different aspects of the same movement. The complexity of measuring human movement has made it extremely difficult to design a system whereby all the parameters involved can be measured simultaneously. Those professionals who need to analyse normal and abnormal human movement are faced with the problem that they need to consider all the factors which together contribute to normal movement, yet there is no practical way of achieving this. One solution is to undertake a subjective assessment of the activity, after which it should become clear which parameters are most important and then a limited number of appropriate measurement tools can be chosen.

KINEMATIC ASSESSMENT OF HUMAN MOVEMENT

Kinematic assessments consider human movement in terms of position and displacement (angular and linear) of body segments, centre of gravity, and acceleration and velocities of the whole body or segments of the body such as a lower limb or the trunk.

A kinematic assessment will provide information on the relationship of parts of the body to each other. This is useful in measuring joint angles during complex movements and has provided the basis for understanding functional activities including rising from a chair or stair climbing: much of the information in Chapter 10 comes from kinematic assessments. Equally valuable is a knowledge of the accelerations and velocities of body segments; for example, in prosthetics it is important that the design of a prosthetic limb ensures that it has the same properties of acceleration and velocity as a normal limb. A knowledge of the displacement patterns of the centre of gravity during movement gives information on efficiency, which is an important factor in most rehabilitation programmes.

Kinematic assessment uses anatomical terminology and a spatial reference system where:

Y = vertical component or direction
X = anterior posterior component or direction
Z = medial lateral component or direction.

Figure 14.1 illustrates the conventions used when measuring angles in the major body planes. Angles in the XY plane are measured from 0° in the X direction with positive angles being anti-clockwise, which makes extension positive and flexion negative. In the YZ plane, angles start at 0° in the Y direction with positive angles being anti-clockwise (Winter 1990).

KINETIC ANALYSIS

Kinetics is the description of human movement in terms of force and these forces can be internal or external (see Ch. 2). Internal forces include those resulting from muscle activity, force generated by stretch on non-contractile and elastic soft tissue and internal friction. Ground reaction forces, forces generated by other people, external loads or wind resistance would all be classed as external.

Apart from the direct information on force production obtained from kinetic analysis of human movement, it is also possible to infer information about muscle function or pain. In addition, it is possible to look at the forces generated by muscles in relationship to task performance or the way external forces impinge on an individual's ability to undertake a functional activity.

ENERGY EXPENDITURE ANALYSIS

There are a number of approaches which can be

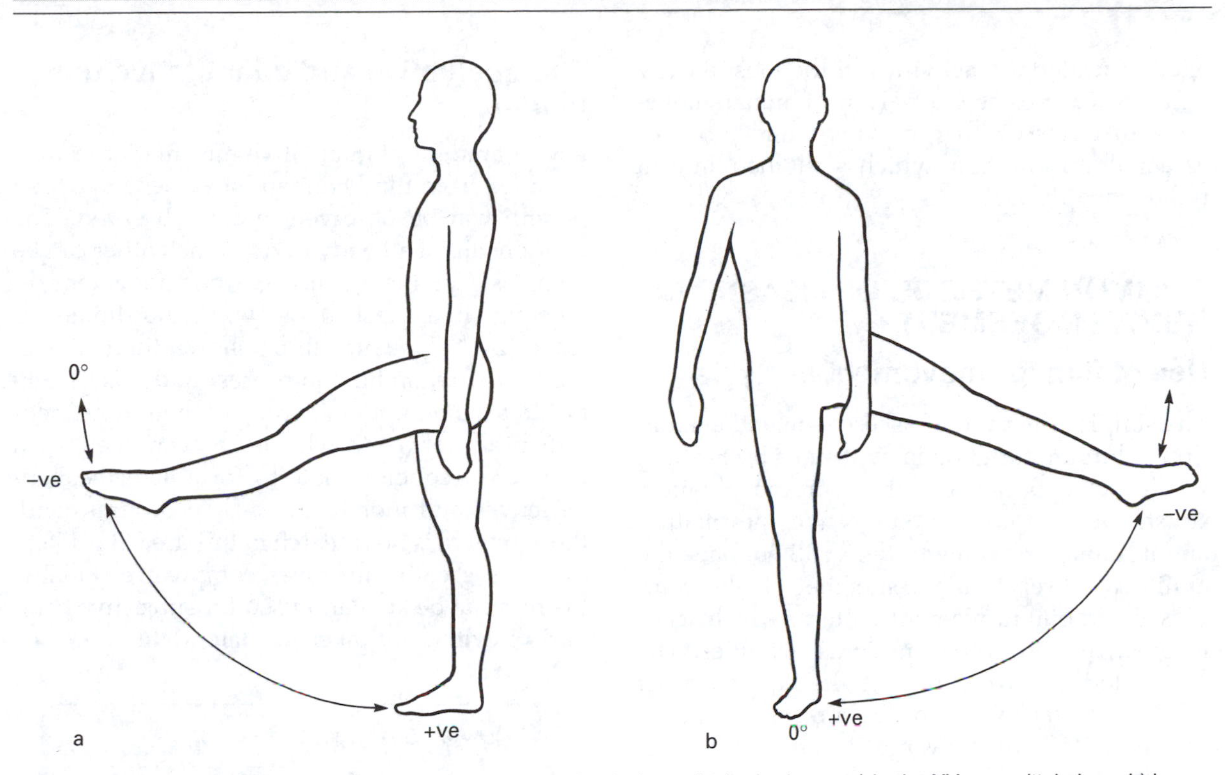

Figure 14.1 The conventions used when measuring angles in the major body planes: a) in the XY or sagittal plane; b) in the YZ or coronal plane.

Box 14.1 Case study 1

Mr G., a keen motor cyclist, was struck from the side by a car which joined a main road without stopping. Apart from the injuries received as a consequence of the impact, Mr G. was then run over by an oncoming bus. He suffered multiple fractures and avulsion of soft tissue, but despite his injuries made a remarkable recovery.

During his rehabilitation, force traces taken from a Kistler force plate showed considerable variation from the ground reaction forces that would be normally expected in walking (a normal trace taken from a force plate is shown later in the chapter, see Fig. 14.6). His trace indicated a marked reduction in initial foot–floor contact force and also in the force that should have been generated on push off. Mr G. was reluctant to approach heel strike at normal velocity because the forces generated caused him considerable pain at the tibial fracture sites. He also experienced difficulty in generating force at push off because of tissue damage to the plantar flexor muscles.

used to calculate energy expenditure during movement; the details are beyond the scope of this book but can be found in any advanced physiology text. Direct calorimetry produces highly accurate and repeatable results but is largely impractical as it requires the use of an airtight insulated chamber in which the activity is performed. Indirect calorimetry most commonly relies on the analysis of the oxygen and carbon dioxide content of expired air and is valuable and reasonably practical but still requires the subject to be attached to some form of device in which expired air can be collected. The need to wear a noseclip and to breathe through a fairly large mouthpiece can be rather daunting and needs quite lengthy acclimatisation. In addition, upper limb activities may be restricted by the tube leaving the mouthpiece (McArdle et al 1991, Whittle 1991).

Consideration of energy expenditure is important in patients who are frail or disabled. When an individual is already functioning at the limit of their ability, it is essential that they are encouraged to undertake activities in the most energy efficient way possible. Analysis of human movement in terms of energy expenditure provides essential information which will then inform treatment approaches.

COMMON METHODS OF MEASURING HUMAN MOVEMENT

Use of film for movement analysis

The simplest method of assessment and evaluation of human function is by visual observation. Whilst an experienced observer can obtain a substantial amount of subjective information about human movement, they will not have the ability to observe and remember all the complex, multijoint movement patterns which occur in even the simplest functional activities. The unassisted eye functions at the equivalent of 1/30th of a second exposure time and can only see the details of slow motion; the brain, too, despite its amazing ability, has a limit on the amount of information it can absorb and remember. As a consequence, when observing complex movement only a limited amount of the detail is actually seen (Terauds 1984). The other major drawback to unaided visual observation is that only subjective information can be obtained and, without baseline data, reliable measurement of change is impossible.

The use of film, whether it is cine, video or still photography, enables movement to be observed in much more detail, permits measurement and provides a permanent record.

Cine and video give a permanent record which can be stored over many years and replayed repeatedly. This is particularly valuable as it allows the analysis of movement to occur after the patient or subject has left when there is time for uninterrrupted observation and analysis. Visual analysis of film is enhanced by the use of freeze-frame, slow motion facilities and computer aided analysis software. Video and cine film of human movement enhances the acquisition of qualitative information, quantitative data and can also be used as feedback to the patient.

The qualitative and quantitative use of film

A vast amount of qualitative information can be obtained from film. Human movement as a total pattern can be observed and re-observed. The relationship of all body parts to each other can be seen, as can the quality of the movement — whether it is fast or slow, incoordinate or smooth. The patient can be shown their film as part of the rehabilitation process and this greatly facilitates their understanding of their movement difficulties. Patients who have seen their own films are frequently able to formulate recovery objectives and monitor their own progress and, though there is no research in this area, it is likely that this greatly improves compliance. Finally, the film can be kept and used for subjective comparison with films taken at a later date.

Box 14.2 Case study 2

Mrs Wilson, 59, had had rheumatoid arthritis for approximately 20 years when she suddenly noticed an increased clumsiness when undertaking functional activities using her hands. She had learned to cope with limited movement and pain at both shoulder and elbow joints and severe ulnar deviation of her metacarpophalangeal joints. For many years she had managed most activities of daily living despite her substantial problems, but now she found she was knocking over items as she went to pick them up and her accuracy at putting down objects like cups or a vase of flowers was seriously compromised. She reported that she had noted no change or deterioration in her physical condition so it was decided to evaluate the total upper limb joint movement patterns by use of video. When viewed in slow motion the tape revealed an inability to fully extend the joints of her index finger beyond the resting position. As she performed grip activities, the index finger became caught on the object she wished to pick up and, because of the lack of extension, she was unable to let go of objects at the end of a task. A physical examination subsequently showed total rupture of the extensor indices tendon.

Quantitative data can be obtained by digitising the video image and subjecting the data to computer processing and analysis. In order to be able to digitise, skin markers must be placed over major joints prior to filming. The film is then generated on the computer screen and the person digitising uses the cursor to indicate the position of the markers. The computer stores the coordinate data for each of the markers for each frame of the film and this information is called upon when calculations are required. Experiments in the author's laboratory have shown that the technique is reliable and accurate and that human error is relatively small, especially with experienced digitisers. Unfortunately, the process is time-consuming as only 15 to 30 points can be digitised per minute.

Whilst computer aided measurements have proved to be accurate and reliable, direct measurements of joint position taken from the video screen are subject to considerable error and this should not be used as a method of quantifying human movement.

Computer aided analysis of video and cine film can give a wide range of information, but unless complex and expensive systems are used, this information is confined to movement in one plane.

From the computer analysis software it is possible to plot body coordinates (centre of gravity, etc.). Knowledge of the position of the centre of gravity is important when comparing the efficiency of movement. Smooth displacements of the centre of gravity tend to indicate a more efficient movement than one where the centre of gravity is raised unnecessarily high.

Stick diagrams can be produced and these are valuable as an initial qualitative analysis of the sequence of movement. A stick diagram of a jump is shown in Figure 14.2.

Data on joint angles in one plane of movement can be collected and the pattern of movement at a joint can be graphically represented and related to other joints or the whole body. Joint angle data are available for any instance in the movement sequence.

The velocity and acceleration of limb segments can be measured and the data give useful information about patterns of movement, for example when comparing the acceleration of the tibia in the swing phase of normal gait, with the acceleration of the shank of an artificial limb in amputee gait.

It is also possible to calculate, from video tape, cadence, stride length and velocity in gait, but it is necessary to provide some form of scaling in the filming area (Whittle 1991).

In conjunction with force plate data, information on joint angles can be used to calculate muscle moment (Winter 1990).

Filming with video is a relatively simple and cheap technique which can be undertaken almost anywhere and, because it does not require any attachments to the subject, it does not disrupt the movement being analysed. (Whether there is a psychological effect on movement patterns brought about by the self-consciousness of being filmed is not known and this should not be discounted.) The value of film as a movement analysis tool has long been recognised by sport scientists, but the health care professions appear to have been less enthusiastic about its use. Film has been used to analyse human movement in a limited range of activities, for example the ability of paraplegic subjects to reach from their wheelchairs (Curtis et al 1995), the measurement of angular velocity of the leg in a patient with cerebral palsy (Winter 1982), the biomechanical analysis of swing through gait (Noreau et al 1995) and the energy transfers of children walking with crutches (McGill & Dainty 1984).

For meaningful data collection, great care must be taken in setting up the filming site and

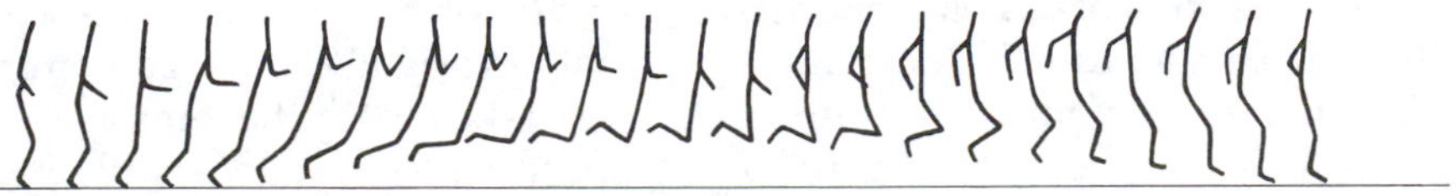

Figure 14.2 A stick diagram of a jump. This was generated from data obtained by digitising a video film and gives an overall, subjective impression of the movement.

arranging the camera. For accurate spatial and temporal measurements to be taken, the camera must be positioned accurately in relation to the subject and timing and scale devices must be included in the field of view.

Frame rate

Video, unlike cine, has a fixed frame rate which in the UK and Europe is 25 frames per second (in the USA it is 30 frames per second). Each frame can be broken down into two fields which in Europe gives 50 fields per second. Some authorities (Winter 1990) feel that this rate is acceptable for all but the fastest athletic movements and is, therefore, acceptable for work with patients. However, a low frame rate will not guarantee capture of specific moments of impact or the instant of change of direction in normal human movement, such as the instant of heel strike, etc. For detailed analysis of specific instants of the movement cycle, it may be more appropriate to use a cine camera with a variable frame rate rather than video, although the comparative difficulty and the cost of using cine has to be a counter argument.

Shutter speed

It is essential to use a camera which has a variable shutter facility. If the shutter speed cannot be controlled, fast movement will give spatial blurring and make the identification of the position of skin markers impossible. A shutter speed of a thousandth of a second will eliminate spatial blurring, but it may require the use of artificial lighting which is sometimes off-putting to the patient.

Skin markers

Computer aided analysis of film requires the placement of skin markers over bony landmarks so that joints can be identified. Unfortunately, the use of skin markers introduces a number of potential errors to which there is no easy solution. Problems arise in designing markers which can be easily seen on screen and some markers will, at times, be obscured by the swing-through of other parts of the body or as the subject moves behind other objects in the filming area. The markers, being attached to skin, will be displaced along with the skin as joint movement occurs and will not retain a fixed position in relation to the underlying joint. To complicate matters further, few joints have fixed axes and, though the skin marker might be aligned directly over the axis at one point in the joint range, as the instantaneous centre of rotation changes it will move away from the static location of the marker. All these factors are likely to introduce error into the data collected.

Filming

The camera should be placed as far from the subject as possible to reduce parallax and perspective error. There is disagreement amongst researchers as to how far the camera should be from the movement, but consensus suggests that 7 m is an acceptable distance which will not require a correction factor to be applied. The axis of the camera should be at right angles to the plane of progression and the zoom facility should be used to enlarge the image so that the markers can be easily seen. If significant movement is occurring in more than one plane, a three-dimensional filming system is needed.

Measurement of joint motion

Traditionally, joint motion has been investigated by measuring the maximum range of movement available at individual joints. These static goniometric measurements, whilst of proven reliability and validity (Gajdosik & Bohannon 1987), have limited value in the description of movement. Despite being used as objective measures for testing the efficacy of therapeutic intervention, they give no indication of the functional range of movement of the joint within an activity. Normally, static goniometric measurements are taken, with the joint in a non-weight bearing position which allows full range of active or passive movement to be measured. However, very few functional activities of the lower limb are performed in a non-weight bearing position, and few upper limb functions are performed without the limb holding a weight, so the results obtained are not an accurate reflection of the subject's capabilities in a functional activity.

With improving technology it has become possible to quantify joint movement within a functional activity. Photographic methods, including cine film and video, provide valuable information, usually in one plane, but they are often time-consuming to use and the filming equipment and computer hardware is costly. A cheaper approach is the electrogoniometer which has made it possible to take measurements of joint position direct from the limb. The electrogoniometer, which was introduced by Karpovitch in the 1950s (Rothstein 1985) can take a number of forms but most commonly it consists of two endblocks joined by an electronic potentiometer which is encased within a protective spring. More sophisticated devices may consist of up to three potentiometers for each joint, thus allowing simultaneous measurement of movement in three planes. The potentiometer consists of a series of strain gauges mounted around the circumference of a wire (Nicol 1987). One endblock is usually fixed to the potentiometer whilst the other allows telescoping when a longitudinal tension or compression is placed upon it. Two different types of goniometer are shown in Figure 14.3. In both cases the way in which they are designed allows measurement to take place when the centre of rotation of the goniometer does not coincide with the centre of rotation of the joint. Figure 14.4 illustrates how this is possible. With these types of electrogoniometers, movement of a joint will result in movement of the potentiometer and the resultant strain on the potentiometer generates electrical signals, i.e. the resistance in the potentiometer is changed. These signals, in the form of voltage and, less commonly, current, are plotted and, after calibration, represent the angular displacement of the joint. Only angular displacements are measured. Linear movements result in telescoping of the spring and therefore no strain occurs, consequently, no voltage is recorded. The joint displacement curves for the hip, knee and ankle joint which are shown in Chapter 10 were taken from data obtained from an electrogoniometer (Fig. 10.3).

Some goniometers have to be aligned over the joint axis and this introduces a source of error if the instantaneous centre of rotation changes throughout the movement or if the goniometer becomes misaligned.

The electrogoniometer itself is light and will not therefore interfere to a large extent with the activity that is being tested. It is designed so that only a light force is needed to bend the potentiometer, making the instrument very sensitive. Error associated with the use of electrogoniometers comes mainly from the means by which they are fixed to the subject and the method by which data are relayed to the computer for analysis. Fixation has proved to be a problem over many years as human limbs are normally conical in shape and, unless the goniometer is stuck to the skin, it is in danger of sliding down or round the limb during movement. When straps or bands are used to hold the goniometer onto the limb they normally need to be so tightly fastened that they are uncomfortable and restrict movement. Sticking the electrogoniometer directly to the skin is a more satisfactory means, although skin movement over the joint, as discussed earlier in this chapter, may present a problem. Some electrogoniometers relay the joint data to the computer via leads, which can be off-putting and restrictive especially on fast movements or when the subject is covering a substantial distance. The swinging of the leads may also introduce movement artifacts. Telemetry is more successful and is being increasingly utilised (Whittle 1991). Some electrogoniometers are supplied with a small data logger, with the capability of recording data during the activity. This information can then be downloaded onto a compatible computer and the results analysed at a later date.

Many researchers have demonstrated the reliability of electrogoniometers (Rowe et al 1989). Cosgrove et al (1991) found that there was a 2° discrepancy with the goniometer and no baseline drift, but more recent researchers such as Myles et al (1995) have found the hysteresis effect to be 3.6° with a residual error of 2.9° for repeated

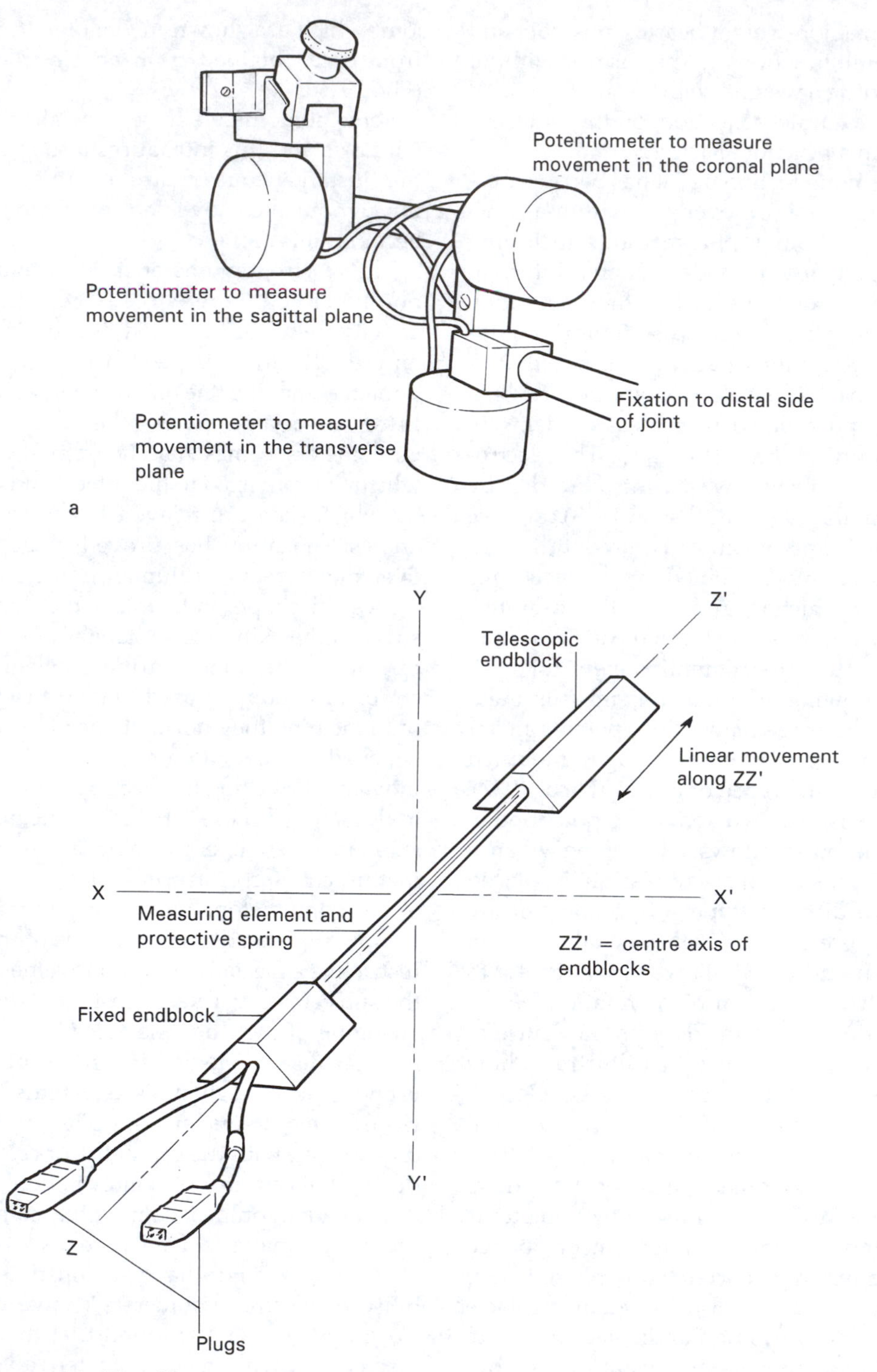

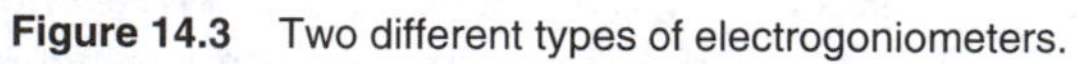

Figure 14.3 Two different types of electrogoniometers.

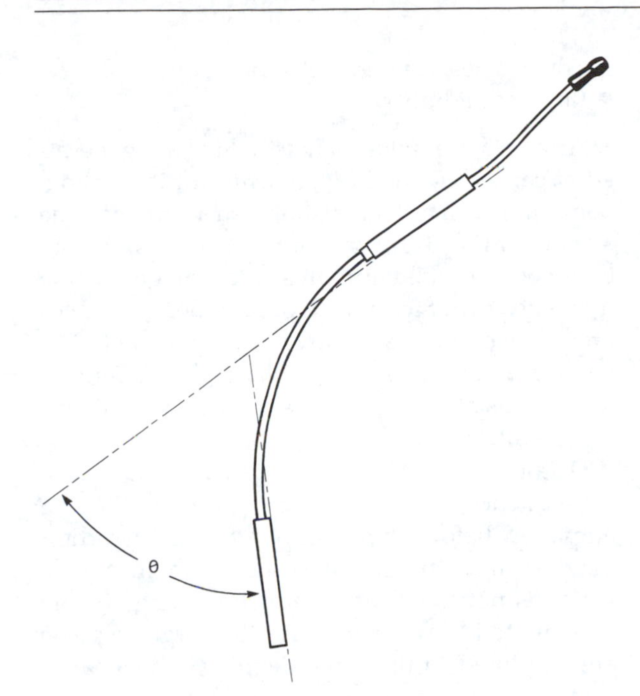

Figure 14.4 This electrogoniometer does not need to be aligned over the axis of the joint. Movement of one endblock in relation to the other enables calculation of the position of the joint.

measurements of large ranges. Smaller joint ranges, however, showed discrepancy only in the order of 1° for hip and knee flexion during walking. Hazelwood et al (1995) tested the construct validity of the electrogoniometer and found the measures to be highly repeatable with little variation. All these errors can be kept to a minimum if the operational definition is implemented with care.

Electrogoniometers have been found to be valid, reliable and easy to use. They help to analyse joint ranges within activity and, therefore, represent a good picture of the functional capabilities of that joint. Although most use is made of the goniometer within the field of research, the costs will inevitably decrease, making it within the price range of clinical therapists.

Other means of measuring joint position exist including optoelectronic devices and polarised light goniometers. These tend to be costly but can produce detailed information (Whittle 1991).

Electromyography (EMG)

EMG is the term used to describe the electrical signals produced as a result of the contraction of a muscle, the method of collecting these signals and the data that are produced.

When a muscle is quiescent there is little electrical activity. However, during muscular activity, electrical signals are produced and can be recorded. Electromyography will show if a muscle is active or not, the duration of that activity and, as the EMG increases in magnitude with tension, the signals will also give an indication of how much torque is being generated.

The basis of EMG

At a cellular level, the muscle fibre or cell is the unit of contraction. During muscle activity there is an electrical potential change and depolarisation and repolarisation of the surface membrane of the cell. There is transmission of impulse across the sarcolemma to the interior of the muscle cell via a complex system of tubules.

When a neural impulse reaches the motor end plate, a wave of depolarisation spreads across the cell resulting in a twitch followed by relaxation. This twitch can last from a few milliseconds to 0.25 s. The depolarisation is followed by a wave of repolarisation.

The muscle electrical potentials, called muscle action potentials (MAPs), will result in a small amount of the electrical current spreading away from the muscle in the direction of the skin, where electrodes can be used to record the electrical activity. The nearer the electrode is to the muscle, the larger the recorded signal. If muscle fibres some distance from the electrodes are conducting electrical current, the MAPs recorded will be smaller than they would have been for similar sized motor units closer to the electrodes.

Values actually obtained from muscle vary between 100 μV (microvolts) to 5 mV (millivolts). The signal can be very small and a problem may exist because electrical activity from sources other than the muscle can overwhelm the desired signal. This unwanted electrical activity is called 'noise' and a number of strategies have to be adopted to eliminate unwanted noise.

Types of electrodes

Either surface or indwelling, needle electrodes can be used, though surface electrode EMG is most common in the analysis of human movement. Both types of electrodes not only pick up electrical activity which passes over their conducting surface but can also register electrical currents nearby.

Surface electrodes are normally small metal discs of about 1 cm diameter, though they can be smaller if tiny muscles are being tested. The electrodes are usually made of silver/silver chloride and are sensitive to electrical signals from superficial muscles. They give a reading corresponding to the average electrical activity.

Needle electrodes are normally fine hypodermic needles containing a conductor which is insulated except for its protruding end. As two electrodes are needed, the outer part of the hypodermic forms one and the conductor inside the hypodermic forms the other.

There is some suggestion that the surface electrodes are more likely to give reliable data and, as a non-invasive technique, are preferable (Winter 1990). Subjective reports indicate that despite the use of local anaesthetics, there is a degree of discomfort from needle electrodes and that the movement of the contracting muscle in relation to the overlying skin, when pierced by the electrode, produces some inhibition to normal movement.

Recording the EMG

In order to be able to use the data collected from muscle, the signal must be 'clean', that is, free from noise, artefacts and distortion (Winter 1990).

Noise may come from a variety of sources. These include:

- other muscles, especially the heart
- nearby electrical machinery including the EMG recording equipment
- radio waves, i.e. ambulance/police/CB radios
- power lines, domestic electrical supply
- fluorescent lights.

Artefacts are 'false signals' which are generated or caused by the EMG machine or its cabling. Some are difficult to distinguish from the true signal coming from the muscle, but others can easily be identified. Into this latter category come movement artefacts which occur when the cables or the electrodes are moved and touched. The movement artefacts are usually at the upper and lower ends of the frequency range and can therefore be filtered out.

Distortion of EMG signals usually results as a consequence of the signals needing to be amplified before they can be of use. Distortion may occur if the signal is amplified in a way which is not linear over the whole range of the system. It is important that the larger signals are amplified to the same degree as the smaller signals.

EMG processing

After the EMG signal has been amplified it has to be processed so that it is in a form which will enable it to be compared or correlated with other physiological and biomechanical signals (Winter 1990). Computers are used for this purpose and it is important to be aware that the original signals will have been subjected to a number of manipulations before the final data are produced. This should not normally be a problem, but where the output is not what was expected then the signal processing should be checked.

Clinical significance of EMG for measuring human movement

A question that has been puzzling researchers for at least 25 years is 'how valuable is EMG in reflecting and predicting muscle function and can findings be extrapolated to total body function?'. EMG surface electrodes are non-invasive and the method is cheap and simple to apply. Various authors have suggested that EMG will provide information on muscle power, muscle sequencing, fatigue, composition of fibre type and metabolism. The simplicity and inexpensive nature of EMG ought to make it an important method of evaluating function, but the fact that the value of EMG is still being questioned after so long indicates its limitations. There are still sufficient doubts about what EMG can reliably do to make it a measurement tool which must be used with caution.

EMG and the phasic activity of muscles

EMG can provide information on whether or not a muscle is active and for how long the period of activity or inactivity continues. There is always a small lag between the onset of electrical activity in a muscle and perceived movement of the limb; this is of the region of 30 ms and it is probably not significant in terms of analysis of the phasic activity of muscles. The lag is partially due to the chemical changes which must take place before the muscle can contract and also due to the need of the muscle to 'take up slack' before joint movement can occur.

A similar lag occurs at the end of muscle activity. With the cessation of electrical activity, the muscle continues to contract for a short period whilst the chemical changes stabilise and the muscle is able to relax. It is likely that the period between the end of electrical activity and the cessation of contraction will vary between muscle groups and will also be dependent on the type of muscle contraction. In the normal human being the quadriceps has been studied most often and it shows a lag period of between 250 ms and 300 ms (Inman et al 1981).

It is valuable to know the duration of involvement of various muscle groups during movement. All current information on the phasing or sequencing of muscles in functional activity is taken from EMG studies. These studies show when the muscles are activated and when they cease activity, but EMG is not able to provide information on whether the activity is isometric, concentric or eccentric. EMG studies of muscles involved in normal walking have shown that the input of individual muscles may only last for a very short time, often in the region of 0.2 s. As this would not give the muscle sufficient time to produce a movement at a joint, it leads to speculation that these muscles are doing no more than producing an isometric action (Inman et al 1981). With such minor involvement of muscles in normal gait, it is no wonder that people are able to walk for many hours before experiencing fatigue.

EMG and force production

There is considerable controversy about whether EMG can give reliable, quantifiable information on the magnitude of muscle activity, the recruitment of contractile elements and muscle metabolism and fatigue.

Whilst it is commonly agreed that EMG can distinguish between a working muscle and a quiescent muscle, there is disagreement about whether the relationship between EMG activity and torque is consistent and linear. If the relationship is linear, EMG can be used to calculate the forces generated by muscles during functional activities. If, however, the EMG signal has an inconsistent, non-linear relationship with muscle torque, then it has little value as a measurement tool.

EMG and isometric tension

Early experiments by Lippold in 1952 showed that for the gastrocnemius muscle, a linear relationship could be shown between the average amplitude of EMG and the tension developed in muscle. This appeared to indicate that EMG could be used to measure the force generated, but unfortunately in subsequent decades other workers found different results. Although the EMG increased, it was not found to do so in direct relationship to the actual amount of force generated by the muscle (Lawrence & De Luca 1983, Rau & Vredenbregt 1973, Zuniga & Simons 1969).

Winter (1990) suggests that there is no more than a 'reasonable relationship' between isometric force generation and EMG activity. If this is true, the results from EMG can only be used to give a general prediction of muscle tension.

EMG and isotonic tension

Much less experimental activity has been undertaken in the field of EMG/isotonic force relationships than EMG/isometric relationships. On low velocity contractions there is an identifiable, but inconsistent, relationship between EMG and muscle torque and the faster the contraction, the more difficult it becomes to see any relationship (Komi 1973).

Mathematical calculations of muscle moment, based on force plate and joint angle data, can be made. Information gained from these calculations and from simultaneous EMG correlates very closely, suggesting that EMG is an accurate method of collecting information about whether a muscle is contracting or not during a functional activity (Olney & Winter 1985). EMG will also give some indication of the magnitude of the force being generated by that muscle, though no detail.

Isokinetic dynamometry

There are no direct methods of measuring the work undertaken by individual muscles during functional movements, though there are a number of mathematical approaches which can be employed to provide data on the net moment at a joint (Winter 1990). Whilst isokinetic dynamometers can record muscle torque throughout a single joint movement, this can only be done with the subject attached to the machine, rendering any movement measured non-functional. The isokinetic dynamometer requires the subject to be positioned with the axis of the joint to be tested aligned with the axis of the machine. The limb must be strapped tightly to the dynamometer chair to ensure that misalignment of these two axes does not occur and movement will only occur through the range permitted by the fixation. Modern isokinetic dynamometers are able to measure muscular torque, work, power, the rate of torque production (explosiveness) and endurance in movements which involve the muscle concentrically and eccentrically. The force generated in isometric muscle activity can also be measured.

Peak torque, which is measured in newton-meters (N.m) is the highest torque output achieved by a muscle as it moves its joint through range of motion. It would appear to be an accurate and reproducible measure (Kannus 1994).

Peak torque is constant up to the velocity of 60°/s but then declines as velocity increases; this is known as the torque–velocity relationship. It declines for several reasons. At low velocities the muscle uses both Type I and Type II fibres, but as the velocity increases, the muscle becomes unable to activate Type I fibres (slow twitch), then the Type IIA fibres until only the fastest fibres are able to be recruited in the time available (Type IIB). With a decreasing number of fibres being recruited, torque production decreases.

Kannus (1994) suggests that with increasing angular velocity, the point at which peak torque is achieved occurs later in the range. This is problematic as there is an optimum part of the range when muscles are able to generate maximum force (mid-range). If the muscle has not recruited all possible fibres by mid-range, the peak force recorded may not be a true representation of actual ability. This delay in recruitment may be due to slow neural recruitment. It may also be that in disuse atrophy, neural recruitment is slower than normal and that this may also result in peak torque occurring later than normal. For these reasons it is suggested that when peak torque at different velocities is compared, the joint position at which peak torque occurred should be taken into consideration.

Torque can be measured anywhere within the range of movement and this is called *angle specific torque*. It would appear that when angle specific torque is measured in inner or outer range, the accuracy of the measurement decreases below an acceptable level and results become inconsistent (Kannus 1994).

Figure 14.5 shows the traces taken from a normal subject who was measured during a concentric contraction of the quadriceps muscle which produced a knee extension followed immediately by eccentric activity in the quadriceps to control knee flexion. Six tests were undertaken and it

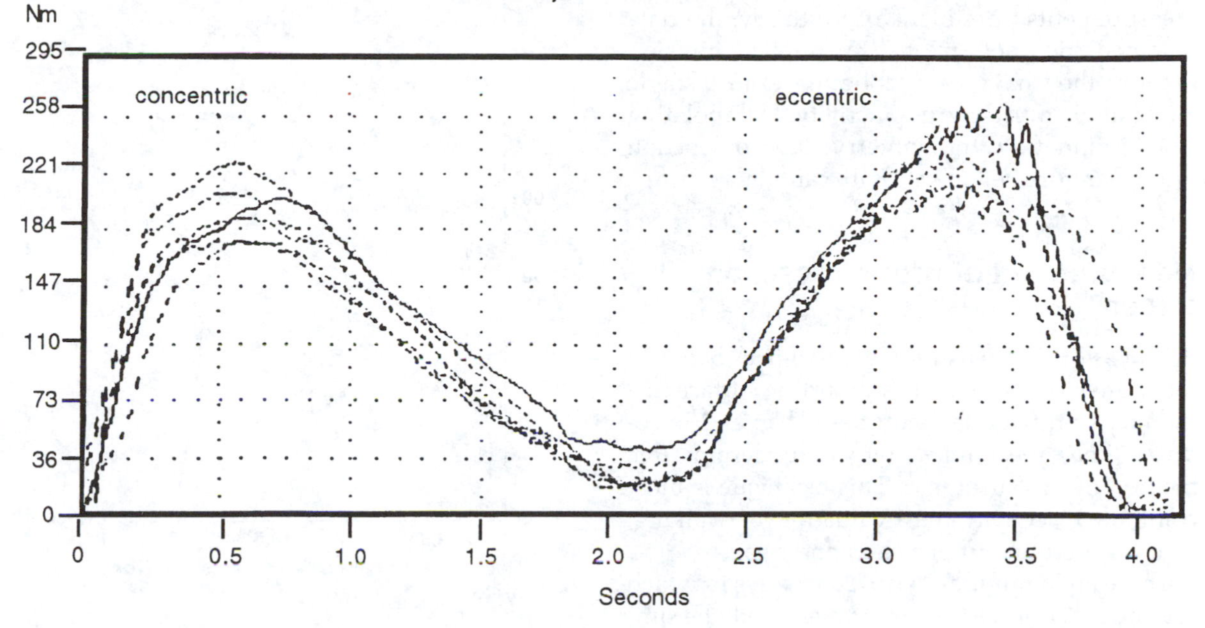

Figure 14.5 Torque curves of repeated concentric/eccentric activity of the quadriceps muscle. Data were obtained by using an isokinetic dynamometer.

is clear that the pattern of muscle activity is fairly constant, though the torque varies between repetitions. This is partially a result of the learning process which invariably means that the first contraction is not as effective as subsequent ones and partially due to fatigue in the later repetitions. The traces clearly indicate that peak torque for both the concentric and eccentric actions is in mid-range, though it is achieved more quickly for concentric activity. As would be expected, the peak torque in eccentric activity is greater than in concentric activity.

Work and power measurements can be obtained from the IKD. In isokinetics, work is defined as the area under the torque curve where the torque curve is torque against angular displacement (work is torque × angular displacement). Work is measured in joules. Power is the rate of muscular work and increases with angular velocity. Average power is total work for a given contraction divided by the time taken and is measured in joules per second or watts.

It is also possible to measure peak torque acceleration energy, which is the greatest amount of work performed in the first 125 ms of a contraction. It is measured in joules and it is suggested that this measurement is indicative of explosive ability as it gives the rate of torque production. There is some doubt about the reliability of these data and their repeatability, especially at low speeds (Kannus 1994).

Endurance indices can be defined as the ability of muscle to perform repeated contractions against a load. An endurance index is supposed to indicate the rate of fatigue. There appears to be no agreement as to the best way to test for endurance, though most tests work on a reduction from peak torque over a specified period of time. Isokinetic dynamometers can provide this information.

The isokinetic dynamometer as a measure of human function

The isokinetic dynamometer is a popular tool because it provides information about muscle groups which may be functioning isometrically, isotonically and isokinetically. It is the only device which makes concentric and eccentric measurements possible. Unfortunately, the data obtained do not relate to human function because they have been collected from a single joint/single muscle group activity. Extrapolating from isokinetic dynamometry data to function must therefore be viewed with caution.

Measurement of ground reaction forces

It is possible to measure reaction forces between the human being and the supporting surface in a number of functional activities; Chapter 2 introduces the mechanics which underpin this method of measurement. This technique is most commonly seen in the evaluation of walking, running, getting out of a chair and sway.

There are a number of different ways in which ground reaction forces can be measured; the simplest usually measure the vertical ground reaction forces or, if a device is placed inside a shoe, the foot-shoe interface forces. The more complex force plates measure both vertical forces and shear forces in the horizontal plane (in an anterior/posterior direction and also in a medial/lateral direction). From these three forces it is possible to calculate a single point about which these forces are said to act and also represent these forces by a single ground reaction vector.

Typically, force plate data are plotted against time, showing the patterns of change of force during the period of time that the foot is in contact with the supporting surface. Figure 14.6 shows the normal vertical reaction forces seen in walking plotted in two different ways.

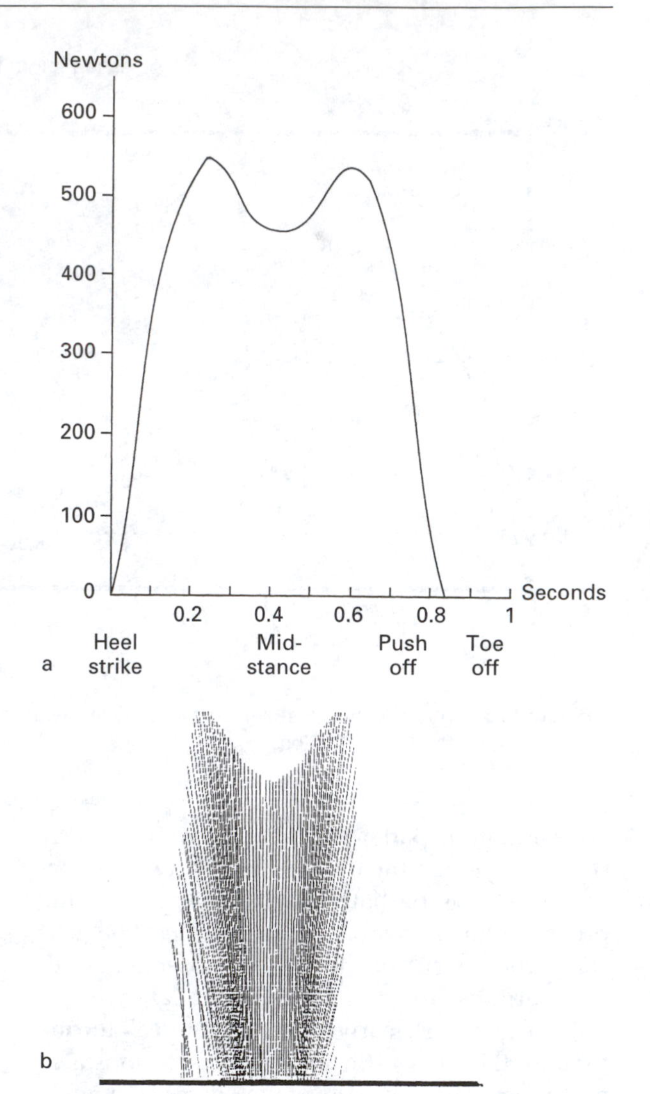

Figure 14.6 Normal force traces taken from the stance phase of walking: a) a force curve; b) a vector diagram.

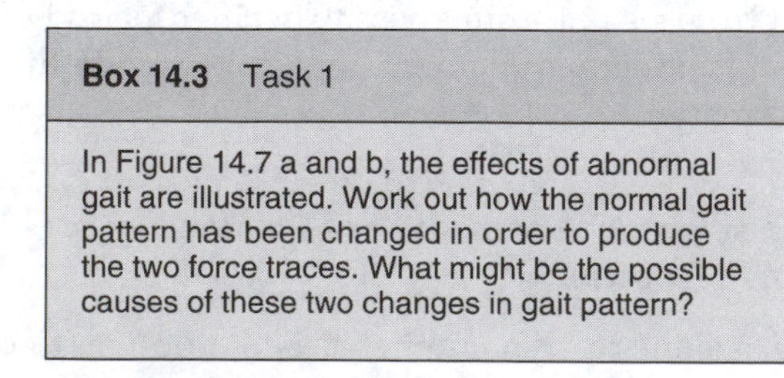

Box 14.3 Task 1

In Figure 14.7 a and b, the effects of abnormal gait are illustrated. Work out how the normal gait pattern has been changed in order to produce the two force traces. What might be the possible causes of these two changes in gait pattern?

The clinical value of force plate data is debatable. Major movement abnormalities are apparent both on visual observation and on examination of a force trace. The advantage that arises from using a force plate is that the data obtained enable quantification of change over time. Force plates are, however, very expensive and their use as a clinical tool is likely to remain limited.

Scales or indices for measuring activities of daily living

A vast number of scales or indices have been

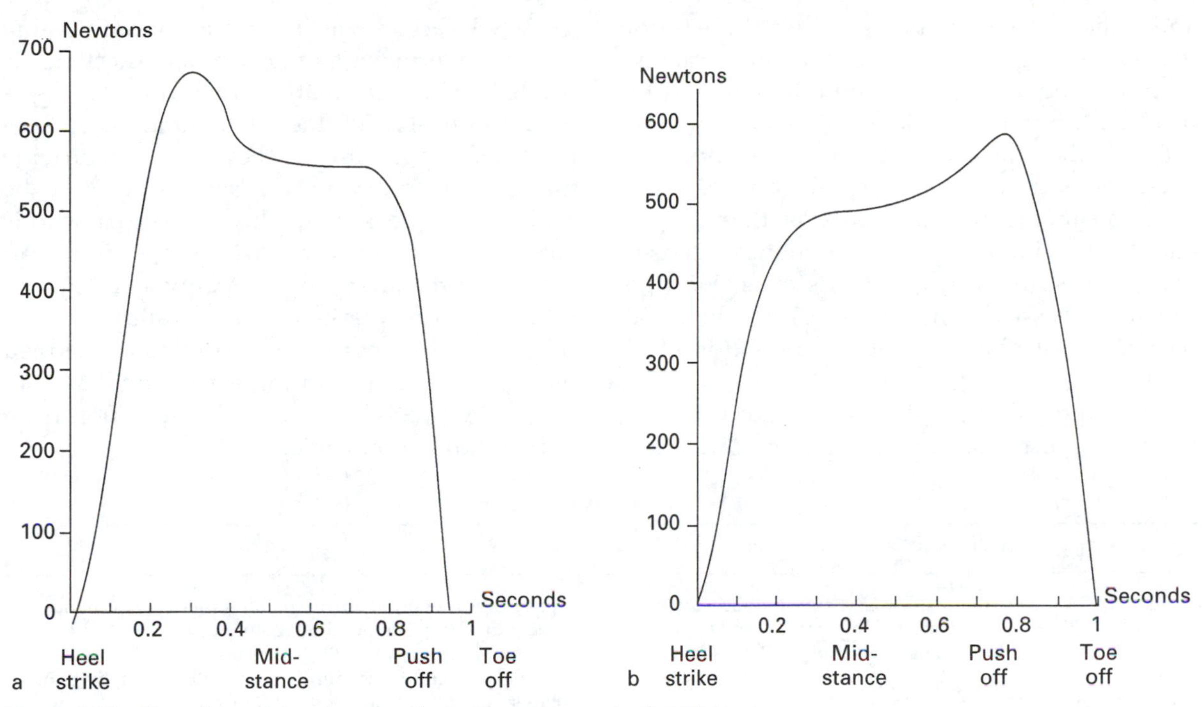

Figure 14.7 Force traces taken from a patient with an abnormal gait pattern.

designed to measure an individual's ability to perform activities of daily living. These would be most correctly regarded as measurements of disability as defined by the World Health Organization (1980). Some of these approaches to measurement are simple, concentrating on only one aspect such as gait and using, for example, the timed walk test. Other approaches are increasingly complex and will evaluate a number of parameters including motor, sensory, cognitive and emotional abilities. Tests commonly in use include the Barthel index, the functional independence measure (FIM) and the Rivermead motor assessment (Wade 1992). Most of these scales have arisen out of the need to move away from measurements which only look at one parameter of human movement such as joint range or muscle torque. These scales ask the more realistic question of whether an activity can actually be performed and, in some instances, they invite judgements on the quality of the performance and encourage the tester to evaluate the patient in relation to his/her environment. The Barthel index, which is one of the better known and simpler scales, illustrates this form of measurement. This index has two categories: scoring depends on whether an individual can perform certain activities of daily living independently or whether they need assistance. The categories, which are indicated in Table 14.1, are awarded marks and an overall score is reached.

Some of these scales and indices are reliable

Table 14.1 Items in the Barthel index

	With help	Independently
1. Feeding	5	10
2. Moving from wheelchair to bed and return	5	15
3. Personal toilet	0	5
4. Getting on and off toilet	5	10
5. Bathing self	0	5
6. Walking on level surface	10	15
(or propelling a wheelchair)	0	5
7. Ascending and descending stairs	5	10
8. Dressing	5	10
9. Controlling bowels	5	10
10. Controlling bladder	5	10

(Mahoney & Barthel 1965)

and valid measurement tools, but the ease of use varies greatly. The Barthel index can be administered in a matter of minutes and may be dealt with purely by asking the patient to indicate whether they can or cannot perform the listed activities. Other systems can be very time-consuming by the nature of their length and this is evidenced by the functional independence measure, which, with its extension (the functional assessment measure) is about 50 pages long (Collen et al 1991, Hamilton et al 1994).

The main value of all these measures is as a means of gross assessment of patient disability. They will record which activities of daily living can be undertaken and how much assistance is needed. The information they provide is a broad indicator of the rehabilitation or care needs of a patient, but they will not provide information on why an activity cannot be achieved or measure quality. A patient who is able to walk unaided but whose pattern of gait requires excessively high energy expenditure may score full marks on these scales, but if a rehabilitation programme needed to be planned then there would need to be additional tests to reveal the precise nature of the deviation from normal human movement.

REFERENCES

Collen F M, Wade D T, Bradshaw C M 1990 Mobility after stroke: reliability of measures of impairment and disability. International Disability Studies 12: 6–9

Cosgrove A P, Graham H K, Mollan R A B 1991 Gait analysis in children with cerebral palsy using electrogoniometers. Presentation, British Orthopaedic Research meeting

Curtis K A, Kindlin C M, Reich K M, White D E 1995 Functional reach in wheelchair users: the effects of trunk and lower extremity stabilisation. Archives of Physical Medicine and Rehabilitation 76: 360–367

Gajdosik R L Bohannon R W 1987 Clinical measurement of range of motion: review of goniometry emphasising reliability and validity. Physical Therapy 67: 1867–1872

Hamilton B B, Laughlin J A, Fielder R C, Granger C V 1994 Inter-rater reliability of the 7-level functional independence measure (FIM). Scandinavian Journal of Rehabilitation Medicine 26: 115–119

Hazelwood M E, Rowe P J, Salter P M 1995 The use of electrogoniometers as a measurement tool for passive movement and gait analysis. Physiotherapy 81(10): 639

Inman V T, Ralston H J, Todd F 1981 Human walking. Williams and Wilkins, Baltimore

Kannus P 1994 Isokinetic evaluation of muscular performance: Implications for muscle testing and rehabilitation. International Journal of Sports Medicine 15: S11–S18

Komi P V 1973 Relationship between muscle tension, EMG and velocity of contraction under concentric and eccentric work. In: Desmedt J E (ed) New Developments in Electromyography and Clinical Neurophysiology 1: 596–606

Lawrence J H, De Luca C J 1983 Myoelectric signal versus force relationship in different human muscles. Journal of Applied Physiology 54(6): 1653–1659

Lippold O C J 1952 The relationship between integrated action potentials in a human muscle and its isometric tension. Journal of Physiology 117: 492–499

McArdle W D, Katch F I, Katch V L 1991 Exercise physiology, energy, nutrition and human performance, 3rd edn. Lea & Febiger, Philadelphia

McGill S M, Dainty D A 1984 Computer analysis of energy transfers in children walking with crutches. Archives of Physical Medicine and Rehabilitation 65: 115–120

Mahoney F I, Barthel D W 1965 Functional evaluation: the Barthel Index. Maryland State Medical Journal 14: 61–65

Myles C, Rowe P J, Salter P, Nicol A 1995 An electrogoniometry system used to investigate the ability of the elderly to ascend and descend stairs. Physiotherapy 81(10): 640

Nicol A C 1987 A flexible electrogoniometer with widespread applications. In: Johnson B (ed) Biomechanics XB. Human Kinetics Publishers, Illinois: pp 1029–1033

Noreau L, Richards C L, Comeau F, Tardif D 1995 Biomechanical analysis of swing-through gait in paraplegic and non-disabled individuals. Journal of Biomechanics 28: 689–700

Olney S J, Winter D A 1985 Predictions of knee and ankle moments of force in walking from EMG and kinematic data. Journal of Biomechanics 18: 9–20

Rau G, Vredenbregt J 1973 EMG force relationship during voluntary static contractions (M. biceps). Medicine & Sport Biomechanics III 8: 270–274

Rothstein J M (ed) 1985 Measurement in physical therapy. Churchill Livingstone, Edinburgh

Rowe P J, Nicol A C, Kelly I G 1989 Flexible goniometer computer system for the assessment of hip function. Clinical Biomechanics 4: 68–72

Terauds J 1984 Sports biomechanics: Proceedings of the International Symposium of Biomechanics in Sport. Academic Publishers, Del Mar, California

Wade D T 1992 Measurement in neurological rehabilitation. Oxford University Press, Oxford

Whittle M 1991 Gait analysis, an introduction. Butterworth-Heinemann, Oxford

Winter D A 1982 Camera speeds for normal and pathological gait analyses. Medical and Biological Engineering & Computing 20: 408–412

Winter D A 1990 Biomechanics and motor control of human movement, 2nd edn. Wiley Inter-science, New York

World Health Organization 1980 The international classification of impairments, disabilities and handicaps. World Health Organization, Geneva

Zuniga E N, Simons D G 1969 Non-linear relationship between averaged electromyogram potential and muscle tension in normal subjects. Archives of Physical Medicine and Rehabilitation 50: 613–620

Index

A

B

C

D

E

F

G

H

I

J

K

L

M

N

O

P

Q

R

S

T

U

V